Testing of Construction Materials

Testing of Construction Materials

Bahurudeen A
PVP Moorthi

CRC Press
Taylor & Francis Group
Boca Raton London New York

CRC Press is an imprint of the
Taylor & Francis Group, an **informa** business

First edition published 2021
by CRC Press
6000 Broken Sound Parkway NW, Suite 300, Boca Raton, FL 33487-2742

and by CRC Press
2 Park Square, Milton Park, Abingdon, Oxon, OX14 4RN

First edition published by CRC Press 2021
CRC Press is an imprint of Taylor & Francis Group, LLC

Library of Congress Cataloging-in-Publication Data

[Insert LoC Data here when available]

ISBN: 9780367644956 (hbk)
ISBN: 9781003124825 (ebk)

Typeset in Times
by Deanta Global Publishing Services, Chennai, India

Authors dedicate to their teachers Prof. Manu Santhanam (IIT Madras) and Prof. Prakash Nanthagopalan (IIT Bombay).

Contents

Preface

The purpose of writing this book is to provide technical details of construction material testing methods in a simple way with high precision. This book conceptualizes the testing of construction materials with a brief technical summary, followed by competitive practice questions to attain scientific understanding of the testing of materials. This book is divided into 11 chapters. Chapters 1 to 10 focus on specific construction materials. In every chapter, only essential test methods are described based on academic university syllabus and real-time quality-control testings. Chapter 11 focuses on the sophisticated analytical techniques and methods used for the analytical investigation of construction materials. Moreover, outdated and unnecessary testings are entirely avoided in this book. The testing procedure described in this book is strictly followed as per Indian standards and international standards including ASTM (American Society for Testing and Materials), South African standards, DIN German standards and European standards. Moreover, this book can also be used in the laboratory classes directly as a "Textbook cum Laboratory Manual". This will be a unique manual for undergraduate students because "Brief Chapter Summary" is given for each material which helps to get a quick technical background of materials before the start of laboratory classes. The testing of construction material is described in a step-by-step procedure. Furthermore, neat illustrations and tables are presented wherever essential. To understand whether the material is in a satisfactory condition, standard values in accordance with Indian and international standards are included. This rich pool of pedagogy includes questions and answers from previous year competitive exams for all test methods. The step-by-step procedure provided in each test guides readers to perform the test in a logical sequence with good clarity. Authors thank the editorial team of CRC press for their constant support and welcome feedback from readers in all aspects.

Bahurudeen A and PVP Moorthi

Authors Biographies

Dr Bahurudeen A is Faculty in Civil Engineering at Birla Institute of Technology and Science (BITS Pilani), Hyderabad Campus, India. He completed his PhD in Civil Engineering, Indian Institute of Technology Madras (IIT Madras). He served as Senior Scientific Officer at IIT Madras. His research focus involves scientific understanding of properties of construction materials. He has published 6 patents and several research articles in reputed journals and conferences on the evaluation of construction materials. Moreover, he is a Primary Developer for outcome-based course content for "Construction Materials and Technology" course, which is an MHRD sponsored national pedagogy project. He has completed several real-time consultancy projects and is actively involved in many educational bodies.

Mr PVP Moorthi is Technical Head of Engineering Delight Academy, Salem, India. He has completed his master's in Infrastructure Engineering and Management from Birla Institute of Technology and Science (BITS Pilani), Rajasthan. He has published several research articles in reputed journals and conferences. His areas of exposure include construction materials, rheology of cementitious materials and non-destructive testing.

List of Symbols

A:	ampere
°C:	degree centigrade or degree Celsius
°F:	degree Fahrenheit
GΩ:	giga ohm
g:	grams
kg:	kilo grams
kN:	kilo newton
kPa:	kilo pascal
MΩ:	mega ohm
MPa:	mega pascal
μm:	micrometer
mA:	milli ampere
ml:	millilitre
mm:	millimetre
mV:	milli voltage
N:	newton
Pa:	pascal
%:	percentage
V:	voltage

List of Abbreviations

AASHTO: American Association of State Highway and Transportation Officials
ACI: American Concrete Institute
ASTM: American Standard for Testing and Materials
BSE: Backscattered electrons
C–A–H: Calcium-aluminate-hydrate
C–S–H: Calcium-silicate-hydrate
CTM: Compression testing machine
DVM: Digital voltmeter
C_2S: Di-calcium silicate or belite
DC: Direct current
FA: Fly ash
GGBS: Ground granulated blast furnace slag
IS: Indian standards
ITZ: Interfacial transition zone
Mk: Meta-kaolin
OPC: Ordinary Portland cement
PCE: Poly carboxylic esters or ethers
PPC: Portland pozzolan cement
pH: Power of hydrogen or potential for hydrogen
PAI: Pozzolanic activity index
RCC: Reinforced cement concrete
RH: Relative humidity
RHA: Rice husk ash
rpm: Rotations per minute
SEM: Scanning electron microscope
SE: Secondary electrons
SCC: Self-compacting concrete
SF: Silica fume
SSA: Specific surface area
SAI: Strength activity index
SCBA: Sugarcane bagasse ash
SMF: Sulphonated melamine formaldehyde
SNF: Sulphonated naphthalene formaldehyde
SCM's: Supplementary cementitious materials
C_4AF: Tetra-calcium alumino ferrites
TGA: Thermogravimetric analysis
C_3A: Tri-calcium aluminates
C_3S: Tri-calcium silicate or alite
UTM: Universal testing machine
VMA: Viscosity modifying admixtures
XRD: X-ray diffraction technique

1 Cement

1.1 INTRODUCTION

Binding materials are extensively used in the construction sector, specifically in concrete. In early age, construction materials were evolved from dry clay. Afterwards, lime was introduced in Egypt as a construction material. Usage of gypsum as a binding material was inadequate due to its limitation with respect to moisture. Fine volcanic ashes were obtained and used in different regions along with lime and sand that not only provided better strength but also contributed higher longevity. A material obtained from Mount Vesuvius named as "Pozzolana" which had an extraordinary binding ability with lime and crushed stone started to harden when it came in contact with water. This process kept on evolving, and around 1824 clayey materials (Argillaceous) that consist of SiO_2, Al_2O_3 and Fe_2O_3 as their major constituent along with limestone (Calcareous), which consists majorly of $CaCO_3$, were burnt at lower temperatures that produced binders of reasonable quality.

John Smeaton planned and constructed Eddystone Lighthouse tower in 1756. During the construction, he discovered that the reactivity of lime, which consists of a high degree of clayey matter, is high. In the long run, such a lime was used along with pozzolana in equal quantities. In the early 1800s, Vicat proposed a method of calcination to achieve high reactivity lime. A mixture of limestone (chalk) and clay was used as the raw material in the calcination process, and this was the principal forerunner to the invention of Ordinary Portland Cement (OPC). Afterwards, Joseph Aspdin attempted further improvement in the process to invent hydraulic cement, which had significant strength attainment, desirable setting properties and durability compared to other binders. Then around 1845, based on different laboratory experiments, the optimum proportion of Argillaceous and Calcareous materials were identified and also higher-temperature calcination was adopted during the process, leading to the best binder.

1.2 CEMENT PRODUCTION

1.2.1 Blending Process of Raw Materials

Limestone and clay are prominent raw materials. Limestone is quarried and crushed to 125 mm size using a primary crusher. Afterwards, it is further ground to 20 mm size with the help of a secondary crusher. Clay is also obtained from a suitable source. Raw materials are ground and blended to achieve better burning in the rotary kiln. Generally, there are two methods for the production of clinker: wet method and dry method. Depending upon the moisture content of the raw material, one among the above methods is preferred. If the moisture content of the raw material is more than 15%, then the wet process is preferred; alternatively, if the moisture content of the raw material is less than 15%, then the dry method is preferred.

1.2.2 Preheating Process

In modern cement plants, a preheater is used to heat the raw materials before the burning process. The preheater is a large tower. After the proper blending of raw materials, the mix is fed into the preheater from the top of the heater. Hot gas from the kiln is sent from the bottom of the preheater to preheat the raw feed. It is interesting to note that a significant amount of calcination can be achieved using this process and also hot gases from the kiln are used effectively.

1.2.3 Burning in a Rotary Kiln

Although cement as a binder works in a highly efficient way for binding different constituents, their production consumes a massive quantity of fuel to de-carbonate $CaCO_3$ (i.e. heating $CaCO_3$ to release CO_2 to form CaO) and to start the sequence of reactions between Argillaceous and Calcareous materials. During the burning process of raw materials in a rotary kiln at higher temperatures, oxides of calcium, silicates and aluminates fuse to form a mixture of special crystalline components called as Bouge's compounds. These components on later stages whenever it comes in contact with water start to set and harden in a controlled manner. The mixture of the crystals consists of tri-calcium silicate in its impure form (alite) C_3S, di-calcium silicate (belite) C_2S, fine crystals of tri-calcium aluminates C_3A and ferrite C_4AF. The overview of the reactions that take place during the process of clinker formation is as follows.

The whole process of clinker production works in broader ranges of temperature.

Water is evaporated, and microstructural change in crystal takes place due to the activation of silicates in argillaceous material. The above two steps take place up to 700°C.

This is followed by de-carbonation of limestone, $CaCO_3$. Combination of CaO with activated silica, alumina and iron oxides takes place in the temperature range of 700°C to 900°C.

Further, the reaction between activated silica and CaO forms belite (C_2S). Belite is di-calcium silicate in its mixed form with 5% of oxides such as Al_2O_3, Fe_2O_3, MgO, SO_3, Na_2O, K_2O, TiO_2 and P_2O_5. This whole reaction takes place in the range of 900–1200°C.

Beyond 1250°C, liquid phases that are formed promote a reaction between free lime (CaO) and belite (an impure form of C_2S) to form alite (C_3S).

The form which consists only of Argillaceous and Calcareous materials is called clinker. Clinker tends to react very quickly in the presence of moisture. This reaction leads to rapid hardening and setting. Then a sequence of reactions takes place between the Argillaceous and Calcareous materials at higher temperatures in a rotary kiln. Further, the percentage variation in the formed Bouge's component can be determined.

1.2.4 Grinding Using a Ball Mill

In order to regulate setting, gypsum is added as an additive during the process of grinding. Gypsum is also called a set regulator. Clinker and approximately 4–5% of Gypsum are ground using a ball mill to obtain the required level of fineness. Large steel balls are used for crushing the clinker. Two different ranges of shots are taken (60–80 mm large balls and 15–40 mm medium-sized balls).

1.2.5 Storage and Packing

After the grinding process, the cement is stored in large silos. Based on order and demand, it is packed and dispatched to the dealers, RMC plants and other construction sites.

1.3 SUPPLEMENTARY CEMENTITIOUS MATERIALS

1.3.1 Need of Pozzolans

Even though cement is considered an excellent binder, clinker production of cement requires a higher temperature to be supplied. Subsequently, clinker production not only consumes a higher amount of energy but also releases a tremendous amount of CO_2 to the atmosphere. Moreover, the use of concrete is increasing rapidly, as concrete is the second most used material in the world after water. Therefore, the release of CO_2 needs to be substantially reduced, taking the environment and global warming as a concern. This leads to intensive search and research for the replacement and higher level of substitution of cement by other alternative materials. These materials that are basically used for the replacement of cement in concrete are called supplementary cementitious materials (SCMs) or pozzolanic materials. By-products obtained from other industries can be used as a replacement material for cement.

1.3.2 Pozzolans and Pozzolanic Reaction

Mineral admixtures also called supplementary cementitious materials (SCMs) can be used as a replacement for cement because of their capability to act as pozzolana. Cement is used as a binder because of its cementitious ability. However, during hydration of cement a major component, $Ca(OH)_2$, is formed. In the presence of moisture, certain materials that possess reactive silicates and aluminates combine with calcium in $Ca(OH)_2$ and form calcium silicate hydrate (C–S–H) or calcium-aluminate-hydrate (C–A–H). Those materials that possess this nature are called pozzolanas, and the reaction is called a pozzolanic reaction. This compound does not possess cementitious nature on its own; nonetheless, in contact with moisture and $Ca(OH)_2$ it forms C–S–H and C–A–H. For instance, the conversion of $Ca(OH)_2$ to C–S–H and C–A–H enhances the resistance against permeability, strength, durability and reduces the heat of hydration.

Pozzolanic materials can be defined as materials that constitute silicates and aluminates as their principal constituents. As and when they combine with compounds of lime in the presence of moisture and at the ordinary temperature they result in the formation of insoluble stable compounds that possess the binding ability. The pozzolanic reaction is expressed as given below:

$$Ca(OH)_2 + \text{Reactive } SiO_2 (\text{or } Al_2O_3) + H_2O \rightarrow \text{C–S–H} \quad (\text{or C–A–H})$$

1.3.3 Pozzolanic Activity Index

In order to identify a suitable material as SCMs, the reactivity of that particular material needs to be assessed, which can be given in terms of an activity index called as **pozzolanic activity index (PAI).**

$$PAI(\%) = \left(\frac{\textit{Strength of OPC mortar}}{\textit{Strength of OPC-Pozzolana mixture}} \right) \times 100$$

1.3.4 Pozzolanic Material

Percentage replacement of different SCMs is primarily based upon their pozzolanic activity and their efficacy towards usage, especially in terms of economy. Different SCMs were used as a replacement for cement in concrete. Some of the commonly used pozzolans are

- Fly ash (FA)
- Ground granulated blast furnace slag (GGBS)
- Silica fume (SF)
- Rice Husk Ash (RHA)
- Meta-kaolin (Mk)
- Sugarcane bagasse ash (SCBA)

The ultimate aim of this replacement is to manufacture a sustainable binding material for the construction. This can be achieved by choosing a material that has relevant and required elemental compounds at the optimum level.

The fundamental reason for the utilisation of a particular by-product as a replacement for cement in concrete is its pozzolanic activity. The extent to which a specific by-product consists of argillaceous materials (i.e. SiO_2, Al_2O_3) in an active state decides its usage and extent of replacement in cement. The present century has led to the introduction and usage of different by-products from various industries, such as granulated blast furnace slag (GGBS) from the iron industry that has a higher amount of CaO and SiO_2, which are similar to the components present in cement. As a result, practically, GGBS can be replaced to a maximum extent of 70% and the preferred optimum level of replacement of GGBS is 50%.

Similarly, fly ash (FA) is considered to be one of the excellent pozzolanic materials obtained during the process of burning different types of coals. Based on the nature of the source and the type of process involved during the process of burning, the chemical composition of the FA varies considerably. There are two types of FA: class F fly ash and class C fly ash. Class F fly ash is pozzolanic in nature; it is mainly obtained from harder and older coals, such as bitumen and anthracite, whereas class C Fly ash is obtained from lignite and sub-bituminous. The underlying reason behind the pozzolanic activity and self-cementing activity of class F and class C fly ash, respectively, is due to the percentage variation in their $\frac{CaO}{SiO_2}$ ratio. In the meantime, the presence of SiO_2 can make them more pozzolanic, but they require some binding agents such as CaO, which tends to bring about dissolution and precipitation in the early stage of the process. If the $\frac{CaO}{SiO_2}$ ratio is low, then the relative presence of SiO_2 is high as compared to CaO and this may lead to high pozzolanic reactions; this is the case of class F fly ash, where the relative presence of SiO_2 is high. Nonetheless, in the case of class C fly ash, along with pozzolanic activity, it possesses a considerable self-cementing property. In general, class F fly ash has CaO less than 7%, which is very meagre when compared to class C fly ash, in which CaO is more than 20%. Generally, 30% mass replacement of cement using fly ash is used in many applications. Silica fume is also used as one among pozzolana material due to its high SiO_2 content. Although obtained as amorphous polymorph from ferrosilicon alloy production, the replacement level of silica fume is very less when compared to other by-products.

1.3.5 Chemical Composition of Cement and SCMs

In order to identify the suitability and utility of a particular alternative material as a replacement for cement during the process of concrete manufacturing, it is necessary to have a scientific insight into the variation of the chemical composition of cement as well as that of SCMs. Moreover, chemical composition has a significant influence on hydration and also on fresh and hardened properties of concrete. In addition to major oxide compounds such as lime, silica and alumina, some other elements such as iron, sulphur trioxide and alkalis are present in cement and SCMs. Table 1.1 depicts the major oxide phases present and their functions during the process of hydration.

TABLE 1.1
Oxide Composition and Their Function

Oxide	Influence	Cement (%)	GGBS (%)	Class F Fly ash (%)	Silica Fume (%)
CaO	Strength and soundness control	60–65	30–50	1–12	< 1
SiO_2	Strength attainment	17–25	28–38	20–60	85–97
Al_2O_3	Setting	3–8	8–24	5–35	–
Fe_2O_3	Gives colour	0.5–6	–	10–40	–

TABLE 1.2
Composition and Functions of Bouge's Components

Component	Function	Composition in cement (%)
Tri-calcium silicate or alite (C_3S)	Cementing and early age strength	55
Di-calcium silicate or belite (C_2S)	Later strength; slow hydration; low heat of hydration	20–25
Tri-calcium aluminate (C_3A)	Initial set; High heat of hydration; Greater affinity towards volume changes causing cracking	5–10 (normally 10.5% of cement)
Tetra-calcium alumino-ferrite (C_4AF)	Less heat	8

The above oxide components, along with some minor oxides and alkalis, are responsible for the development of Bouge's component after the clinkering process. Bouge's component can be engineered to get a definite proportion that can change the type of cement based on applicability and usage. Bouge's components, their composition and their action on hydration are detailed in Table 1.2.

1.3.6 Advantages of Pozzolanic Reaction

The pozzolanic reaction has several advantages because of the changes brought about by the reaction. Some of them are given below.

Compared to ordinary cement concrete, the strength of concrete increases due to the additional formation of C–S–H due to consumption of $Ca(OH)_2$.

Formation of C–S–H refines pores and the interfacial transition zone (ITZ) found between the transitions of aggregates and cement matrix.

The heat of hydration is reduced significantly due to the low initial reaction.

$Ca(OH)_2$, considered vulnerable because of its reaction with the environment in the presence of moisture, is reduced in pozzolanic reaction. This results in reduced permeability and increased durability.

Apart from these technical advantages, the consumption of SCMs is sustainable. Subsequently, usage of SCMs reduces the release of CO_2 produced during the production of OPC that rapidly consumes limited natural resources.

1.3.7 Salient Features of Pozzolanic Materials

Usage of a particular pozzolana as a replacement material is based on its reaction in the activity index test. Some of the materials used as pozzolanas are fly ash (FA), ground granulated blast furnace slag (GGBS) and silica fume (SF). Other pozzolanic materials are sugarcane bagasse ash (SCBA), rice husk ash (RHA) and meta-kaolin (Mk).

1.3.7.1 Fly Ash

Usage of fly ash as an SCM gives twofold benefits by reducing disposal problems and enhancing the properties of concrete. Fly ash can be classified as class F and class C. Class F fly ash can act only as pozzolana, while class C fly ash can act as pozzolana as well as a binder on its own due to relatively higher amount of alkalis and sulphates. As fly ash is the by-product of the coal industry, the above variation in the chemical composition of fly ash is mostly due to variation in the source of coal. Electrostatic precipitator and bag-house precipitator are two methods through which fly ash is generally collected.

The microstructure of fly ash shows that fly ash is a solid spherical particle. A set of hollow spherical fly ash particles with entrapped gas in it is called cenospheres. Another set of hollow spherical fly ash particles with solid particles inside them are called plerospheres.

Fresh concrete properties such as viscosity and yield stress are lowered by the usage of fly ash as a replacement for cement. Since they are spherical in nature, they can slide among each other during the process of casting of concrete, an effect termed as a ball-bearing effect that results in the easy flow and higher workability that in turn reduces segregation and bleeding. Hardened concrete properties, such as the chloride diffusion rate, are reduced compared to OPC. Further, the reduced heat of hydration results in less thermal cracking. Even though early strength gain of fly ash concrete is less, the ultimate strength of concrete with fly ash is more.

1.3.7.2 Ground Granulated Blast Furnace Slag

GGBS can act as both pozzolana and a binder, so they are generally called cementitious and pozzolanic materials. Obtained as a by-product of the iron-ore industry, their influence on the fresh concrete property is less, except for a small delay in initial setting time. The ultimate strength of slag cement is more when compared to OPC, and the durability of concrete is also increased by their replacement.

1.3.7.3 Silica Fume

Silica fume is obtained as a by-product from the silicon or silicon alloy industry. Silica fume contains a high amount of amorphous silica (90% and above) as a major oxide composition. Consequently, their higher fineness and high powder volume result in more water demand, leading to early loss of workability and higher consequences, especially during the early age of concrete, such as rapid slump loss. Moreover, a significant reduction in bleeding can be achieved.

Handling and transportation of silica fume are tedious because of the bulk in volume. To achieve an effective way of handling, silica fume is supplied in three different forms: (1) Bulk powder (tedious to transport and handle), (2) Dry densified silica fume (relatively easy to transport and handle) and (3) Slurry (50% water + 47% Silica fume + 3% chemical agents, occupies less storage place).

Usage of silica fume significantly enhances hardened concrete properties, a substantial increase in initial strength due to high pozzolanic reactivity as well as micro filling effect. Micro filling effect is due to the relatively smaller particle size of silica fume compared to other SCMs, which reduces permeability and increases the strength of ITZ by enhancing the interface between aggregates and cement matrix. Flexural strength and resistance to chemical attack are found to be increased. Silica fume further reduces the rate of corrosion, although plastic shrinkage is another concern in concrete blended with silica fume.

1.3.7.4 Meta-kaolin

Meta-kaolin is produced by calcination of kaolinite at 740–840°C. Calcination of kaolinite results in loss of bound water and crystallisation. Even though meta-kaolin can be used as an excellent mineral admixture, the production process of meta-kaolin makes it less economical. It means that fly ash, slag and silica fume are by-products, whereas meta-kaolin needs to be produced.

1.4 HYDRATION OF CEMENT

Hydration of cement is a complex process. As and when cement comes in contact with water, it starts to dissolve and precipitate due to supersaturation of the solution by different ions that are released from different Bouge's components. Experimental investigations and numerous studies confirm that the mechanism of cement hydration is a two-step process that involves solution hydration followed by solid-state hydration reaction.

1.4.1 Through Solution Hydration

Through the solution, hydration reaction involves the dissolution of anhydrous components in the presence of water, resulting in their corresponding ionic constituents. These ionic constituents, due to the increase in their relative concentrations, result in supersaturation of the solution. After supersaturation, these ionic constituents combine to form their respective hydrates. Subsequently, these hydrates formed are less soluble in nature, which leads to eventual precipitation. Further, precipitation leads to a decrease in the saturation level of particular ionic constituents that tend to be re-released by the dissolution of a specific component and this cycle of dissolution, supersaturation, the formation of hydrates and the eventual precipitation of formed hydrates due to their low solubility goes on. This initial state of hydration is dominant until there is no restriction on the mobility of the ions in the solution.

1.4.2 Solid-State Hydration

In later stages, the restriction of the mobility of the ions in the solution leads to reactions that take place directly on the surface of the anhydrous cement. This especially takes place on residual cement particles. The study of hydration of cement is imperative for the reason that it is the main constituent of the concrete that binds all the constituents together. Further, the nature of their binding influences the properties of the concrete. Hydration of any component involves the absorption of water by the compound, resulting in the dissolution of the component. Subsequently, after the process of dissolution, the solution supersaturates with respect to the particular element, supersaturation results in the gaining of enough amount of energy to form new seeds on the surface of the components already present in the solution. This seed formation is called nucleation. When these seeds are formed, they tend to grow in size and start to aggregate, forming new products from the old components. A similar kind of process takes place during the hydration of cement.

Hydration of cement is a complex process, as cement consists of different components. Consequently, the hydration rate is controlled by different significant parameters other than state parameters such as temperature and concentration. Some of the most critical parameters that control hydration are cement fineness, crystal defects in various components of the cement obtained from kiln and surface constituents. As mentioned earlier in previous sections, the principal constituent of cement is tri-calcium silicate (C_3S); therefore, hydration of C_3S is of utmost importance. That can also act as a good model for cement reaction with water. In a similar way, hydration of C_2S is also identical to C_3S. The reaction of C_3S and C_2S with water is schematically given in Equations 1.1 and 1.2, respectively. Tri-calcium silicate and di-calcium silicate react with water and form calcium silicate hydrate (C–S–H) and calcium hydroxide (CH). CH is generally developed as pore solutions. The formation of CH plays a vital role in the case of the usage of mineral admixtures such as class F fly ash, as reactive silicates in the case of a fly ash–like material tend to react with CH and fill the pores formed by this CH.

$$2C_3S + 6H_2O \rightarrow C_3S_2H_3 + 3CH \tag{1.1}$$

$$2C_2S + 4H_2O \rightarrow C_3S_2H_3 + CH \tag{1.2}$$

Equations 1.1 and 1.2 have significant implications – 2 moles of C_3S requires 6 moles of H_2O for the formation of $C_3S_2H_3$, and it forms 3 moles of CH. Similarly, the equivalent mole of C_2S requires 2 moles less, although producing similar moles of $C_3S_2H_3$ and a further reduced mole of CH. Reducing the mole of CH has different advantages.

Similar to tri-calcium and di-calcium silicates' reaction with water, aluminates react with water. In fact, C_3A is the first component to react with water in the cement system due to its higher reactivity. In order to restrict this, gypsum ($CaSO_4{\cdot}2H_2O$) is added, as mentioned in the earlier section. The reaction of C_3A without gypsum and with gypsum is given by Equations 1.3 and 1.4, respectively.

$$2C_3A + 21H_2O \rightarrow C_3AH_{13} + C_2AH_8\left(\text{without gypsum}\right) \tag{1.3}$$

$$C_3A + 3C\bar{S}H_2 \rightarrow C_6A\bar{S}_3H_{32}\left(\text{with gypsum-formation of ettringite}\right) \tag{1.4}$$

As stated previously, the initial hydration reaction takes place due to the ability of the ions to move in the solution. As, the reaction proceeds, it leads to the formation of hydrated products. Further, the reaction takes place on the surface of the formed hydrated products by diffusion. During any stage of hydration, hydration products consist of fine-grained products with a collectively larger surface area. Hydration of different components in cement leads to different effects. Hydration of aluminates

TABLE 1.3
Hydration Products, Their Shapes and Some Useful Implications

Hydrated products	Crystal shape/ amorphous form	Useful implication from the study
Ettringite (Tri-calcium aluminate + gypsum)	Prismatic needles	High rate of reaction; formed with high heat liberation; high rate of reaction; setting
Mono sulpho aluminates	Platelet crystals	Formed during later stages due to conversion of ettringite to mono-sulphate
C–S–H (Tri-calcium silicate + water)	Gel-like	Moderate rate of reaction; high heat liberation; high strength
C–S–H (di-calcium silicate + water)	Gel-like	Deficient reaction; low heat liberation; later strength
$Ca(OH)_2$ or CH (Portlandite)	Large plane hexagonal crystal	–

leads to the setting process, while hydration with silicate leads to strength development. The initial stage of the reaction of aluminates with gypsum leads to the formation of calcium sulpho aluminates (ettringite), while the later stage reaction leads to the formation of calcium mono sulpho aluminate. Each crystal formed during the process of hydration looks unique when observed through a scanning electron microscope (SEM). Table 1.3 gives different products formed during the process of hydration, their crystal shapes and some implications that are observed in regular studies.

Hydration of cement is an exothermic process that results in the liberation of heat. The process of hydration can be divided into different periods based on heat liberation. The initial reaction of cement with water releases a high amount of heat due to the wetting of cement. Afterwards, highly reactive components such as C_3A and C_3S liberate a considerable amount of heat, which can be measured by calorimetric techniques. Based on the evolution of heat, there are four stages during the hydration of cement.

- Stage 1. Rapid heat evolution due to wetting is known as the heat of wetting. Generally, it lasts for the first 15 minutes.
- Stage 2. This stage lasts for 2–4 hours. This stage is called a dormant or induction period.
- Stage 3. In this stage rapid hydration of C_3S accelerates the reaction and the heat curve reaches its peak generally between 8 and 10 hours. Also, after 48 hours of the final set, the hardening process is started.
- Stage 4. Reaction slows down after the acceleration period (i.e. after stage 3). Steady-state is reached in 12–24 hours.
- Stage 5. Steady-state of heat evolution.

1.5 TESTING OF CEMENT

As described in previous sections, testing of cement is of utmost importance before using it as a binding material in any kind of construction activity. Subsequently, cement and supplementary cementitious materials are chemically active, which brings about significant changes in the system and can influence the process of hydration. The dissolution process that starts as and when cement comes in contact with water can be measured and quantified by the heat evolved. As already stated, the reaction of water with cement is an exothermic reaction. This reaction can be measured by means of a calorimeter that can be isothermal, adiabatic or semi-adiabatic. Comparative results obtained by measuring evolved heat for different samples can effectively show the parameters that need to be taken care of during this process. This unit covers some of the general tests that are usually performed in a laboratory in order to identify the quality of cement. Some of the properties conducted with respect to practical relevance in the field are fineness, consistency, soundness and strength. Even though these tests are not directly related to the heat of hydration, the rate of reaction defines the degree at which cement products are formed. This is a direct consequence that is related to how consistent the material is with respect to time, which helps in altering the workable nature of the concrete, how fast the concrete is gaining its strength and how durable the concrete is against environmental actions. Recommendations are given in IS 4031 for conducting and evaluating certain physical properties of the cement. Some of the major tests conducted are as follows.

1.5.1 Fineness

The value of the cement as a binder can be improved by reducing the mean size of grains by grinding. The increase in its cementing nature can be due to two reasons: first, as cement becomes finer in size it results in good coverage of sand and inert material grains than can be achieved with coarser grains. Second, the later stage of hydration in cementitious suspension is a result of the action of water on the surface of hydrated products directly. This process can be increased only if the relative number of grains increases with respect to its mass. Coarser particles have less number of grains when compared to finer particles which are similar in mass. Consequently, the measurement of the mean size of grains is of utmost importance. This can be performed by three methods: (1) Air permeability method – Blaine permeability method; (2) Sieve method (using 90 μm sieve); and (3) Sedimentation method – Wagner Turbidity method.

1.5.2 Consistency

Consistency defines the viscous nature of the cement paste. Consistency is determined by Vicat's apparatus. When water is mixed with cement, cement paste of certain viscosity is formed. Normal consistency is defined as the amount of water required to form a cement paste through which Vicat's plunger can penetrate to a point, where measured penetration is 5–7 mm from the bottom of the Vicat's mould.

1.5.3 Soundness

Soundness is the undesirable expansion that cement can undergo after gauging it with water. The study of the soundness of cement is vital since the expansion of cement after setting may result in disruption of the hardened concrete, ultimately leading to cracks in the constructed facility. Expansion of cement is mostly due to slow hydration of certain cement constituents present. Unsoundness in cement can be due to excess amount of lime, which may react with oxides of acid, excess amount of magnesia and an excess amount of sulphates. Since, unsoundness of cement cannot be noticed until a considerable amount of time, accelerated tests are required. One of the most crucial tests is done by using Le Chatelier's apparatus.

1.5.4 Compressive Strength

Cement is used in concrete that binds together different constituents and gains the assured designated strength. The gain in strength is again related to the type of cement used in the concrete. Therefore, strength gain needs to be tested in terms of compressive strength and tensile strength. Strength is tested by casting test specimens of standard size as prescribed by IS: 4031 (Part 6). Compressive strength can be tested by a compression testing machine and tensile strength can be identified by the Briquette test or by split tensile strength test.

1.6 TESTING PROCEDURES

1.6.1 Consistency of Cement

This test is used to determine the consistency of cement.

Test summary

- Normal consistency of cement is defined as the percentage of water which produces a consistency that permits a plunger of 10 mm diameter to penetrate up to a depth of 5–7 mm above the bottom of Vicat's mould.
- This test method is used to determine the consistency of cement and SCMs by Vicat's apparatus, according to IS: 4031 (Part 4) (IS 4031 Part 4 2014).

Apparatus Required

- Vicat's apparatus
- 10 mm plunger
- Trowel

Procedure

Step 1. Adjust the needle in Vicat's apparatus to show zero reading when the plunger touches the non-porous surface below, as shown in Figures 1.1 and 1.2.

Step 2. Take 400 grams of cement in a dry tray.

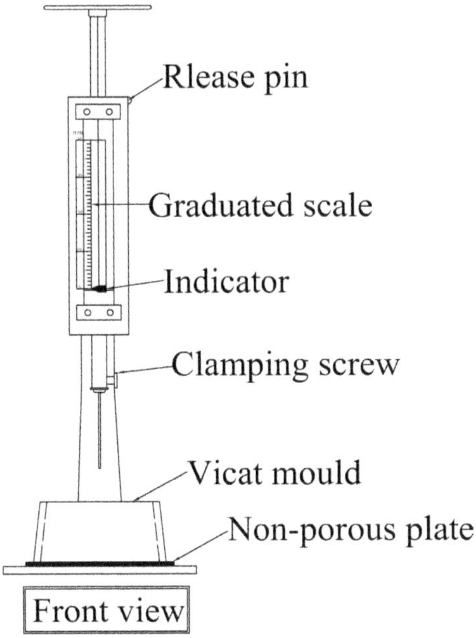

FIGURE 1.1 Front view of Vicat's apparatus.

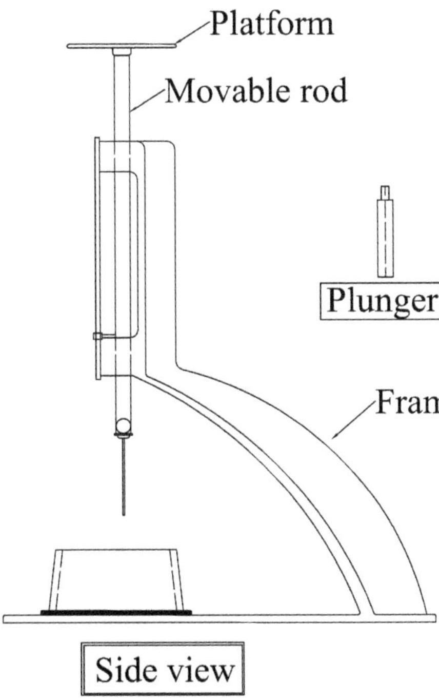

FIGURE 1.2 Side view of Vicat's apparatus.

Step 3. Measure clean water by weight of the cement (approximately start with 25% and increase depending on the given cement). Start the stopwatch when water is mixed with the cement.

Step 4. After mixing thoroughly, cement is filled in the Vicat's mould placed on a glass plate.

Step 5. Tamp the mould gently to expel out any entrapped air and level the surface with a trowel. Place the prepared mould under the plunger.

Step 6. The screw is adjusted to lower the plunger until the plunger is just in contact with the top surface of the cement paste. Quickly release the plunger to allow it to sink.

Step 7. Measure the reading from the scale; it represents the penetration of plunger inside the prepared cement paste. For instance, if the penetration reading is 5 then it represents the plunger is 5 mm from the bottom of the Vicat's mould. In case the plunger reading is more than 5–7 mm, extra water is added. If the reading is less, then reduce the water content.

Step 8. The above operation of mixing, filling and levelling is repeated until a reading of 5–7 mm is obtained.

Step 9. The entire operation of mixing and filling of the mould including testing should be completed within 3–5 minutes from the moment the water is added to the sample.

$$\text{Normal consistency of cement, } P = \frac{\text{Weight of water}}{\text{Weight of cement}} \times 100$$

Report

Normal consistency of cement (P) is ....

Student Remark

....
....
....
....

$$\frac{\textbf{\textit{Marks obtained}}}{\textbf{\textit{Total marks}}} = -$$

Instructor signature

Instructor Remark

....
....
....
....

1.6.2 Initial and Final Setting Time

This test is used to determine the setting time of cement.

Test Summary

Initial setting time

- The initial setting time of cement is defined as the period between the time when the water is added to the cement and the time at which the needle of 1 mm^2 section fails to pierce the test block to a depth of 5 ± 0.5 mm from the bottom of the mould.

Final setting time

- Final setting time is defined as the period between the time when the water is added to the cement and the time at which the needle (annular attachment of 5 mm diameter) makes an impression on the test block while the attachment fails to make it.
- This test method is used for cement and SCMs by Vicat's apparatus, according to IS: 4031 (Part 5) (IS 4031 Part 5 2014).

Use and Significance

- The importance of this property of the cement concrete is that the mortar should be mould in its final position before initial setting time and no movement should be made until the final setting.
- For instance, time is needed to transport the concrete, and immediate setting after mixing with water is not preferred.
- On the other hand, the removal of formwork helps to use it for the casting of other concrete elements and final setting should not extend more than the desirable limits.

Apparatus Required

- Vicat's apparatus
- A needle of 1 mm^2
- Annular attachment of 5 mm diameter
- Trowel

Procedure

Initial and final setting time of cement

Step 1. Setting refers to the phenomenon of the start of chemical reaction once cement comes in contact with water and the loss of plasticity of cement paste. It is also referred to as stiffening.

Step 2. Initial setting time of cement is defined as the period between the time when the water is added to the cement and the time at which the needle of 1 mm^2 section fails to pierce the test block to a depth of 5 ± 0.5 mm from the bottom of mould.

Step 3. Take 400 grams of cement and place it on a dry tray.

Step 4. Measure a quantity of water equal to 0.85 P, where P is the normal consistency of cement.

Step 5. A needle of 1 mm^2 is fixed in the Vicat's apparatus. Adjust the needle indicator to zero mark when the tip of the needle just touches the glass plate.

Step 6. Mix the measured quantity of water with the cement and simultaneously note down the time. Fill the Vicat's mould with the prepared cement paste. Afterwards, expel the air by properly placing cement paste and level the top surface of cement paste.

Step 7. The test block is placed under the needle, as shown in Figure 1.3. Now, the needle is lowered until it just touches the surface of the cement paste. The needle is gently released and allowed to penetrate into the test block.

Step 8. Repeat the test until the needle penetrates 5 ± 0.5 mm from the bottom of the mould. When the needle has reached this penetration, note the time.

Step 9. Time for mixing and filling of the mould should be between 3 and 5 minutes.

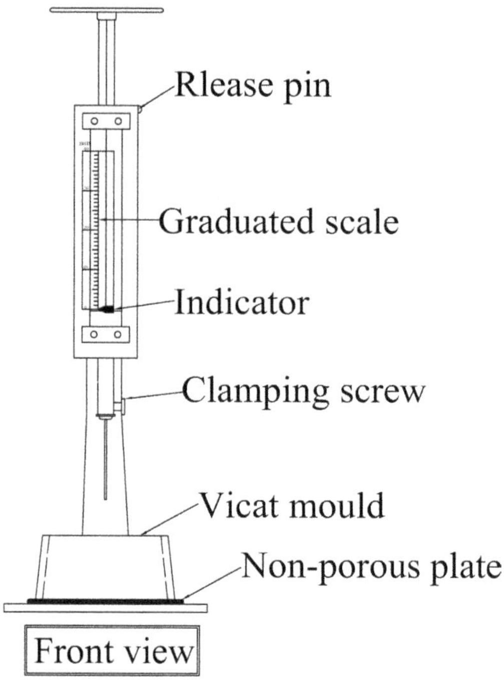

FIGURE 1.3 Front view of Vicat's apparatus.

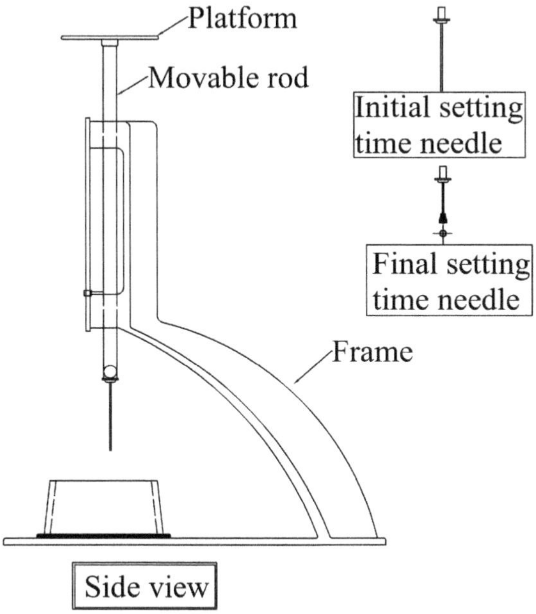

FIGURE 1.4 Side view of Vicat's apparatus with initial and final setting time needles.

Step 10. The final setting time is determined on a similar cement block. In this test the 1 mm^2 needle is replaced by a needle with an annular attachment of 5 mm diameter.

Step 11. Final setting time is defined as the period between the time when the water is added to the cement and the time at which the needle makes an impression on the test block while the attachment fails to make it.

Step 12. Initial and final setting times are reported and rounded off to the next 5 minutes.

Step 13. Final setting time indicates the complete loss of plasticity of cement paste. Vicat's apparatus with initial and final setting time needles is shown in Figure 1.4.

Report

Initial setting time ....

Final setting time ....

Student Remark

....

....

....

....

$$\frac{\textit{Marks obtained}}{\textit{Total marks}} = -$$

Instructor signature

Instructor Remark

....
....
....
....

1.6.3 Soundness

Soundness test is used for checking the quality of the cement.

Test Summary

This test method is used to determine the soundness of cement according to IS: 4031 (Part 3)(IS 4031 Part 3 2014).

Use and Significance

- When cement is mixed with water, hydration of cement and formation of initial hydrated products occur. Moreover, this reaction subsequently leads to setting/stiffening.
- After it sets and hardens, it should not undergo significant changes in volume, particularly under restraint conditions, the amount of expansion should be limited.

Required Apparatus

- Le Chatelier's apparatus
- Trowels
- Enamel tray
- Measuring jar

Notes

- Expansion may take place due to slow hydration or delayed hydration of certain components.
- Reaction of some components such as free lime, magnesia and calcium sulphate after hardening results in expansion.

Procedure

Step 1. Measure 100 grams of cement sample and place it on an enamel tray.

Step 2. Now measure 0.78 P of clean water in a graduated measuring jar, where P is the normal consistency of the cement sample.

Step 3. The stopwatch is started once the water comes in contact with cement. Mixing is completed within 3–5 minutes until cement paste of uniform consistency is formed.

Step 4. Fill Le Chatelier's mould with cement paste which is placed in a bottom glass plate, as shown in Figures 1.5 and 1.6. After filling the mould with cement paste, cover the top of the mould with another glass plate. Moreover, place a 50-gram weight above the top glass plate and tie up the whole assembly with a thread.

Step 5. Measure the distance between the tips of the two indicators. A similar procedure is repeated for other samples. All samples are then placed under water at a temperature of 27–32°C for 24 hours, during which slacking of lime and magnesia takes place.

Step 6. At the end of 24 hours, heat the water with the assembly submerged at the rate so that the water starts boiling within 25–30 minutes. This is done to accelerate the slacking of magnesia with water.

Step 7. The boiling of water is continued for 3 hours. At the end of 3 hours, remove the assembly from the heat source and keep it to cool to room temperature. Take out the samples from the water and measure the distance between the two indicator ends.

Step 8. Initial and final readings of the indicator are noted down (i.e. before boiling and after boiling). The final reading is subtracted from the initial reading in order to find the respective expansion in both the samples. Average expansion can be calculated for all the samples and reported as the final expansion of given cement.

Step 9. If the obtained mean value is less than 10 mm, then the cement is said to be sound. In the case of unsound cement, aerated cement test is carried out, which involves spreading out the cement in a 75 mm thick layer at a relative humidity of 50–80%, maintaining a temperature of 27 ± 3°C. A similar procedure is followed, but the aerated cement is kept for seven days and retested.

Step 10. If the cement sample fails to meet the IS requirements even after aeration test, the cement lot may be rejected, especially if the cement needs to be used for water retaining structures.

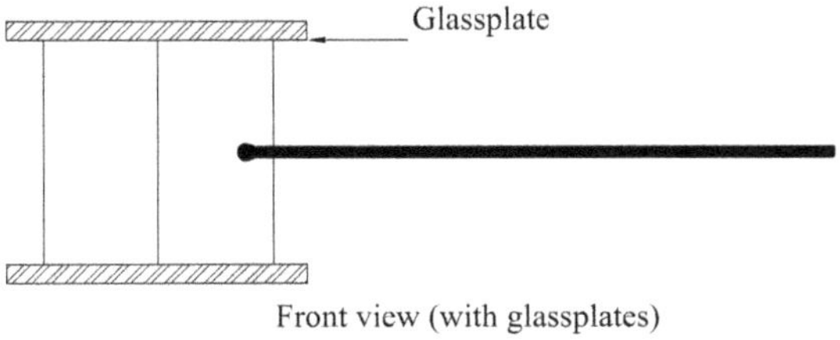

FIGURE 1.5 Le Chatelier's apparatus.

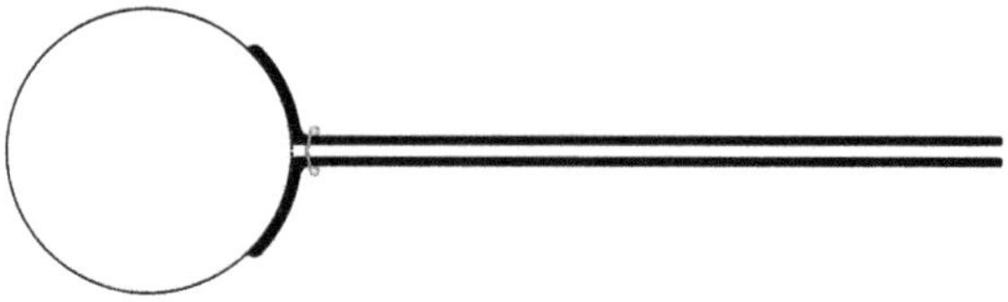

FIGURE 1.6 Le Chatelier's apparatus.

Report

Is the cement (sound/unsound) ....

Student Remark

....
....
....
....

$$\frac{\textbf{\textit{Marks obtained}}}{\textbf{\textit{Total marks}}} = -$$

Instructor signature

Instructor Remark

....
....
....
....

1.6.4 Compressive Strength

Compressive strength test is used for checking the quality of the cement.

Test Summary

This test method is used to determine compressive strength according to IS: 4031 (Part 6) (IS 4031 Part 6 2019).

Use and Significance

- Ordinary Portland Cement (OPC) is manufactured in three different grades, 33 Grade, 43 Grade and 53 Grade, which corresponds to their compressive strength after 28 days of curing per specified standard.
- In the same way, cement used in construction which binds different constituents to form homogeneous mass should be of the required compressive strength to give proper strength to the concrete.

Required Apparatus

- Trowels
- Enamel tray
- Measuring jar
- Standard moulds $70.6 \times 70.6 \times 70.6$ mm
- Compression testing machine

Notes

- Even though different grades of cement are manufactured, all types of cement can be used for the mix proportion of a particular grade of concrete based on the requirement. One of the differences is that the rate at which the cement can gain strength is more for 53-grade cement when compared to other grades.

Caution

- During the preparation of cube specimens for compression testing, vibration given to the cement mortar need to be properly taken care of. Irregularity in vibration may result in air voids that may significantly decrease the compressive strength of cement mortar.
- Moulds need to thoroughly oiled or wetted with a dry cloth in order to ensure or reduce the absorption of water from the mixture by the moulds. Moreover, it helps in easy de-moulding.

Procedure

Step 1. Compressive strength of cement is carried out on cement mortar cube specimens made up of one part of cement and three parts of standard sand.

Step 2. Standard sand comprises three different grades of sand. The particle size of grade 1 sand varies from 2 mm to 1 mm. Grade 2 sand varies in size from 1 mm to 500 μm, and grade 3 sand varies from 500 μm to 90 μm.

Step 3. 200 grams of each grade 1, grade 2 and grade 3 sands are measured and placed on an enamel tray. The total weight of 600 grams of sand is thoroughly mixed. 200 grams of cement is measured and added to the sand. Dry mixing is done thoroughly.

Step 4. Water is taken as $\left(\frac{P}{4}+3\right)$ times the weight of the combined mixture of sand and cement, where P is the normal consistency of the cement. Before adding a measured quantity of water to the dry cement–sand mixture, apply oil to all the interior faces of the mould and also each of the edges that measure 70.6 mm.

Step 5. After oiling inner portions of the mould, including the base plate, place the assembled mould on the standard vibrating machine. Firmly hold the mould by suitable means.

Step 6. Add the measured quantity of water to the dry mixture and simultaneously start a stopwatch. Total time of mixing and filling the mould needs to be within 3–5 minutes from the addition of water.

Step 7. Place the prepared mortar mix in the mould by use of a hopper and vibrate at 1200 ± 400 vibrations per minute. Cement mortar is vibrated for 2 minutes. Ensure that there is no leakage of mortar through the joints of the mould.

Step 8. Remove the mould and place it on a levelled surface and finish the surface with a trowel. Similarly, the other two mortar cubes are placed with cement mortar and levelled.

Step 9. After the elapse of initial setting time of cement (generally after 30 minutes), mortar cubes are covered with wet gunny bags to maintain relative humidity above 90% and a temperature of $27 \pm 2°C$.

Step 10. At the end of 24 hours, the gunny bags are removed and the mould is dismantled with the help of gentle tapping by a hammer without damaging the edges.

Step 11. Mortar specimens are then placed in a water curing tank at $27 \pm 2°C$. At the prescribed day of testing, take the mortar specimens not more than half an hour prior to the testing. Before testing, measure the dimension of the mortar specimens (perpendicular to the loading direction) and record it.

Step 12. Place the mortar specimen under the platens of the compression testing machine. Switch on the machine so that a uniform hydraulic load is applied on the cubes during the test.

Step 13. Once the load reaches a constant value (indicated by a constant dial gauge movement), switch off the machine and note down the crushing load. The same procedure is followed for the other two mortar specimens to know the crushing load of the other two cubes.

Step 14. Compressive strength can be calculated by dividing the crushing load by the area of the specimen. The average compressive strength is reported on respective days.

Step 15. This value can be compared with specified values of different cement according to IS code.

Step 16. For example, compressive strength at three days for OPC (IS: 269) is 160 kg/cm^2 and for PPC (IS: 1489) no standard values are specified. At seven days, compressive strength is 220 kg/cm^2 and 175 kg/cm^2 for OPC and PPC, respectively.

Report

Average compressive strength of cement mortar ....

Standard deviation ....

Student Remark

....
....
....
....

$$\frac{\textit{Marks obtained}}{\textit{Total marks}} = -$$

Instructor signature

Instructor Remark

....
....
....
....

1.6.5 Blaine's Air Permeability Test

This test is used to characterise the surface area of the cement, which describes the fineness of the cement, also called as a specific surface area (SSA). Understanding the specific surface area of the cement is essential to determine its initial reactivity.

Test Summary

This test method is used to determine the specific surface area (SSA) of cement and SCMs using Blaine's air permeability apparatus, according to IS: 4031 (Part 2) (IS 4031 Part 2 2013).

Use and Significance

- Greater the surface area of cement particles, higher will be its contact with water and greater its reactivity.
- The fineness of the cement can be measured by IS 90 µm sieve test and air permeability test.

Required Apparatus

- 90 µm sieve.
- Blaine's air permeability apparatus.

Caution

Care should be taken to maintain the temperature and RH during testing.

Procedure

Fineness of cement by IS 90 µm sieve

Step 1. Measure 200 grams of cement sample and place it in 90 µm sieve with a pan fitted at the bottom. Let it be w_1.

Step 2. Place the lid and sieve it for 15 minutes in clockwise, anti-clockwise, forward and backward directions.

Step 3. The retained weight indicates a particle size larger than 90 μm. Residual cement particles retained after 15 minutes of sieving is measured. Let the weight retained be w_2.

Step 4. The percentage of weight retained can be calculated as $\frac{w_2}{w_1} \times 100$.

Step 5. The percentage of weight retained should be less than 10% for ordinary Portland cement. For rapid hardening cement percentage retained should be less than 5%.

Fineness of cement by Blaine's air permeability apparatus

Step 1. Blaine's air permeability apparatus consists of a permeability cell with a plunger at the top. The diameter of the air permeability cell is 12.5 ± 1.0 mm, and the height is 50 ± 1.5 mm.

Step 2. The bottom portion of the air permeability cell consists of a conical-shaped ledge that helps in connecting it with a U-tube manometer through a rubber coupling.

Step 3. The junction of the top portion and bottom portion (conical-shaped ledge) consists of a perforated metal disc of 1 mm perforation.

Step 4. Filter paper disc of No. 40 Whatman is used in conjunction with the perforated metal disc in the cell. It means that the perforated metal disc and the filter paper are placed one above another. It helps to avoid mixing of the mercury with cement.

Step 5. U-tube manometer consists of a stopcock provided with a pressure bulb with rubber tubing. This pressure bulb can be used to produce air pressure differences between the two arms. The other end of the manometer is open to atmospheric pressure.

Step 6. Before starting the test, perforated metal disc is placed in the air permeability cell and lowered down until it is rested well at the junction by the help of a steel rod.

Step 7. Similar to perforated metal disc, the filter paper disc is placed in the air permeability cell and lowered down so that the filter paper disc covers and rests on the perforated metal disc.

Step 8. Calculated quantity of cement (2.8 grams) is placed in the cell, and the plunger is inserted and compressed until the plunger collar comes in contact with the top of the cell to ensure the required volume and hence the porosity.

Step 9. Permeability cell is connected to the U-tube manometer through rubber coupling. A stopcock is opened and pressure is given by pressure bulb. The stopcock is closed once the liquid level reaches the topmost mark in the manometer, say AA.

Step 10. Once the pressure level just touches the mark BB (intermediate mark), a stopwatch is switched on. Once the liquid level just touches the mark, say CC, the stopwatch is switched off.

Step 11. The time taken for the liquid level to drop from the mark BB to CC is noted in seconds. Let us denote the time taken in seconds as "T".

Step 12. The specific surface area of the cement is determined by the following equation.

$$S = k\sqrt{T}$$

Where constant k in the present case is 486, therefore

$$S = 486\sqrt{T}$$

Where specific surface area S is reported in cm^2/g.

Step 13. A minimum of three observations needs to be taken in order to determine the specific surface area of cement.

Step 14. The value of constant k can be determined by using standard cement whose specific surface area (SSA) is known (3400 cm^2/g).

Step 15. Time is taken (T) for the liquid to rise from the mark BB to the mark CC is generally considered for a standard weight of cement, which is usually 2.8 grams.

Step 16. In general, standard cement has a specific surface area (SSA) of 3400 cm^2/g and time T is measured as 49 seconds.

Step 17. To determine the fineness of any given cement, the following steps are carried out. Fill the cell with mercury (Hg) until the top and level with the help of a plastic plate. Remove the Hg from the cell into a dish. The weight of the dish along with Hg is noted down as W_A.

Step 18. Measure 2.8 grams of cement sample and place it in the cell. Place filter paper disc on the top and insert the plunger until the collar of the plunger touches the top surface of the cell.

Step 19. Remove the plunger and fill the space with Hg until it reaches the top, level the surface of Hg with a plastic plate. Remove Hg from the cell to the same dish that was previously used to weight W_A. The weight of the dish with Hg is noted down as before as W_B.

Step 20. The bulk volume occupied by the cement is then calculated as

$$V = \frac{W_A - W_B}{D}$$

Where, D is the density of Hg 13.53 g/cm^3 at 28°C.

Step 21. The weight of the cement sample required for the test can be calculated as

$$W = \rho V(1-e)$$

Where ρ is the density of the cement = 3.15 g/cm^3, V is the bulk volume of cement, which is equal to 1.86 cm^3, and e is the required porosity of cement bed, which is equal to 0.500 ± 0.005 for OPC and PPC and 0.530 ± 0.005 for early strength cement. Therefore, the weight of cement is

$$W = 3.15 \times 1.86(1 - 0.500) = 2.93\,\text{g}$$

Step 22. Once the weight of cement is calculated, the experiment is performed and SSA is calculated.

Step 23. Blaine's air permeability apparatus along with the cell and the plunger is shown in Figures 1.7 and 1.8.

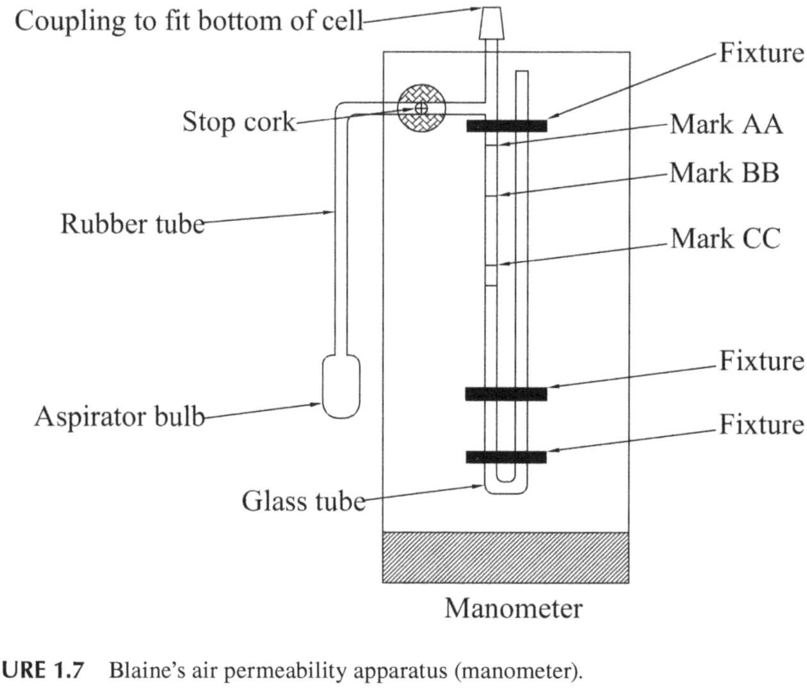

FIGURE 1.7 Blaine's air permeability apparatus (manometer).

Report

Specific surface area of the given sample is ....

Student Remark

....

....

....

....

$$\frac{\textit{Marks obtained}}{\textit{Total marks}} = -$$

Instructor signature

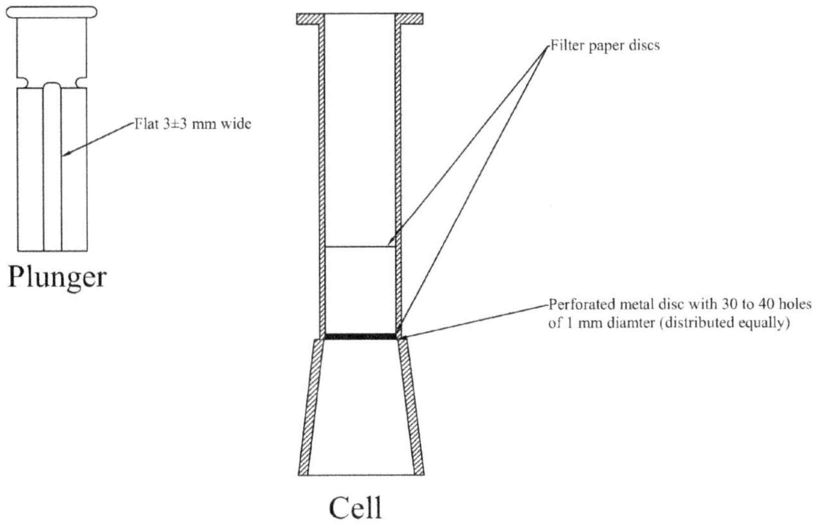

FIGURE 1.8 Blaine's air permeability cell and plunger.

Instructor Remark

....
....
....
....

1.6.6 Strength Activity Index Test for Pozzolanic Material

To reduce the production of energy-intensive ordinary Portland cement, green alternative cementitious materials are used, and these materials can react with the $Ca(OH)_2$ formed during the process of hydration and form additional C–S–H. Materials that possess the ability to form hydration products on reaction with $Ca(OH)_2$ are called as pozzolanic materials.

Test Summary

This test method is used to sample and investigate the reactivity of supplementary cementitious materials for use in Portland cement concrete. Under specific conditions, standard mortar cube specimens are cast and tested according to ASTM: C311M-17(ASTM C311/C311M-18 2019).

Use and Significance

- The reactivity of a material that can be used as a pozzolana can be identified.
- The suitability of using a pozzolana as a blend in Portland pozzolana cement can be identified.

Required Apparatus

- Weighing balance
- Non-absorbent mixing bowl and planetary mixer for mixing mortar
- Flow table with flow mould
- Cube moulds of 50 mm dimension with a non-absorbent base plate
- Scrapper
- Graduated cylinder
- Trowel
- Tamping rod of 12×25 mm cross-section and a length of 125–150 mm
- Compression testing machine of capacity 100 kN

Procedure

Step 1. Control mortar specimens and SCM (supplementary cementitious material) blended specimens are prepared in order to investigate the level of reactivity of a pozzolanic material in terms of strength.

Step 2. Control mortar specimens are prepared by cement: graded sand at a ratio of 1:2.75. As per standard, 500 g of ordinary Portland cement, 1,375 g of Indian standard-graded sand and 242 ml water need to be used to cast six control mortar specimens of size 50×50×50 mm. The amount of water for the control specimen is directly given in the standard. However, the amount of water for blended mortar specimens is decided based on the required flow.

Step 3. The flow of the control mortar specimen is calculated from the flow table experiment.

Flow Table Experiment

Step 4. The mixing bowl is cleaned and dried before starting the mixing procedure. During the preparation of control as well as test specimens, the amount of water required is added first in the mixing bowl.

Step 5. After the addition of water, ordinary Portland cement is added with water and mixed at a speed of 140±5 rpm for 30 seconds.

Step 6. The proportioned quantity of sand is slowly added during the next 30 seconds of mixing. Mixing should be continued at 140±5 rpm.

Step 7. The mixer is stopped and the speed is increased to 285±10 rpm. The mixing is continued for 30 seconds.

***Caution*:** Mixer needs to be stopped while the speed is changed from one speed to the other by a gear system.

Step 8. After mixing for a period of 1 minute and 30 seconds, the mixer is stopped and manual homogenisation is made by scraping the sides of mixing bowl for 15 seconds.

Step 9. Once manual homogenisation is completed, the mixing bowl is closed with the lid and made to stand still for a period of 1 minute and 15 seconds.

Step 10. The final mixing is done for a period of 1 minute at a speed of 285 ± 10 rpm. Once, the final mixing is completed mixing paddle is removed, and scrapped for the removal of mortar.

Step 11. Flow table mould is cleaned, dried and placed at the centre of the flow table.

Step 12. The mixed mortar is placed in the flow mould in layers. Each layer is approximately 25 mm in thickness and tamped 20 times with the help of a tamping rod to spread the mortar uniformly through the mould.

Step 13. The final layer is filled with excess mortar. Once the tamping is completed, a straight edge is used to cut off the excess of mortar to form a plane surface.

Step 14. The top of the table is cleaned and wiped, and care should be taken specially to remove the excess water from the side of the flow mould.

Step 15. The flow mould is lifted after 1 minute of filling the mortar. The flow table is dropped 10 times through a height of 12.5 mm within 6 seconds.

Step 16. The average flow value of the mortar due to the resulting drops is measured by calculating the average diameter at four places equidistant from each other. It means that the diameter is measured from four different directions and the average value is taken. The average flow spread of control mortar is noted as the flow value of the control specimen.

Step 17. Now, 20% of ordinary Portland cement is replaced with the pozzolanic material. Graded sand is taken similar to the control mortar (1375 g).

Step 18. A similar procedure of mixing and flow table testing is followed for each new mix with different amounts of water. The flow value of the control is considered as 100%. The above trials are followed until a flow of $\pm 5\%$ of the control mix is obtained. It means that different trials are adopted until a flow value of 95–105% of the control flow is obtained. The amount of water required to attain this flow is considered as the water for SCM blended mortar specimen preparation.

Step 19. The quantity of material required for casting 6 test SCM blended mortar specimen is calculated (OPC + 20% SCM, 1375 g graded sand, water from flow test). Mixing is done as prescribed in the previous steps of the control mix.

Step 20. Moulds and base plates are coated with low viscous mineral oil or grease. Moulds and base plates are placed on a level surface.

Step 21. The mixed mortar is placed in the moulds in layers. Each layer is tamped for 25 times to ensure that the mixed mortar is spread through the mould, with each layer of thickness 25 mm.

Step 22. The mould is filled to overflow. Once tamping is completed with the help of a straight edge, the moulds are levelled to get a plain and smooth surface.

Step 23. Specimens with the moulds are covered with cover plates and placed under wet gunny bags for 24 hours. Specimens are removed from the moulds and placed under saturated lime water curing to reduce the risk of leaching of $Ca(OH)_2$ by the specimens.

Step 24. The specimen is removed after curing for 7–28 days underwater and tested for compressive strength. Before testing the specimen, ensure that the specimen is cleaned and wiped and the loose material from the surface of the specimen is removed.

***Caution*:** During testing, the specimen is placed in such a way that the surfaces other than the cast face are under examination.

Step 25. The load is applied at a uniform rate of 3.5 MPa/minute until failure. The load at which the failure occurs is noted down.

Step 26. The ratio between the load at failure and the cross-sectional area at which the load is applied is the compressive strength of the specimen. Compressive strength is calculated to the nearest 0.05 MPa. The average strength of three mortar cube specimens is reported as compressive strength.

Step 27. The ratio of the strength of blended mortar specimen (OPC+20% SCM) to the strength of control specimens (only OPC) is represented as strength activity index (SAI) or pozzolanic activity index (PAI).

Observations and Calculations

Step 1. Flow obtained for the control mix is .... mm.

Step 2. Trial mixes for blended mortar specimens to identify the amount of water required for a flow of ± 5% of control mix flow is identified from Table 1.4.

Step 3. The weight of water corresponding to ± 5% of control mix is taken for casting blended test specimens.

Step 4. The compressive strength of control and test specimen can be identified as shown in Table 1.5.

Report

***Caution*:** The fineness of the pozzolana to be used for blending needs to be identified. As reactivity is measured in terms of strength, fineness is one of the major physical properties that influence the reactivity of the pozzolana.

TABLE 1.4
Trial Mixes to Identify the Amount of Water for the Required Flow of ±5% of Control Mix

	Trail 1	Trial 2	Trail 3	Trial 4	Trial 5
Weight of Portland cement (g)					
Weight of pozzolana (g)					
Weight of sand (g)					
Weight of water (g)					
Average flow difference (%)					

TABLE 1.5
Compressive Strength of Specimens

	Control specimen			Test specimen		
Specimen No	1	2	3	1	2	3
Temperature						
Load at failure (kN)						
Stress at failure (MPa)						
Average compressive strength (MPa)						
SAI	[Strength (OPC-pozzolan mix)/Strength of control mix] × 100 =					
Quality of SCM						
Qualified (≥75%)						
Disqualified (less than 75%)						

***Specification*:** According to ASTM: C311M-17, the ratio of compressive strength of the test specimen to that of the control specimen should be more than 75% for the material to be satisfied as a pozzolanic material.

Student Remark

....
....
....
....

$$\frac{\textit{Marks obtained}}{\textit{Total marks}} = -$$

Instructor signature

Instructor Remark

....
....

....
....

1.6.7 Lime Reactivity of Pozzolanic Material

Due to the pozzolanic reaction, additional strength gain and reduction in permeability can be achieved for blended cement concrete. Reactive silica present in the alternative cementitious materials reacts with $Ca(OH)_2$ formed during the process of cement hydration and form additional C–S–H. This reaction is known as pozzolanic reaction and materials that possess the ability to form hydration products on reaction with $Ca(OH)_2$ are called pozzolanic materials.

Test Summary

This test method is used to determine the reactivity in terms of compressive strength of a material that can be used as a pozzolana when mixed with hydraulic lime. Under specific conditions, standard mortar cubes are prepared and tested according to IS: 1727 (IS 1727 2013).

Use and Significance

- The reactivity of a material that can be used as a pozzolana can be identified.
- The suitability of using a pozzolana as a blend in Portland Pozzolana cement can be identified.

Required Apparatus

- Weighing balance
- Non-absorbent mixing bowl and planetary mixer for mixing mortar
- Flow table with flow mould
- Cube moulds of 50 mm dimension with a non-absorbent base plate
- Scrapper
- Graduated cylinder
- Trowel
- Tamping rod of 12×25 mm cross-section and a length of 125–150 mm (non-absorbent)
- Compression testing machine of capacity 100 kN

Procedure

Step 1. To prepare test specimens, dry materials of lime: pozzolana: standard sand is proportioned in the following ratio 1: 2*m*: 9 by weight.

$$m = \frac{\text{Specific gravity of pozzolana}}{\text{Specific gravity of lime}}$$

Step 2. The amount of water required to gauge the specimen is added based on a trial-and-error basis.

Step 3. The quantity of water to be added can be determined in terms of the flow value. Water needed to obtain a flow value of 70 ± 5% (to that of flow table) for 10 drops within 6 seconds is measured based on trials, and it is considered to be the amount of water required to gauge the specimen.

Caution: Each trial mixes should be prepared with new mortar. For the first trial, the amount of water required is taken as 60% of the weight of lime and pozzolana. For subsequent trials, the weight of water is adjusted to obtain a flow of 70 ± 5%.

Step 4. The mixing bowl needs to be cleaned and dried before starting the mixing procedure. During each trial, the amount of water required is added first in the mixing bowl.

Step 5. The calculated amount of lime and pozzolana are added along with water and mixed at a speed of 140 ± 5 rpm for 30 seconds.

Step 6. The proportioned quantity of sand is added slowly during the next 30 seconds of mixing. Mixing should be continued at 140 ± 5 rpm.

Step 7. The mixer is stopped as well as the speed is increased to 285 ± 10 rpm and mixing is continued for 30 seconds.

Caution: The mixer needs to be stopped while the speed is changed from one to another by a gear system.

Step 8. After mixing for a period of 1 minute and 30 seconds, the mixer is stopped and manual homogenisation is made by scraping the sides of the mixing bowl for 15 seconds.

Step 9. Once manual homogenisation is completed, the mixing bowl is closed with the lid and made to stand still for a period of 1 minute and 15 seconds.

Step 10. Final mixing is done for a period of 1 minute at a speed of 285 ± 10 rpm. As soon as, the final mixing is completed mixing paddle is removed and scrapped for the removal of mortar.

Step 11. Flow mould is cleaned, dried and placed in the centre of the flow table.

Step 12. Mixed mortar is placed in the flow mould in four layers. Each layer is of approximately 25 mm thickness and tamped 20 times by the use of a tamping rod to spread the mortar uniformly through the mould.

Step 13. The final layer is filled with excess mortar. Once the tamping is completed in the final layer, a straight edge is used to cut off the excess of mortar to finish the top surface.

Step 14. The top of the table is cleaned and wiped and care should be taken especially to remove the excess water from the side of the flow mould.

Step 15. Flow mould is lifted after 1 minute of filling the mortar. The flow table is dropped 10 times with the height of fall 12.5 mm within 6 seconds.

Step 16. The average flow spread of the mortar due to the resulting drops is measured by evaluating the average diameter in four places equidistant from each other.

Step 17. The increase in flow diameter is expressed as a percentage with respect to the original base diameter of the flow table.

Step 18. A similar procedure is followed for each new mix with different weights of water. The above procedure is repeated until the amount of water required to obtain a flow of $70 \pm 5\%$ is determined.

Step 19. The quantity of material required for casting six cube specimens is calculated. Mixing is done as prescribed in the previous steps.

Step 20. Moulds and base plates are coated with low viscous mineral oil or grease. Moreover, cube moulds and base plates are placed on a level surface.

Step 21. The mixed mortar is placed in the cube moulds in layers. Each layer is tamped for 25 times to ensure that the mixed mortar is spread through the mould, with each layer of thickness 25 mm.

Step 22. The mould is filled to overflow. Once tamping is completed with the help of a straight edge the moulds are levelled to get a plain and smooth surface.

Step 23. The specimens with the moulds are covered with cover plates and placed under wet gunny bags for 48 hours. Specimens are removed from the moulds and cured at $50 \pm 2°C$ at a relative humidity of 90–100% for eight days.

***Note*:** The specimens should not be cured underwater.

Step 24. The specimens are removed after curing for eight days and tested for compressive strength once the specimen reaches room temperature. Before testing the specimen, ensure that the specimen is cleaned and wiped and the loose material from the surface of the specimen is removed.

***Caution*:** During testing, the specimen is placed in such a way that the surface opposite to the surface of the casted faces is under examination.

Step 25. The load is applied at a uniform rate of 3.5 MPa/minute until failure. The load at which the failure occurs is noted down. Average compressive strength of three mortar specimens is reported as lime reactivity.

Step 26. The ratio between the load at failure and the cross-sectional area at which the load is applied is the compressive strength of the specimen. Compressive strength is calculated to the nearest 0.05 MPa.

Observations and Calculations

- Trial mixes to identify the amount of water required for a flow of 70 ± 5% is identified from Table 1.6.
- The compressive strength of the test specimen can be identified as shown in Table 1.7.

TABLE 1.6
Trial Mixes to Identify the Amount of Water for the Required Flow of 70 ± 5%

	Trail 1	Trial 2	Trail 3	Trial 4	Trial 5
Weight of lime (g)					
Weight of pozzolana (g)					
Weight of sand (g)					
Amount of water (%)					
Average flow (%)					

TABLE 1.7
Compressive Strength of Test Specimens

Specimen No	1	2	3	4	5	6
Temperature						
Load at failure (kN)						
Stress at failure (MPa)						
Average compressive strength (MPa)						

Report

***Caution*:** The fineness of the pozzolana to be used for blending needs to be identified. One of the significant physical properties that influence the reactivity of the pozzolana is its fineness.

***Specification*:** According to IS: 1727 compressive strength of the test specimen as well as fineness, the material considered to be pozzolana should have a minimum compressive strength of 4.0 MPa and fineness of at least 320 m^2/kg.

Student Remark

....
....
....
....

$$\frac{\textit{Marks obtained}}{\textit{Total marks}} = -$$

Instructor signature

Instructor Remark

....
....
....
....

1.7 PRACTICE QUESTIONS AND ANSWERS FROM COMPETITIVE EXAMS

1. The fineness of cement is determined by
   a) Blaine's air permeability apparatus
   b) Le Chatelier's apparatus
   c) Vicat's apparatus
   d) None of the above

2. Soundness of cement is determined by
   a) Le Chatelier's apparatus
   b) Vicat's apparatus
   c) Compression testing machine
   d) None of the above

3. If the grade of cement is increased from 33 to 53 grade, what parameters increase?
   a) Strength
   b) Fineness
   c) Heat of hydration
   d) All the above

4. What is the purpose behind the addition of gypsum in cement?
   a) To accelerate the initial setting time
   b) To reduce the heat of hydration
   c) To regulate the setting time
   d) None of the above

5. What happens during the hydration of cement if no gypsum is available in cement?
   a) Flash set
   b) False set
   c) Perfect set
   d) All the above

6. In which of the following condition false set occurs
   a) Excess presence of gypsum
   b) Gypsum in un-hydrated form

c) Gypsum in hydrated form
d) All the above

7. When fineness of cement is higher, what characteristics does the cement possess?
   a) Higher heat of hydration
   b) Lower heat of hydration and higher strength
   c) Higher heat of hydration and high soundness
   d) None of the above

8. During a site test, it was identified that the soil possesses higher content of sulphate. Moreover, the site is located nearby a beach. What kind of cement will you recommend for the construction?
   a) Ordinary Portland cement
   b) Cement with a higher proportion of C_3A
   c) Cement with the higher replacement of slag
   d) None of the above

9. One of the early hydration product formed is $Ca(OH)_2$, which of the following statement (s) is/are correct?
   1. It is a crystalline phase which gives very high cohesion and bonding strength.
   2. It is an amorphous phase and has very low strength.
   3. It is a crystalline phase with very poor strength.
   4. It acts as an entry point for the deterioration.
      a) All the statements are correct
      b) 1 and 2 are correct
      c) 2 and 3 are correct
      d) 3 and 4 are correct
10. Aft (ettringite) phases are formed during the hydration of cement. What are the compounds involved in the formation of ettringite?
    a) C_3S, gypsum and water
    b) C_3A, gypsum and water
    c) C_3S, C_3A and water
    d) All the above

11. The compound which is mainly responsible for initial setting and early strength gain of ordinary Portland cement is
    a) C_3A
    b) C_3S
    c) C_2S
    d) C_4AF

[GATE 2016 FN; 1 mark]

12. Group I gives a list of test methods and test apparatus for evaluating some of the properties of ordinary Portland cement (OPC) and concrete group II gives the list of these properties.

Group I	Group II
P. Le Chatelier test	1. Soundness of OPC
Q. Vee-Bee test	2. Consistency and setting time of OPC
R. Blaine air permeability test	3. Consistency of workability of concrete
S. The Vicat apparatus	4. Fineness of OPC

The correct match of the items in group I with items in group II is

a) P-1, Q-3, R-4, S-2
b) P-2, Q-3, R-1, S-4
c) P-4, Q-2, R-4, S-1
d) P-1, Q-4, R-2, S-3

[GATE 2017 AN; 2 marks]

13. The Le Chatelier apparatus is used to determine
    a) Compressive strength of cement
    b) Fineness of cement
    c) Setting time of cement
    d) Soundness of cement

[GATE 2018 FN; 1 mark]

14. The setting time of cement is determined using
    a) Le Chatelier apparatus
    b) Briquette testing apparatus
    c) Vicat apparatus
    d) Casagrande's apparatus

[GATE 2018 AN; 1 mark]

15. During the process of hydration of cement, due to increase in dicalcium silicate (C_2S) content in the cement clinker, the heat of hydration
    a) Increases
    b) Decreases
    c) Initially decreases and then increases
    d) Does not change

[GATE 2020 FN; 1 mark]

16. Match list I with list II. Select the correct answer using the codes given below the lists:

	List I		List II
A	Fineness of cement	1	Le Chatelier apparatus
B	Setting time	2	Vicat's needle
C	Soundness	3	Air permeability apparatus
D	Workability	4	Slump cone

Codes:

	A	B	C	D
a)	1	2	3	4
b)	3	1	4	2
c)	3	2	1	4
d)	1	4	3	2

[IES 1995]

17. If "p" is the standard consistency of cement, the amount of water used in conducting the initial setting time test on cement is
    a) 0.65 p
    b) 0.85 p
    c) 0.6 p
    d) 0.8 p

[IES 1995]

18. Assertion (A): Pozzolana is added to cement to increase early strength.
    Reason (R): It offers greater resistance to the attack of aggressive water.
    a) Both A and R are true and R is the correct explanation of A.
    b) Both A and R are true and R is not the correct explanation of A.
    c) A is true but R is false.
    d) A is false but R is true.

[IES 1995]

19. For complete hydration of cement the water–cement ratio needed is
    a) Less than 0.25
    b) More than 0.25 but less than 0.35
    c) More than 0.35 but less than 0.45
    d) More than 0.45 but less than 0.60

[IES 1996]

20. Blast furnace slag has approximately
    a) 45% calcium oxide and about 35% silica
    b) 50% alumina and 20% calcium oxide
    c) 25% magnesia and 15% silica
    d) 25% calcium sulphate and 15% alumina

[IES 1996]

21. Gypsum is used as an admixture in cement grouts for
    a) Accelerating the setting time
    b) Retarding the setting time
    c) Increasing the plasticity
    d) Reducing the grout shrinkage

[IES 1996]

22. Which of the following pairs with respect to ordinary Portland cement are correctly matched?
    1. Initial setting time – 30 minutes
    2. Final setting time – 10 hours
    3. Normal consistency – 10%

Select the correct answer using the codes given below:
- a) 1, 2 and 3
- b) 2 and 3
- c) 1 and 2
- d) 1 and 3

[IES 1997, AE 2015 and AE 2018]

23. Consider the following statements:

    High early strength of cement is obtained as a result of
    1. Fine grinding
    2. Decreasing the lime content
    3. Burning at higher temperatures
    4. Increasing the quantity of gypsum

Which of these statements are correct?
- a) 1 and 2
- b) 1 and 3
- c) 2, 3 and 4
- d) 1, 3 and 4

[IES 1997]

24. The temperature range in a cement kiln is
    - a) 500–1000°C
    - b) 1000–1200°C
    - c) 1300–1500°C
    - d) 1600–2000°C

[IES 1998]

25. Before testing setting time of cement, one should test for
    - a) Soundness
    - b) Strength
    - c) Fineness
    - d) Consistency

[IES 1998]

26. Match list I (property of cement) with list II (testing apparatus) and select the correct answer using the codes given below:

	List I		List II
A	Specific gravity	1	Blain's apparatus
B	Setting time	2	Le Chatelier's flask
C	Soundness	3	Compress meter
D	Fineness	4	Autoclave
		5	Vicat's apparatus

Codes:

	A	B	C	D
a)	3	5	1	2
b)	2	5	1	4
c)	2	5	4	1
d)	5	3	4	1

[IES 1999]

27. The role of superplasticiser in a cement paste is to
    a) Disperse the particles
    b) Disperse the particles and to remove air bubbles
    c) Disperse the particles, remove air bubbles and to retard setting
    d) Retard setting

[IES 1999]

28. Consider the following oxides:
    1. Al_2O_3
    2. CaO
    3. SiO_2

The correct sequence in increasing order of their percentage in an ordinary Portland cement is
    a) 2, 1, 3
    b) 1, 3, 2
    c) 3, 1, 2
    d) 1, 2, 3

[IES 1999]

29. Increase in fineness of cement
    a) Reduces the rate of strength development and leads to higher shrinkage
    b) Increases die rate of strength development and reduces the rate of deterioration
    c) Decreases the rate of strength development and increases the bleeding of cement

d) Increases the rate of strength development and leads to higher shrinkage

[IES 1999]

30. Consider the following statements:
    1. Masonry in rich cement mortar though having good strength with high shrinkage is much liable for surface cracks.
    2. Lime mortar possesses poor workability and poor water retentivity and also suffers high shrinkage.
    3. Masonry in lime mortar has a better resistance against rain penetration and is less liable to crack when compared to masonry in cement mortar.

Which of these statements are correct?

a) 1, 2 and 3
b) 1 and 2
c) 2 and 3
d) 1 and 3

[IES 1999]

31. Consider the following statements:
    1. Addition of a small quantity of slaked lime to Portland cement in cement mortar increases the plasticity of the mortar.
    2. Light-weight mortar is prepared by mixing cement and finely crushed fire bricks with water.
    3. Fire resistance mortar is prepared by mixing cement and finely crushed fire bricks with water.

Which of these statements are correct?

a) 1 and 2
b) 1 and 3
c) 2 and 3
d) 1, 2 and 3

[IES 2000]

32. Guniting is the application of mortar
    a) On a surface under pneumatic pressure
    b) On a vertical surface
    c) On brickwork by manual method
    d) Of fluid consistency for repair works

[IES 2001]

33. Consider the following statements:

High Alumina Cement (HAC)

1. Has high early compressive strength and high heat of hydration than OPC-43 grade
2. Is not suitable to be used in cold regions

Which of these statements is/are correct?
a) 1 alone
b) 2 alone
c) Both 1 and 2
d) Neither 1 nor 2

[IES 2001]
34. Which one of the following statements regarding the cement fineness is NOT correct?
a) Fine cement is more liable to suffer from shrinkage cracking than coarse cement.
b) Fine cement will show faster rate of hardening than coarse cement.
c) Fine cement shows faster rate of heat evolution and total quantity of heat evolved is much large than coarse cement.
d) Fine cement shows the same setting time as coarse cement.

[IES 2001]
35. For marine works, the best suited cement is
a) Low heat Portland cement
b) Rapid hardening cement
c) Ordinary Portland cement
d) Blast furnace slag cement

[IES 2001]
36. Assertion (A): Flash set is the stiffening of the cement paste within a few minutes after mixing.
Reason (R): Flash set occurs due to insufficient gypsum to control the rapid reaction of C_3A with water.
a) Both A and R are true and R is the correct explanation of A.
b) Both A and R are true and R is not the correct explanation of A.
c) A is true but R is false.
d) A is false but R is true.

[IES 2002]
37. Assertion (A): The amount of cement paste should be sufficient to cover the surface of all particles for proper workability and bond.
Reason (R): The water–cement ratio is fixed accordingly.
a) Both A and R are true and R is the correct explanation of A.
b) Both A and R are true and R is not the correct explanation of A.
c) A is true but R is false.
d) A is false but R is true.

[IES 2002]
38. Assertion (A): The higher percentage of tri-calcium silicate in cement results in rapid hardening with an early gain in strength at a higher heat of hydration.

Reason (R): A higher percentage of dicalcium silicate in cement results in slow hardening and less heat of hydration and greater resistance to chemical attack.

a) Both A and R are true and R is the correct explanation of A.
b) Both A and R are true and R is not the correct explanation of A.
c) A is true but R is false.
d) A is false but R is true.

[IES 2002]

39. Assertion (A): For a given composition, finer cement will develop strength and generate heat more quickly than a coarse cement.
Reason(R): The reaction between water and cement starts on the surface of the cement particles and in consequence the greater the surface area of a given volume of cement, the greater the hydration.
a) Both A and R are true and R is the correct explanation of A.
b) Both A and R are true and R is not the correct explanation of A.
c) A is true but R is false.
d) A is false but R is true.

[IES 2002]

40. Four main oxides present in ordinary Portland cement are: CaO, Al_2O_3, SiO_2 and Fe_2O_3. Identify the correct ascending order of their proportions in a typical composition of OPC.
a) Al_2O_3, Fe_2O_3, CaO, SiO_2
b) Al_2O_3, CaO, Fe_2O_3, SiO_2
c) Fe_2O_3, Al_2O_3, SiO_2, CaO
d) Fe_2O_3, SiO_2, Al_2O_3, CaO

[IES 2002]

41. Assertion (A): The greater the surface area of a given volume of cement the greater the hydration.
Reason (R): The reaction between the water and cement starts from the surface of the cement particles.
a) Both A and R are true and R is the correct explanation of A.
b) Both A and R are true and R is not the correct explanation of A.
c) A is true but R is false.
d) A is false but R is true.

[IES 2003]

42. The proper size of mould for testing compressive strength of cement is
a) 7.05 cm cube
b) 10.05 cm cube
c) 15 cm cube
d) 12.05 cm cube

[IES 2003]

43. The specific gravity of commonly available ordinary Portland cement is
    a) 4.92
    b) 3.15
    c) 2.05
    d) 1.83

[IES 2003]

44. A quick-setting cement has an initial setting time of about
    a) 50 minutes
    b) 40 minutes
    c) 15 minutes
    d) 5 minutes

[IES 2003]

45. Match list I (cement mortar for different work) with list II (proportion of cement: sand in mortar) and select the correct answer using the codes given below the lists:

	List I		List II
A	Normal brick work	1	1: 4
B	Plastering works	2	1: 3
C	Grouting works	3	1: 6
D	Guniting	4	1: 15

Codes:

	A	B	C	D
a)	3	4	2	1
b)	1	2	3	4
c)	3	1	4	2
d)	1	4	2	3

[IES 2003]

46. What is the quantity of cement (in kg) and of dry sand (in cubic metre) respectively required for preparing 1 cubic metre of wet cement mortar of 1:5 proportion?
    a) 270 and 1.00
    b) 290 and 1.05
    c) 290 and 1.00
    d) 310 and 1.05

[IES 2004]

47. Consider the following statements regarding cement mortar:
    1. Silicate type chemical-resistant mortar has good resistance to hydrofluoric acid.
    2. Sulphur-type chemical-resistant mortar has poor resistance to alkalis.
    3. The interior surface of a building may be plastered with cement mortar containing cement and sand in the ratio 1: 6.

    Which of these statements are correct?
    a) 1, 2 and 3
    b) 1 and 2
    c) 2 and 3
    d) 1 and 3

[IES 2004]

48. Consider the following statements:
    Low percentage of C_3S and high percentage of C_2S in cement will result in
    1. Higher ultimate strength with less heat generation
    2. Rapid hardening
    3. Better resistance to chemical attack

    Which of these statements are correct?
    a) 1 and 2
    b) 2 and 3
    c) 1 and 3
    d) 1, 2 and 3

[IES 2004]

49. Match list I (type of cement) with list II (characteristics) and select the correct answer using the codes given below the lists:

	List I		List II
A	Rapid hardening cement	1	Lower C_3A content than that in OPC
B	Low heat Portland cement	2	Contains pulverised fly ash
C	Portland pozzolana cement	3	Higher C_3S content than that in OPC
D	Sulphate resisting cement	4	Lower C_3S and C_3A contents than that in OPC

Codes:

	A	B	C	D
a)	1	2	4	3
b)	3	4	2	1
c)	1	4	2	3
d)	3	2	4	1

[IES 2005]

50. Assertion (A): Low heat Portland cement is used in dam construction.
Reason(R): Low heat Portland cement attains higher 28 days strength than ordinary Portland cements.
    a) Both A and R are true and R is the correct explanation of A.
    b) Both A and R are true and R is not the correct explanation of A.
    c) A is true but R is false.
    d) A is false but R is true.

[IES 2006]

51. As per specifications, the initial setting time of ordinary Portland cement should not be less than
    a) 10 minutes
    b) 20 minutes
    c) 30 minutes
    d) 60 minutes

[IES 2006]

52. In cements, generally, the increase in strength during a period of 14 –28 days is primarily due to
    a) C_3A
    b) C_2S
    c) C_3S
    d) C_4AF

[IES 2006]

53. Consider the following types of cements:
    1. Portland pulverised fly ash cement
    2. High alumina cement
    3. Ordinary Portland cement
    4. Rapid hardening cement

Which one of the following is the correct sequence of the above cements in terms of their increasing rate of strength gain?
    a) 2-3-4-1
    b) 1-3-4-2
    c) 2-1-3-4
    d) 3-1-2-4

[IES 2007]

54. The ultimate strength of cement is influenced by which one of the following?
    a) Tri-calcium silicate
    b) Di-calcium silicate
    c) Tri-calcium aluminate
    d) Tetra-calcium alumino-ferrite

[IES 2007]

55. Consider the following statements:
    1. Setting and hardening of cement takes place after the addition of water.
    2. Water causes hydration and hydrolysis of the constituent compounds of cement which Act as binders.

Which of these statements is/are correct?

a) 1 only
b) 2 only
c) Both 1 and 2
d) Neither 1 nor 2

[IES 2007]

56. Assertion (A): The rate of hydration is faster in finer cements.
Reason (R): The surface area is more in case of finer cement.
    a) Both A and R are true and R is the correct explanation of A.
    b) Both A and R are true and R is not the correct explanation of A.
    c) A is true but R is false.
    d) A is false but R is true.

[IES 2009]

57. What is the requirement of water (expressed as % of cement w/c) for the completion of chemical reactions in the process of hydration of OPC?
    a) 10–15%
    b) 15–20%
    c) 20–25%
    d) 25–30%

[IES 2009]

58. If "W" is the percentage of water required for normal consistency of cement, water to be added for determination of initial setting time is
    a) 0.50 W
    b) 0.62 W
    c) 0.75 W
    d) 0.85 W

[IES 2011]

59. Match list I (grade of cement and age) with list II (compressive strength in N/mm^2) and select the correct answer using the code given below the lists:

	List I		List II
A	Grade 33 (7 days)	1	27
B	Grade 43 (28 days)	2	43
C	Grade 53 (3 days)	3	22
D	Grade 43 (7 days)	4	33

Codes:

	A	B	C	D
a)	4	2	1	3
b)	3	2	1	4
c)	4	1	2	3
d)	3	1	2	4

[IES 2011]

60. Consider the following statements:
Cement mortars richer than 1:3 are not used in masonry work because
   1. There is no gain in strength of masonry
   2. There is high shrinkage
   3. They are prone to segregation

Which of these statements are correct?
   a) 1, 2 and 3
   b) 1 and 2 only
   c) 2 and 3 only
   d) 1 and 3 only

[IES 2011]

61. Consider the following statements:
More than 6% magnesium oxide by weight in cement results in
   1. High early strength and high heat generation
   2. Less tendency towards volume change and formation of cracks

Which of these statements is/are correct?
   a) 1 Only
   b) 2 only
   c) Neither 1 nor 2
   d) Both 1 and 2

[IES 2012]

62. Fineness of cement is measured in the units of
   a) Volume/mass
   b) Mass/volume
   c) Area/mass
   d) Mass/area

[IES 2012]

63. The initial setting time of cement depends most on
   a) Tri-calcium aluminate
   b) Tri-calcium silicate
   c) Tri-calcium alumino-ferrite

d) Di-calcium silicate

[IES 2012]

64. Gypsum is added into the raw materials during the manufacture of cement so that the final product exhibits
    a) Retarded initial setting time
    b) Improved modulability for cornices, etc.
    c) Increased compressive strength
    d) Augmented bond strength

[IES 2012]

65. The objectives of producing cement include:
    1. Incorporating industrial wastes
    2. Increasing free lime
    3. Increasing fineness
    4. Saving energy needed in the process

    a) 1, 2, 3 and 4
    b) 1, 3 and 4 only
    c) 1, 2 and 4 only
    d) 2, 3 and 4 only

[IES 2013]

66. Which compound of cement is responsible for the strength of cement?
    a) Magnesium oxide
    b) Silica
    c) Alumina
    d) Calcium sulphate

[IES 2014]

67. Which type of cement is recommended in large mass concrete works such as a dam?
    a) Ordinary Portland
    b) High alumina
    c) Low heat Portland
    d) Portland pozzolona

[IES 2014]

68. Consider the following statements regarding "setting of cement":
    1. Low heat cement sets faster than OPC.
    2. Final setting time decides the strength of cement.
    3. Initial setting time of Portland pozzolona is 30 minutes.
    4. Air induced setting is observed when stored under damp conditions.
    5. Addition of gypsum retards the setting time.

Which of the above statements are correct?

a) 1, 2 and 3

b) 2, 3 and 4
c) 3, 4 and 5
d) 2, 3 and 5

[IES 2014]

69. Consider the following statements regarding "strength of cement":
    1. Strength test on cement is made on cubes of cement–sand mix.
    2. Water to be used for the paste is $0.25P$, where P is the water needed for normal consistency.
    3. The normal consistency is determined on Le Chatelier's apparatus.
    4. Cubes are cast in two layers in leak-proof moulds further compacted in each layer by vibration on a machine.

    Which of the above statements are correct?
    a) 1 and 2
    b) 2 and 3
    c) 1 and 4
    d) 3 and 4

[IES 2014]

70. Which of the following ingredients refer to binding materials of mortar?
    1. Cement
    2. Lime
    3. Sand
    4. Ashes

    Select the correct answer using the code given below
    a) 1 and 4
    b) 3 and 4
    c) 1 and 2
    d) 2 and 3

[IES 2014]

71. Consider the following related to sand in mortars:
    1. It increases the volume of the mortar mix.
    2. It increases the strength of masonry.
    3. The cost of the mortar is reduced.
    4. Shrinkage of the mortar is almost prevented.
    5. Surkhi can replace sand in cement mortar used in plastering, and this modified mortar is more durable.

    Which of the above are relevant to "sand" in mortar?
    a) 1, 2, 4 and 5
    b) 1, 3 and 4 only
    c) 3, 4 and 5 only
    d) 2, 3 and 4 only

[IES 2015]

72. Consider the following statements related to "composite mortar":
    1. Addition of lime to cement mortar improves its workability.
    2. Composite mortar is obtained by adding 10% by weight of cement and mixing with water.
    3. Composite mortar is not preferred in tall buildings.
    4. Mechanical grinding is essential for developing composite mortar.

Which of the above statements are true in this case?

a) 1, 2 and 3 only
b) 1, 3 and 4 only
c) 2, 3 and 4 only
d) 1, 2, 3 and 4

[IES 2015]

73. One bag of Portland cement, 50 kg in weight, would normally have a bulk volume of
    a) 30 l
    b) 35 l
    c) 40 l
    d) 45 l

[IES 2015]

74. Consider the following forms of water in a hydrated cement paste:
    1. Capillary water
    2. Chemically combined water
    3. Interlayer water
    4. Adsorbed water

Which of the above forms of water will, on their removal, cause shrinkage of the paste?

a) 1, 2 and 3
b) 1, 2 and 4
c) 2, 3 and 4
d) 1, 3 and 4

[IES 2015]

75. Consider the following statements:
    High early strength of cement is obtained as a result of
    1. Fine grinding
    2. Decreasing the lime content
    3. Burning at higher temperature
    4. Increasing the quantity of gypsum

Which of the above statements are correct?

a) 1 and 2

b) 1 and 3
c) 2 and 3
d) 3 and 4

[IES 2015]

76. Assertion (A): Air-entraining cement has a higher initial setting time than OPC and resists frost action better.
Reason (R): Air-entraining cement has a longer final setting time compared to OPC.
a) Both A and R are true and R is the correct explanation of A.
b) Both A and R are true and R is not the correct explanation of A.
c) A is true but R is false.
d) A is false but R is true.

[IES 2015]

77. Which of the following statements is correct regarding the strength of cement?
1. Particle sizes less than 3 μm increase the viscous nature of the cement.
2. Finer particles in cement can be replaced by fly ash to improve the strength.
a) 1 only
b) 2 only
c) Both 1 and 2
d) Neither 1 nor 2

[IES 2016]

78. The constituent compound in Portland cement which reacts immediately with water and also sets earliest is
a) Tri-calcium silicate
b) Di-calcium silicate
c) Tri-calcium aluminate
d) Tetra-calcium alumino-ferrite

[IES 2016]

79. Statement (I): the finer the cement, the greater is the need for water for hydration and workability.
Statement (II): Bleeding of a mix occurs due to low water–cement ratio.
a) Both statement (I) and statement (II) are individually true and statement (II) is the correct explanation of statement (I).
b) Both statement (I) and statement (II) are individually true and statement (II) is not the correct explanation of statement (I).
c) Statement (I) is true but statement (II) is false.
d) Statement (I) is false but statement (II) is true.

[IES 2016]

80. Which of the following statements are correct with regard to cement mortar?
    1. Workability of cement mortar can be improved by the addition of lime.
    2. Fly ash cement is economical in plastering jobs.
    3. Addition of saw dust improves workability.
    4. Sand in mortar can be replaced by finely crushed fire bricks.
    a) 1, 2, 3 and 4
    b) 1, 2 and 3 only
    c) 3 and 4 only
    d) 1, 2 and 4 only

[IES 2016]

81. Statement (I): Fire resistance of plastering can be achieved by mixing surkhi to the cement mortar.
    Statement (II): Insulation against sound and fire can be achieved by adding sufficient water in situ just before applying the mortar.
    a) Both statement (I) and statement (II) are individually true and statement (II) is the correct explanation of statement (I).
    b) Both statement (I) and statement (II) are individually true and statement (II) is not the correct explanation of statement (I).
    c) Statement (I) is true but statement (II) is false.
    d) Statement (I) is false but statement (II) is true.

[IES 2016]

82. Consider the following statements:
    1. Hydrophobic cement grains possesses low wetting ability.
    2. Rapid hardening cement is useful in concreting under static or running water.
    3. Quick-setting cement helps concrete to attain high strength in the initial period.
    4. White cement is just a variety of ordinary cement free of colouring oxides.

    Which of the above statements are correct?
    a) 1 and 4 only
    b) 1 and 3 only
    c) 2 and 4 only
    d) 2 and 3 only

[IES 2018]

83. Which one of the following statements is not correct with respect to fly ash?
    a) As a part replacement of cement in the range of 15–30%, fly ash reduces the strength in the initial period, but once the pozzolanic process sets in, higher strength can be obtained.

b) Fly ash as a part replacement of sand has a beneficial effect on strength even at an early age.
c) Fly ash as a part replacement of sand is economical.
d) A simultaneous replacement of cement and fine aggregates enables the strength at a specified age to be equalled depending upon the water content.

[IES 2019]

84. Which one of the following statements is not correct with respect to the properties of cement?
    a) Highly reactive pozzolanas enhance the early age strength of the composite cement.
    b) Pozzolanic activity refines pore structure, which decreases electrolytic resistance of concrete.
    c) The expansion due to alkali silica reaction can be controlled by the replacement of as high as 60% of OPC with high calcium pozzolana.
    d) Such high amounts of replacement cements result in higher accelerated carbonation depths compared to the pure use of OPC only.

[IES 2019]

85. Hydration of which compound is responsible for an increase in the strength of cement in later age?
    a) Tri-calcium aluminate
    b) Tetra-calcium alumina ferrite
    c) Tri-calcium silicate
    d) Di-calcium silicate

[IES 2019]

86. The creep strain of cement attains its terminal value by
    a) 1 year
    b) 2 years
    c) 5 years
    d) 6 months

[IES 2019]

87. Statement (I): Expansive cement is used in repair work for opened up joints.
    Statement (II): Expansive cement expands while hardening.
    a) Both statement (I) and statement (II) are individually true and statement (II) is the correct explanation of statement (I).
    b) Both statement (I) and statement (II) are individually true and statement (II) is not the correct explanation of statement (I).
    c) Statement (I) is true but statement (II) is false.
    d) Statement (I) is false but statement (II) is true.

[IES 2019]

88. Which of the following conditions are recommended for using sulphate resisting cement?
    1. Concrete to be used in the foundation and basement, where the soil is not infested with sulphates
    2. Concrete used for fabrication of pipes which are likely to be buried in a marshy region or sulphate-bearing soils
    3. Concrete to be used in the construction of sewage treatment works
        a) 2 and 3 only
        b) 1 and 2 only
        c) 1 and 3 only
        d) 1, 2 and 3

[IES 2020]

89. Which one of the following cements is a deliquescent?
    a) Quick-setting Portland cement
    b) White and coloured cement
    c) Calcium chloride cement
    d) Water repellent cement

[IES 2020]

90. The cement and water slurry coming on the top and setting on the surface is called
    a) Crazing
    b) Efflorescence
    c) Sulphate deterioration
    d) Laitance

[IES 2020]

91. Which one of the following lightweight element will be added to enhance the protective properties for X-ray shielding mortars?
    a) Sodium
    b) Potassium
    c) Lithium
    d) Calcium

[IES 2020]

92. Statement (I): Finer grinding of cement results in early development of strength.
    Statement (II): Finer the cement, higher is the rate of hydration.
    a) Both statement (I) and statement (II) are individually true and statement (II) is the correct explanation of statement (I).
    b) Both statement (I) and statement (II) are individually true and statement (II) is not the correct explanation of statement (I).
    c) Statement (I) is true but statement (II) is false.
    d) Statement (I) is false but statement (II) is true.

[IES 2020]

93. Statement (I): Pozzolana is added to cement to increase early strength.
    Statement (II): It reduces the heat of hydration.
    a) Both statement (I) and statement (II) are individually true and statement (II) is the correct explanation of statement (I).
    b) Both statement (I) and statement (II) are individually true and statement (II) is not the correct explanation of statement (I).
    c) Statement (I) is true but statement (II) is false.
    d) Statement (I) is false but statement (II) is true.

[IES 2020]

94. For the manufacture of Portland cement, the proportions of raw materials used are
    a) Lime 63%, silica 22% and others 15%
    b) Lime 70%, silica 20% and others 10%
    c) Lime 40%, silica 40% and others 20%
    d) None of the above

[SSC JE 2007]

95. Compound of cement which reacts immediately with water and sets first is
    a) Tri-calcium silicate
    b) Tri-calcium aluminate
    c) Di-calcium silicate
    d) All of the above

[SSC JE 2007]

96. The proper size of cube mould for testing the compressive strength of cement is
    a) 7.05 cm
    b) 10.05 cm
    c) 10 cm
    d) 15 cm

[SSC JE 2007]

97. Specific gravity of OPC is generally
    a) 4.92
    b) 3.15
    c) 2.10
    d) 1.75

[SSC JE 2007]

98. The compressive strength of ordinary Portland cement after 3 days should not be less than
    a) 50 kg/cm^2
    b) 100 kg/cm^2
    c) 115 kg/cm^2
    d) 150 kg/cm^2

[SSC JE 2009]

99. The constituent of cement which is responsible for the initial setting time of cement is
    a) Di-calcium silicate
    b) Tri-calcium silicate
    c) Tri-calcium aluminate
    d) All the above

[SSC JE 2009]

100. The main ingredients of Portland cement are
    a) Lime and silica
    b) Lime and alumina
    c) Silica and alumina
    d) All the above

[SSC JE 2010]

101. Good-quality cement contains a higher percentage of
    a) Tri-calcium silicate
    b) Tri-calcium aluminate
    c) Di-calcium silicate
    d) None of the above

[SSC JE 2010]

102. Water required per bag of cement is
    a) 7 kg
    b) 14 kg
    c) 28 kg
    d) 35 kg

[SSC JE 2010]

103. In ordinary Portland cement, the first one to react with water is:
    a) C_3A
    b) C_2S
    c) C_3S
    d) C_4AF

[SSC JE 2010]

104. The volume of one bag of cement is
    a) 0.0214 cu.m
    b) 0.0347 cu.m
    c) 0.0434 cu.m
    d) 0.0606 cu.m

[SSC JE 2011]

105. The standard consistency test is done in a

a) Blaine's apparatus
b) Le Chatelier's apparatus
c) Vane apparatus
d) Vicat's apparatus

[SSC JE 2011]
106. The 28-day compressive strength of cement is tested on 70.7 mm size cubes of mortar having cement to sand proportion of
a) 1:5
b) 1:2
c) 1:3
d) 1:4

[SSC JE 2012]
107. For Portland cement of 43 grades, 28-day mean compressive strength should be
a) 43 MPa
b) 43.5 MPa
c) 33 MPa
d) 38.5 MPa

[SSC JE 2012]
108. Out of the constituents of cement, namely, tri-calcium silicate (C_3S), di-calcium silicate (C_2S), tri-calcium aluminate (C_3A) and tetra-calcium aluminoferrite (C_4AF), the first one to set and harden is
a) C_3A
b) C_4AF
c) C_3S
d) C_2S

[SSC JE 2012]
109. The compound first to settle in cement is
a) Tri-calcium silicate
b) Tetra-calcium alumino-ferrite
c) Tri-calcium aluminate
d) Di-calcium silicate

[SSC JE 2012]
110. Which of the following Bogue's compounds of cement liberates maximum heat of hydration?
a) C_3S
b) C_4AF
c) C_3A
d) C_2S

[SSC JE 2012]

111. The amount of water used in performing setting time test of cement is (assuming p=standard consistency of cement)
  a) 0.60 p
  b) 0.65 p
  c) 0.80 p
  d) 0.85 p

[SSC JE 2013]

112. Gypsum used in cement manufacturing acts as
  a) Accelerator
  b) Air-entraining agent
  c) Plasticiser
  d) Retarder

[SSC JE 2013]

113. Which of the following cement is suitable for use in urgent repairs of existing massive concrete structures such as large dams?
  a) Ordinary Portland cement
  b) Low heat cement
  c) Rapid hardening cement
  d) Sulphate resisting cement

[SSC JE 2013]

114. You are asked to construct a massive concrete dam. The type of cement you will use is
  a) Ordinary Portland cement
  b) Rapid hardening Portland cement
  c) Low heat cement
  d) Blast furnace slag cement

[SSC JE 2013]

115. The initial setting time of ordinary Portland cement is
  a) 10 min
  b) 45 min
  c) 30 min
  d) 60 min

[SSC JE 2013]

116. During the manufacture of Portland cement, gypsum or Plaster of Paris is added to
  a) Increase the strength of cement
  b) Modify the colour of cement
  c) Reduce heat of hydration of cement
  d) Adjust setting time of cement

[SSC JE 2013]

117. Which of the following is added for quick setting of cement?
   a) Gypsum
   b) Alum
   c) Zinc sulphate
   d) Aluminium sulphate

[SSC JE 2013]

118. The density of cement is taken to be
   a) 1000 kg/m^3
   b) 1250 kg/m^3
   c) 1440 kg/m^3
   d) 1800 kg/m^3

[SSC JE 2014]

119. The high early strength of rapid hardening cement is due to its
   a) Increased content of gypsum
   b) Burning at high temperature
   c) Increased content of cement
   d) Higher content of tri-calcium silicate

[SSC JE 2014]

120. Dicalcium silicate
   a) Hydrates rapidly
   b) Generates less heat of hydration
   c) Hardens rapidly
   d) Has less resistance to sulphate attack

[SSC JE 2014]

121. The fineness of cement can be found out by sieve analysis using IS sieve number
   a) 20
   b) 10
   c) 9
   d) 6

[SSC JE 2014]

122. As the cement sets and hardens, it generates heat. This is called as
   a) Heat of hydration
   b) Latent heat
   c) Heat of vaporisation
   d) Sensible heat

[SSC JE 2014]

123. As a cheap alternative, the fineness of cement is tested by using

a) IS 100 μ Sieve, where at least 90% (by weight) should be retained
b) IS 90 μ Sieve, where at least 90% (by weight) should pass
c) IS 90 μ Sieve, where at least 95% (by weight) should pass
d) IS 100 μ Sieve, where at least 90% (by weight) should pass

[SSC JE 2014]
124. The most important constituents of cement are
a) C_3A and C_2S
b) C_3S and C_3A
c) C_3S and C_2S
d) C_3A and C_4AF

[SSC JE 2014]
125. Weight of one bag of cement is
a) 70 kg
b) 50 kg
c) 60 kg
d) 65 kg

[SSC JE 2014]
126. Fineness test of cement gives us an estimate of
a) Workability of concrete
b) Heat of hydration
c) Rate of hydration
d) Durability of concrete

[SSC JE 2014]
127. The soundness of cement is tested by
a) Vicat's apparatus
b) Le Chatelier's apparatus
c) Compression testing machine
d) Standard Briquette test

[SSC JE 2015]
128. Gypsum is added to cement in small quantity to
a) Control initial setting time
b) Control final setting time
c) Give colour to the cement
d) Make cement hydrophobic

[SSC JE 2015]
129. Snowcem is
a) Coloured cement
b) Powdered lime
c) Chalk powder
d) Mixture of chalk powder and lime

[SSC JE 2015]

130. White cement should have the least percentage of
   a) Aluminium oxide
   b) Iron oxide
   c) Silica
   d) Magnesium oxide

[SSC JE 2015]

131. The purpose of the soundness test of cement is
   a) To determine the presence of free lime
   b) To determine the setting time
   c) To determine the sound-proof quality of cement
   d) To determine the fineness

[SSC JE 2017]

132. The development of strength of cement and its fineness are
   a) Directly proportional
   b) Inversely proportional
   c) Not related
   d) Randomly related

[SSC JE 2017]

133. Flash set of ordinary Portland cement paste is
   a) Premature hardening
   b) Surface hardening only
   c) Hardening without the development of heat of hydration
   d) All the options are correct

[SSC JE 2017]

134. The specific surface expressed in square cm/gm of a good Portland cement should not be less than
   a) 1,750
   b) 2,000
   c) 2,250
   d) 2,500

[SSC JE 2017]

135. The compound of Portland cement which reacts immediately with water and also sets first is
   a) Tri-calcium silicate
   b) Di-calcium silicate
   c) Tri-calcium aluminate
   d) Tetra-calcium alumino-ferrite

[SSC JE 2017]

136. Rapid hardening cement attains early strength due to
 a) Larger proportion of lime grounded finer than normal cement
 b) Lesser proportion of lime grounded coarser than normal cement
 c) Lesser proportion of lime grounded finer than normal cement
 d) Larger proportion of lime grounded coarser than normal cement

[SSCJE 2017]

137. The percentage of water for normal consistency is:
 a) 5–15%
 b) 10–25%
 c) 15–25%
 d) 20–30%

[SSC JE 2017]

138. Soundness test of cement determines
 a) Quantity of free lime
 b) Ultimate strength
 c) Durability
 d) Initial setting

[SSC JE 2017]

139. For a 50 kg cement bag, the amount of water required is
 a) 16.5 litres
 b) 18.5 litres
 c) 20.5 litres
 d) 22.5 litres

[SSC JE 2017]

140. For the construction of thin R.C.C. structures the type of cement to be avoided is
 a) Ordinary Portland cement
 b) Rapid hardening cement
 c) Low heat cement
 d) Blast furnace slag cement

[SSC JE 2017]

141. Percentage of pozzolanic material containing clay up to 80% used for the manufacture of pozzolana cement is
 a) 30%
 b) 40%
 c) 50%
 d) 60%

[SSC JE 2017]
142. Pick up the incorrect statement applicable to the field test of good cement.
a) When one thrusts one's hand into a bag of cement, one should feel warm.
b) The colour of the cement is bluish.
c) A handful of cement thrown into a bucket of water should sink immediately.
d) All options are correct.

[SSC JE 2017]
143. An ordinary Portland cement when tested for its fineness should not leave any residue on I.S Sieve No.9, more than
a) 5%
b) 10%
c) 15%
d) 20%

[SSC JE 2017]
144. Hardening of cement occurs at a
a) Rapid rate during the first few days and afterwards it continues to increase at a decreased rate
b) Slow rate during the first few days and afterwards it continues to increase at a rapid rate
c) Uniform rate throughout its age
d) None of these

[SSC JE 2017]
145. For quality control of Portland cement, the test essentially done is
a) Setting time
b) Soundness
c) Tensile strength
d) All options are correct

[SSC JE 2017]
146. If 1500 g of water is required to have 1875 g cement paste of normal consistency, the percentage of water is
a) 20%
b) 25%
c) 30%
d) 35%

[SSC JE 2017]
147. The mixture of different ingredients of cement is burnt at
a) 1000°C
b) 1200°C

c) 1400°C
d) 1600°C

[SSC JE 2017]
148. To obtain cement dry powder, limestones and shales or their slurry is burnt in a rotary kiln at a temperature between
a) 1100°C and 1200°C
b) 1200°C and 1300°C
c) 1300°C and 1400°C
d) 1400°C and 1500°C

[SSC JE 2017]
149. Hydration of cement is due to chemical action of water with
a) Tri-calcium silicate and di-calcium silicate
b) Di-calcium silicate and tri-calcium aluminate
c) Tri-calcium aluminate and tri-calcium alumino-ferrite
d) All options are correct

[SSC JE 2017]
150. Which constituent of the cement, upon addition of water, sets and hardens first?
a) Tri-calcium silicate
b) Tri-calcium aluminate
c) Di-calcium silicate
d) Free lime

[SSC JE 2017]
151. The cementing property of cement is mainly due to
a) Lime
b) Alumina
c) Silica
d) Gypsum

[SSC JE 2017]
152. The setting and hardening of cement paste is mainly due to the hydration and hydrolysis of
a) Tri-calcium silicate
b) Tetra-calcium alumino-ferrite
c) Di-calcium silicate
d) Tri-calcium aluminate

[SSC JE 2017]
153. The rapid hardening Portland cement is obtained by
a) Grinding the clinker to a high degree of fineness
b) Adding calcium sulphate to the mixture

c) Adding gypsum after grinding
d) Burning the mixture at a lower temperature

[SSC JE 2017]
154. As per IS specifications, what should be the maximum final setting time for ordinary Portland cement?
a) 30 minutes
b) 10 hours
c) 1 hour
d) 6 hours

[SSC JE 2017]
155. Air permeability test of cement is conducted to find the
a) Unsoundness
b) Ignition loss
c) Specific gravity
d) Fineness

[SSC JE 2017]
156. When water is added to cement
a) Heat is generated
b) Heat is absorbed
c) Chemical reaction is initiated
d) Both heat is generated and chemical reaction is initiated

[SSC JE 2017]
157. To hydrate 500 kg cement fully, water needed is
a) 100 kg
b) 110 kg
c) 120 kg
d) 130 kg

[SSC JE 2017]
158. The maximum percentage of chemical ingredient of cement is
a) Magnesium oxide
b) Iron oxide
c) Aluminium
d) Lime

[SSC JE 2017]
159. Efflorescence in cement is caused due to an excess of
a) Alumina
b) Iron oxide
c) Silica
d) Alkalis

[SSC JE 2017]

160. Pick up the incorrect statement from the following.
    a) The degree of grinding of cement is called fineness.
    b) The process of changing cement paste into hard mass is known as setting of cement.
    c) The phenomenon by virtue of which cement does not allow transmission of sound is known as soundness of cement.
    d) The heat generated during chemical reaction of cement with water is known as heat of hydration.

[SSC JE 2017]

161. Pick up the incorrect statement from the following.
    a) Cement and standard sand mortar are used in the ratio of 1: 3.
    b) Water is added to the rate of (P/4 + 3) percentage of water, where P is the percentage of water for standard consistency.
    c) A cube mould of 10 cm × 10 cm × 10 cm is used.
    d) The prepared moulds are kept in an atmosphere of 50% relative humidity.

[SSC JE 2017]

162. Pozzolana cement is used with confidence for the construction of
    a) Dams
    b) Massive foundations
    c) Abutments
    d) All options are correct

[SSC JE 2017]

163. For the manufacture of Portland cement the proportions of raw material used are
    a) Lime 63%; silica 22%; other ingredients 15%
    b) Silica 2%; lime 63%; other ingredients 15%
    c) Silica 40%; lime 40%; other ingredients 40%
    d) Silica 70%; lime 20%; other ingredients 20%

[SSC JE 2017]

164. To retard the initial setting time of cement, the compound responsible is
    a) Tri-calcium silicate
    b) Gypsum
    c) Di-calcium silicate
    d) Tri-calcium aluminate

[SSC JE 2017]

165. Quick-setting cement is produced by adding
    a) Less amount of gypsum in very fine powdered form
    b) More amount of gypsum in very fine powdered form

c) Aluminium sulphate in very fine powdered form
d) Pozzolana in very fine powdered form

[SSC JE 2017]
166. If P is the percentage of water required for normal consistency, water to be added for determination of initial setting time is
a) 0.70 P
b) 0.75 P
c) 0.80 P
d) 0.85 P

[SSC JE 2017]
167. Pick up the correct statement from the following.
a) Adding 5–6% moisture content by weight increases the volume of dry sand from 18% to 38%.
b) The bulking of fine sand is more than that of coarse sand.
c) If the percentage content of moisture exceeds 10%, increase in bulk of sand starts increasing.
d) All options are correct.

[SSC JE 2017]
168. The minimum percentage of the chemical ingredient of cement is
a) Magnesium oxide
b) Iron oxide
c) Alumina
d) Lime

[SSC JE 2017]
169. Pick up the correct proportions of the chemical ingredients of cement.
a) Lime: Silica: Alumina: Iron oxide = 63: 22: 6: 3
b) Silica: Lime: Alumina: Iron oxide = 63: 22: 6: 3
c) Alumina: Silica: Lime: Iron oxide = 63: 22: 6: 3
d) Iron oxide: Alumina: Silica: Lime = 63: 22: 6: 3

[SSC JE 2017]
170. Pick up the correct statement from the following.
a) Lime in excess causes the cement to expand and disintegrate.
b) Silica in excess causes the cement to set slowly.
c) Alumina in excess reduces the strength of the cement.
d) All options are correct.

[SSC JE 2017]
171. For an ordinary Portland cement
a) Residual does not exceed 10% when sieved through IS Sieve No.9
b) Soundness varies from 5 mm to 10 mm

c) Initial setting time is not less than 30 minutes
d) Compressive stress after 7 days is not less than 175 kg/cm^2

[SSC JE 2017]
172. The commercial name of white and coloured cement in India is
a) Colocrete
b) Rainbow cement
c) Silvicrete
d) All options are correct

[SSC JE 2017]
173. Which IS code gives specifications about cement plaster?
a) IS 1500
b) IS 1221
c) IS 1400
d) IS 1661

[SSC JE 2018]
174. In a lime cement plaster, ratio 1: 1: 6 corresponds to
a) Lime: Cement: Sand
b) Cement: Lime: Sand
c) Lime: Sand: Gravel
d) Cement: Sand: Gravel

[SSC JE 2018]
175. Which of the following statement is false?
a) With passage of time, the strength of cement increases.
b) With passage of time, the strength of cement decreases.
c) After a period of 24 months, the strength of cement reduces to 50%.
d) The concrete made with storage deteriorated cement gains strength with time.

[SSC JE 2018]
176. In the process of hydration of OPC, to complete all chemical reaction, the water requirement (expressed as the percentage of cement) is
a) 5–8%
b) 8–16%
c) 20–25%
d) 35–45%

[SSC JE 2018]
177. Which of the following is determined with the help of a Le Chatelier's device?
a) Abrasion resistance
b) Chemical resistance

c) Soundness
d) Strength

[SSC JE 2018]
178. Initial setting time of rapid hardening Portland cement is nearly
a) Half a minute
b) 5 minutes
c) 30 minutes
d) 45 minutes

[SSC JE 2018]
179. The centre needle of the attachment of the Vicat plunger projects the circular cutting edge by
a) 0.2 mm
b) 0.5 mm
c) 1 mm
d) 5 mm

[SSC JE 2018]
180. The field test for the quality of cement consists in putting a small quantity of cement in a bucket containing water. Good-quality cement will:
a) Immediately dissolve in the water
b) Float on the water surface
c) Sink to the bottom of the bucket
d) Produce this steam

[SSC JE 2018]
181. Which of the following is an important factor that affects the shrinkage of cement concrete?
a) Quantity of cement
b) Size of coarse aggregate
c) Size of fine aggregate
d) Amount of water added during mixing of concrete

[SSC JE 2018]
182. For the repair of roads
a) Low heat cement is used
b) Rapid hardening cement is used
c) High alumina cement is used
d) Sulphate resisting cement is used

[SSC JE 2018]
183. Which of the following proportion of cement and standard sand is in cement mortar while testing the compressive and tensile strength of cement?
a) 1: 2
b) 1: 3

c) 1: 4
d) 1: 6

[SSC JE 2018]
184. Which of the following shows the correct decreasing order of rate of hydration of Portland cement compounds?
a) $C_3A > C_4AF > C_3S > C_2S$
b) $C_3A > C_4AF > C_2S > C_3S$
c) $C_3A > C_3S > C_2S > C_4AF$
d) $C_4AF > C_3A > C_3S > C_2S$

[SSC JE 2018]
185. Argillaceous material are those
a) Which have alumina as the main constituent
b) Which have alumina as the main constituent
c) Which evolve heat on the addition of water
d) Which easily break when hammered lightly

[SSC JE 2018]
186. Which of the following constituents are present in the blast furnace slag?
a) 50% alumina and 20% calcium oxide
b) 45% calcium oxide and 35% silica
c) 25% calcium oxide and 15% silica
d) 25% magnesia and 15% silica

[SSC JE 2018]
187. Which of the below is not a plaster type based on material?
a) Cement
b) Gypsum
c) Pozzolana
d) Lime

[SSC JE 2018]
188. Which of the below is not a plaster finish?
a) Rough-cast
b) Pebbel dash
c) Sand faced
d) Wooden

[SSC JE 2018]
189. The cement becomes useless if its absorbed moisture content exceeds
a) 0.01
b) 0.02
c) 0.03
d) 0.05

[SSC JE 2018]

190. According to the IS specifications, initial setting time of the ordinary Portland cement should be
   a) 10 minutes
   b) 30 minutes
   c) 6 hours
   d) 10 hours

[SSC JE 2018]

191. If one cement bag has 0.035 m^3 volume of cement, the number of bags required for one tonne of cement is
   a) 10
   b) 12
   c) 15
   d) 20

[SSC JE 2018]

192. Which of the following statements are true?
   A. A gap of 0.3 m is to be maintained between cement bag and wall while storing cement.
   B. A gap 100 cm is to be maintained between cement bag and wall while storing cement.
   a) Only A
   b) Only B
   c) Both A and B
   d) None of these

[SSC JE 2018]

193. How many methods of Ferro cementing are there?
   a) 3
   b) 2
   c) 4
   d) 6

[SSC JE 2018]

194. Identify the correct statements
   a) White cement is unsuitable for ordinary work.
   b) Pozzolana cement is grey in colour.
   c) C_3S is tri-calcium silicate.
   d) All options are correct.

[SSC JE 2018]

195. Pick up the incorrect statement from the following
   a) The bottom and top ends of slump mould are parallel to each other.
   b) The axis of the mould is perpendicular to the end faces.

c) The internal surface of the mould is kept clean and free from set cement.
d) The mould is in the form of a frustum of hexagonal pyramid.

[SSC JE 2018]
196. Under normal conditions using ordinary cement the period of removal of the form work is
a) 7 days for beam soffits
b) 14 days for bottom slabs of span 4.6 m and more
c) 21 days for bottom beams over 6 m spans
d) All options are correct

[SSC JE 2018]
197. Pick up the correct statement from the following.
a) Water enables a chemical reaction to take place with cement.
b) Water lubricates the mixture of gravel, sand and cement.
c) Only a small quantity of water is required for hydration of cement.
d) All options are correct.

[SSC JE 2018]
198. Identify the correct statement
a) Expanding cement is used for filling the cracks.
b) White cement is mostly used for decorative works.
c) Portland pozzolana cement produces less heat of hydration.
d) High-strength Portland cement is produced from special materials.

[SSC JE 2018]
199. What do you mean by "warehouse pack" of cement?
a) Full capacity of the warehouse
b) Pressure excretion of the bags of upper layers
c) Pressure compaction of the bags of lower layers
d) Packing the warehouse

[SSC JE 2018]
200. Which of the below is not a property of Ferro cement?
a) Impervious nature
b) Capacity to resist shock
c) No need of formwork
d) Strength per unit mass is low

[SSC JE 2018]
201. Which of the following is a calcareous raw material used in the cement production?
a) Cement rock
b) Limestone

c) Marine shells
d) All options are correct

[SSC JE 2018]
202. Which of the following is the most common alternative to cement in concrete?
a) Slag
b) Fly ash
c) Asphalt
d) Lime

[SSC JE 2018]
203. Water–cement ratio is usually expressed in
a) Litres of water required per bag of cement
b) Litres of water required per kg of cement
c) Both A and B
d) None of these

[AE 2008]
204. The levelling operation that removes humps and hollows and gives a true, uniform concrete surface is called
a) Screening
b) Floating
c) Trowelling
d) Compacting

[AE 2008]
205. The constituents of cement which act as binder are
a) Sand and silica
b) Carbon and silica
c) Tri-calcium silicate, di-calcium silicate and carbon dioxide
d) Tri-calcium silicate, di-calcium silicate and tri-calcium aluminate

[AE 2008]
206. Le Chatelier's apparatus is used to carry out
a) Consistency test
b) Tensile test
c) Soundness test
d) Compressive strength

[AE 2008]
207. Efflorescence in cement is caused due to an excess of
a) Alumina
b) Iron oxide
c) Silica
d) Alkalis

[AE 2010]

208. Setting time of cement is decreased by adding
    a) Gypsum
    b) Calcium chloride
    c) Sodium oxide
    d) All of these

[AE 2010]

209. Gypsum is added in cement to
    a) Increase its initial setting time
    b) Decrease its initial setting time
    c) Increase its compressive strength
    d) Increase its bond strength

[AE 2010]

210. Plaster of Paris can be obtained from the calcination of
    a) Limestone
    b) Gypsum
    c) Dolomite
    d) Bauxite

[AE 2010]

211. Which constituent of cement upon addition of water, sets and hardens first?
    a) Tri-calcium silicate
    b) Tri-calcium aluminate
    c) Di-calcium silicate
    d) Free lime

[AE 2010]

212. The heat of hydration of cement is expressed in terms of
    a) Grams
    b) k-cal
    c) $\frac{\text{calories}}{\text{gram}}$
    d) $\frac{\text{calories}}{\text{cm}^3}$

[AE 2010]

213. As the cement sets and hardens it generates heat which is called
    a) Latent heat
    b) Sensible heat
    c) Heat of humidity
    d) Heat of hydration

[AE 2010]

214. The minimum percentage of ingredients in cement is that of
   a) Lime
   b) Aluminium
   c) Silica
   d) Magnesium oxide

[AE 2010]

215. The test conducted by Vicat's apparatus is for
   a) Fineness
   b) Compression strength
   c) Tensile strength
   d) Consistency

[AE 2010]

216. Le Chatelier apparatus is used to perform
   a) Fineness test
   b) Soundness test
   c) Consistency test
   d) Compressive strength test

[AE 2010]

217. For the repair of road, the cement used is
   a) Low heat cement
   b) Sulphate resistant cement
   c) High alumina cement
   d) Rapid hardening cement

[AE 2013]

218. Increase in fineness of cement
   a) Reduces the rate of strength development and leads to higher shrinkage
   b) Increases the rate of strength development and reduces the rate of deterioration
   c) Decreases the rate of strength development and increase the bleeding of cement
   d) Increases the rate of strength development and leads to higher shrinkage

[AE 2014]

219. Before testing setting time of cement one should test for
   a) Soundness
   b) Strength
   c) Fineness
   d) Consistency

[AE 2014]

220. The quantity of water required for setting test of cement is _____ times the quantity of water required for normal consistency test
   a) 0.85
   b) 0.78
   c) 0.65
   d) 0.52

[AE 2014]

221. The Bogue compound tri-calcium aluminate in clinkering process of cement is named as
   a) Alite
   b) Belite
   c) Celite
   d) Felite

[AE 2015]

222. Setting time of cement is determined using
   a) Vicat apparatus
   b) Casagrande apparatus
   c) Le Chatelier apparatus
   d) Slump cone apparatus

[AE 2015]

223. The average tensile strength of cement, after 3 days and 7 days, should not be less than
   a) 2.0 N/mm^2 and 2.5 N/mm^2, respectively
   b) 2.5 N/mm^2 and 3.0 N/mm^2, respectively
   c) 3.0 N/mm^2 and 3.5 N/mm^2, respectively
   d) 4.0 N/mm^2 and 5.0 N/mm^2, respectively

[AE 2015]

224. Increase in fineness of cement
   a) Reduces the rate of strength development and leads to higher shrinkage
   b) Increases the rate of strength development and leads to higher shrinkage
   c) Increases the rate of strength development and reduces the rate of deterioration
   d) Decrease the rate of strength development and increase the bleeding of cement

[AE 2015]

225. The undesirable properties of cement are due to the formation of
   a) Di-calcium silicate
   b) Tri-calcium silicate
   c) Tri-calcium aluminate
   d) Tetra-calcium aluminate

[AE 2015]
226. As per IS specification, the minimum time for initial setting of ordinary Portland cement is
 a) 20 minutes
 b) 30 minutes
 c) 60 minutes
 d) 120 minutes

[AE 2015]
227. Compressive strength of cement is determined using the cube of size
 a) 15 cm
 b) 50 cm
 c) 7.6 cm
 d) 7.06 cm

[AE 2015]
228. The cement becomes unsound if the percentage of free magnesia exceeds __________ times by weight of cement
 a) 1.5
 b) 2.5
 c) 3.5
 d) 6.0

[AE 2016]
229. The average tensile strength of cement, after 3 days and 7 days, should not be less than
 a) 2.0 N/mm^2 and 2.5 N/mm^2, respectively
 b) 2.5 N/mm^2 and 3.0 N/mm^2, respectively
 c) 3.0 N/mm^2 and 3.5 N/mm^2, respectively
 d) 3.5 N/mm^2 and 5.0 N/mm^2, respectively

[AE 2018]
230. A supplementary cementitious material (SCM) is selected for use in a construction project. To confirm its reactivity, assessment of pozzolanic activity test was conducted. OPC 43 grade cement was used for pozzolanic activity test for casting control cement mortar specimens. Moreover, 20% replacement was adopted using SCM with cement and specimens were cast as per standard guidelines. Average compressive strength of SCM blended mortar specimens at 20% replacement level was found to be 38 MPa. Strength activity index was observed as 98%. The ratio of obtained strength of control specimens (only cement) to the expected strength of control specimens after 28 days of curing as per standards is
 a) 0.9017
 b) 0.9200
 c) 0.8917
 d) 0.9000

231. Write Bogue's compounds of cement in ascending order in terms of percentage available in cement.
   a) C_3S, C_2S, C_3A, C_4AF
   b) C_4AF, C_3A, C_2S, C_3S
   c) C_2S, C_3S, C_3A, C_4AF
   d) C_3A, C_2S, C_3S, C_4AF

232. Suggestions to reduce the heat of hydration only by changing cement composition are
   a) Reducing C_3A
   b) Reducing C_3S
   c) Increasing C_2S
   d) All the above

233. Argillaceous material is the source of ....and .... oxides in cement
   a) Silica
   b) Alumina
   c) Silica and Alumina
   d) None of the above

234. In the rotary kiln, around 1300°C, the liquid phase appears and promotes the reaction between belite and free lime to form
   a) C_3S
   b) C_2S
   c) C_3A
   d) C_4AF

235. In cement hydration, set regulator reacts with
   a) C_3S
   b) C_2S
   c) C_3A
   d) C_4AF

236. In optical microscopy of cement, large crystals are .... and small rounded crystals are ...., respectively.
   a) C_3S and C_2S
   b) C_3A and C_2S
   c) C_3A and C_3S
   d) C_3A and C_2AF

237. If you arrange Bogue compounds in terms of heat of hydration, the compounds which has the third-highest heat of hydration is
   a) C_3A
   b) C_2S

c) C_3S
d) C_4AF

238. Coloured cement consists of the Portland cement with the .... % of the pigment.
a) 5–10
b) 10–15
c) 15–20
d) 20–25

239. The average particle size of fly ash
a) 20 μm
b) 40 μm
c) 60 μm
d) None of the above

240. Large hollow spheres with solid spheres inside fly ash are called
a) Cenospheres
b) Plerospheres
c) Ketospheres
d) None of the above

241. Slurry form of silica fume is mainly used to reduce
a) Working cost
b) Transportation cost
c) Capital cost
d) Investment cost

242. In addition, pozzolanic reaction, silica fume has a unique characteristic of .... which is not available with other pozzolanic materials?
a) Enhances workability
b) Wall effect
c) Physical/filling effect
d) All the above

243. Calcination of kaolinite clay should be in the range of .... Fahrenheit.
a) 1,364–1,544
b) 1,464–1,664
c) 1,264–1,464
d) None of the above

244. .... fibres are damaged in alkaline media in case of FRP sheets.
a) Poly-ethylene
b) Wood

c) Glass
d) None of the above

245. Soft glass is obtained by fusing mixture of silica, lime and soda. Percentage of Na_2O in the proportion of the overall mixture is
a) 15%
b) 16%
c) 17%
d) 18%

246. .... glass can withstand high temperature.
a) Borosilicate
b) Tempered
c) Silicate
d) None of the above

247. You are in a situation to use a high level of replacement of pozzolanic material with cement. Then most preferred pozzolanic material is
a) Fly ash
b) Carbon black
c) Rice hush ash
d) Slag

1.	a	2.	a	3.	d	4.	c	5.	a	6.	b	7.	a	8.	c	9.	d	10.	b
11.	b	12.	a	13.	d	14.	c	15.	b	16.	c	17.	b	18.	d	19.	c	20.	a
21.	b	22.	c	23.	b	24.	c	25.	d	26.	c	27.	c	28.	b	29.	d	30.	d
31.	b	32.	a	33.	a	34.	c	35.	d	36.	a	37.	b	38.	b	39.	a	40.	c
41.	a	42.	a	43.	b	44.	d	45.	c	46.	d	47.	b	48.	c	49.	b	50.	c
51.	c	52.	c	53.	b	54.	b	55.	c	56.	a	57.	c	58.	d	59.	b	60.	b
61.	c	62.	c	63.	a	64.	a	65.	b	66.	b	67.	c	68.	c	69.	c	70.	c
71.	b	72.	b	73.	b	74.	d	75.	b	76.	c	77.	d	78.	c	79.	c	80.	d
81.	c	82.	a	83.	c	84.	a	85.	d	86.	c	87.	a	88.	a	89.	c	90.	d
91.	c	92.	a	93.	d	94.	a	95.	b	96.	a	97.	b	98.	d	99.	c	100.	a
101.	a	102.	b	103.	a	104.	b	105.	d	106.	c	107.	a	108.	a	109.	c	110.	c
111.	d	112.	d	113.	c	114.	c	115.	c	116.	d	117.	d	118.	c	119.	d	120.	b
121.	c	122.	a	123.	b	124.	b	125.	b	126.	c	127.	b	128.	a	129.	a	130.	b
131.	a	132.	a	133.	a	134.	c	135.	c	136.	a	137.	d	138.	a	139.	b	140.	b
141.	a	142.	d	143.	b	144.	a	145.	d	146.	b	147.	c	148.	d	149.	d	150.	b
151.	a	152.	d	153.	a	154.	b	155.	d	156.	d	157.	c	158.	d	159.	d	160.	c
161.	c	162.	d	163.	a	164.	b	165.	a	166.	d	167.	d	168.	a	169.	a	170.	d
171.	d	172.	d	173.	d	174.	b	175.	a	176.	d	177.	c	178.	c	179.	b	180.	b
181.	d	182.	b	183.	b	184.	d	185.	a	186.	b	187.	c	188.	d	189.	d	190.	b
191.	d	192.	d	193.	a	194.	d	195.	d	196.	d	197.	d	198.	c	199.	c	200.	d
201.	d	202.	b	203.	a	204.	a	205.	d	206.	c	207.	d	208.	a	209.	a	210.	b
211.	b	212.	c	213.	d	214.	a	215.	d	216.	b	217.	d	218.	d	219.	d	220.	a
221.	c	222.	a	223.	a	224.	b	225.	c	226.	b	227.	d	228.	d	229.	a	230.	a
231.	b	232.	d	233.	c	234.	a	235.	c	236.	a	237.	d	238.	a	239.	a	240.	b
241.	b	242.	c	243.	a	244.	c	245.	c	246.	a	247.	d						

REFERENCES

Assistant Engineer (AE). 2020. Tamil Nadu Public Service Commission (TNPSC), Tamil Nadu, India. http://www.tnpsc.gov.in/previous-questions.html (accessed September 18, 2020).

ASTM C311/C311M-18. (2019). *Standard test methods for sampling and testing fly ash or natural pozzolans for use in Portland cement concrete*. American Society for Testing and Materials. West Conshohocken, Pennsylvania.

Engineering Service Exam (ESE). 2020. Union Public Service Commission, New Delhi, India. https://www.upsc.gov.in/examinations/previous-question-papers (accessed September 18, 2020).

Graduate Aptitude Test in Engineering (GATE). 2020. GATE Office, Chennai, India. http://gate.iitm.ac.in/gate2019/previousqp18.php (accessed September 18, 2020).

IS 1727. (2013). *Methods of test for Pozzolanic materials*. Bureau of Indian Standards. New Delhi.

IS 4031 Part 2. (2013). *Determination of fineness by Blaine air permeability method*. Bureau of Indian Standards. New Delhi.

IS 4031 Part 3. (2014). *Determination of soundness*. Bureau of Indian Standards. New Delhi.

IS 4031 Part 4. (2014). *Determination of consistency of standard cement paste*. Bureau of Indian Standards. New Delhi.

IS 4031 Part 5. (2014). *Determination of initial and final setting times*. Bureau of Indian Standards. New Delhi.

IS 4031 Part 6. (2019). *Determination of compressive strength of hydraulic cement other than masonry cement*. Bureau of Indian Standards. New Delhi.

Junior Engineer (JE), Staff Selection Commission (SSC), New Delhi, India. https://ssc.nic.in/Portal/SchemeExamination (accessed September 18, 2020).

2 Chemical Admixtures

2.1 INTRODUCTION

Chemical admixtures are chemicals that are added with the concrete mixture to improve its fresh as well as hardened properties. ACI 116R defines chemical admixture as a material added, especially during or before mixing of concrete (ACI 116R 2000). Based on the targeted enhancement in properties, concrete chemical admixtures can be classified as follows:

1. Water-reducing agents
2. Set regulators
3. Air entrainers
4. Special admixtures
   a. Viscosity-modifying admixtures
   b. Corrosion inhibitors
   c. Shrinkage-reducing admixtures

2.2 WATER-REDUCING AGENTS

Water reducers generally work on the principle that they release the entrapped water between cement particles by specific mechanisms. Water reducers act as workability enhancers, i.e. by producing concrete of given workability at a lower water-to-cement ratio as compared to that of control concrete with no admixtures.

Based on the amount of water reduced, water reducers can be classified as

1. Normal-range water reducers (5–8%)
2. Mid-range water reducers (8–15%)
3. High-range water reducers (15–25%).

Some of the normal-range water reducers are lignosulfonates, hydroxyl-carboxylic acids and carbohydrates.

High-range water reducers can be further classified as

1. 1G-First generation (modified lignosulfonates)
2. 2G-Second generation (sulphonated naphthalene formaldehyde (SNF), sulphonated melamine formaldehyde (SMF))
3. 3G-Third generation (PCE based admixtures poly carboxylates and poly acrylates).

2.2.1 Mechanism

Water reducers generally work on two mechanisms. (1) Electrostatic repulsion: This makes use of the electrical charges developed on the surface of the cement components whenever dissolved in water. (2) Steric hindrance: Generally, it occurs in water reducers of the third generation where long side chains in the main chain of the polymer result in the formation of a separation between cement particles.

2.2.2 Compatibility Criteria

Lignosulfonates are obtained as lignin during the process of paper making. The waste liquor obtained from the process undergoes neutralisation, precipitation and fermentation processes, thus producing lingo-sulphonates of various ranges with varying composition and purity. The molecular weight of the admixture needs to be appropriately tuned to get the required performance; low molecular weight constituents are removed by ultra-centrifuging. Lignosulphonates constitute hydroxyl, carboxyl, methoxy and sulphonic groups. For better performance and compatibility, sulphonic functional group ($-HSO_3$) of lingo-sulphonate should be in the β position of naphthalene ring.

Identification of the compatibility of a particular superplasticiser with cement is of utmost importance, since this may result in immediate slump loss, bleeding and segregation, an increase in setting time and a loss in the strength of concrete in the hardened state. In general, optimum dosage and compatibility can be checked by the mini-slump and Marsh cone tests. Some of the factors that influence compatibility are the chemical and crystallographic nature of cement, fineness of cement, temperature, type of superplasticiser and its dosage, time and method of mixing.

2.3 SET REGULATORS

These kinds of admixtures are generally used in order to regulate setting time. In cold climatic conditions the setting is delayed; hence accelerators are used. In a similar way, in hot climatic conditions, retarders are used. Generally, retarders are organic and inorganic in nature and they interfere in the process of hydration, where hydration products that need to be formed will be delayed/accelerated. Moreover, the dissolution of ions either speeds up (accelerator) or slows down (retarders) in the cementitious system. Some examples of set regulators are gypsum, calcium chloride, potassium and sodium carbonate, sodium silicate, gluconates, lingo-sulphonates, sugars, etc. Concentration during addition of set regulators should be taken care of since a small variation in their concentrations can bring adverse effects.

2.4 AIR-ENTRAINING AGENTS

This chemical admixtures generate air bubbles and stabilise them by adequately changing the surface tension of water. They are generally used in structures where freeze-thaw action is pre-dominant. Some of the widely used air-entraining agents are vinsol resin and pine resin.

2.5 SPECIAL ADMIXTURES

There are other sets of chemical admixtures that can act as viscosity modifiers, shrinkage reducers and corrosion inhibitors.

2.5.1 Viscosity-Modifying Admixtures (VMAs)

Viscosity modifiers are generally used in highly flowable and workable concrete, such as SCC, where chances of segregation and bleeding are high. They can also be used as "anti-washout admixtures" since they can prevent washout of concrete in underwater applications. Some of the chemicals used as VMA's are xanthan gum, diutan gum, hydroxypropyl methylcellulose, etc.

2.5.2 Shrinkage-Reducing Admixtures

Shrinkage-reducing admixtures generally act in such a way that they reduce the surface tension of water in capillary pores. Subsequently, surface tension in water is responsible for the formation of a meniscus in capillary pores. As and when capillary water dries, it results in an inward pull from the capillary walls. This can be reduced by shrinkage-reducing admixtures and corresponding shrinkage cracks can be prevented.

2.5.3 Corrosion Inhibitors

In reinforced cement concrete (RCC) structures high alkalinity of the concrete prevents the steel from corrosion by the formation of a passive layer around it. As this alkaline environment changes to acidic in the presence of enough water and oxygen, the iron oxide (FeO) passive layer formed on the surface of the steel starts to deteriorate, resulting in the formation of rust ($Fe(OH)_3$), thus resulting in severe corrosion, spalling and deterioration. Moreover, a significant budget is needed to repair the structures.

Corrosion inhibitors act in different ways: they (1) act as oxidising and non-oxidising passivators of steel, (2) act as a scavenger as they consume oxygen around steel, (3) act as a hindrance to the steel surface access by the formation of an adsorbed layer such as films, and (4) reduce the available water by making cement paste hydrophobic, rendering cathodic effect.

Some of the examples include calcium nitrate, which can act as inorganic inhibitor and amines, and esters and alkanol-amines as organic inhibitors. In the meantime, they are expensive and their addition is restrained to small dosages, not more than 2%.

2.6 TESTING PROCEDURES

2.6.1 Determination of Ash Content

Admixtures are considered to be inevitable in today's concrete, notably, in case of concrete that requires a good amount of slump. During the synthesis of admixture, several inorganic reagents are used. The presence of inorganic materials in an admixture may result in the inhibition of admixture efficiency due to the intervention of the inorganic materials from the admixture in dissolution and hydration of cementitious materials.

Test Summary

In this test method, a pre-determined mass of admixture is ignited in a furnace to identify inorganic materials in the admixture (IS 9103 2018).

Use and Significance

- The presence of inorganic materials in an admixture may result in the intervention of the hydration process of cementitious materials, consequently resulting in setting as well as workability related problems. This test is useful to understand the inorganic material in the form of ash content in admixture.

Required Apparatus

- Weighing balance
- A silica or porcelain crucible with lid
- Air oven or steam water bath
- A desiccator
- A furnace

Procedure

Step 1. Take the initial weight of pre-ignited, cleaned crucible along with the lid. Let the weight be W_1.

Step 2. Place a sample of 1 g in the crucible, weigh to the nearest of 0.001 g. Let the weight of the sample along with the crucible and lid be W_2. Place the crucible in air oven or water bath at 90°C to remove the liquid portion by evaporation.

Step 3. Once the liquid portion evaporates, place the sample in a furnace and set the temperature so that within a period of 1 hour temperature increases to 300°C and within a period of 2–3 hours temperature increases to 600°C.

Step 4. Maintain the temperature at 600 ± 25°C for a period of 16–24 hours.

Step 5. After the specified period, remove the crucibles from the furnace and place it in a desiccator along with a lid until the crucible reaches room temperature for 30 minutes.

Step 6. Weigh the crucible with sample along with the lid to the nearest of 0.001 g. Let the weight be W_3.

Observations and Calculations

Initial mass and final mass after weighing is recorded in Table 2.1.

TABLE 2.1
Determination of Ash Content

	Sample 1	Sample 2	Sample 3
Weight of crucible and lid (W_1)			
Weight of crucible, lid and sample (W_2)			
Weight of crucible, lid and ash (W_3)			
Ash content $\frac{W_3 - W_1}{W_2 - W_1} \times 100$			

Report

Ash content ....

Student Remark

....
....
....
....

$$\frac{\textit{Marks obtained}}{\textit{Total marks}} = -$$

Instructor signature

Instructor Remark

....
....
....
....

2.6.2 Determination of Dry Material Content or Solid Content

Admixtures are considered to be inevitable in today's concrete, especially in case of concrete that requires a good amount of slump with less water content. Admixtures are added in concrete either in liquid or in solid form. In case if the admixtures are added in liquid form, a corresponding water reduction needs to be carried out. Water correction can be done by identifying the solid content in plasticiser.

Test Summary

In this test method, a pre-determined mass of admixture is heated with completely dry sand in a wide mouth glass bottle in an electric oven (IS 9103 2018).

Use and Significance

- Solid content of admixture can be determined using this test method. While using admixed concrete, it is important to make water correction to maintain water-to-cement ratio.

Required Apparatus

- Wide mouth glass bottle with 60 mm diameter and 30 mm height with a glass stopper
- Desiccator
- Pipette
- Electric oven
- Weighing balance

Procedure

Step 1. In the wide mouth glass bottle place 25–30 grams of dry sand. The bottle with sand along with stopper is placed in an oven for 17 ± ¼ hours at 105 ± 2°C. The period of drying can be reduced if the sand is already dry.

Step 2. When the glass bottle achieves a constant mass, the glass bottle along with a stopper is transferred into the desiccator to cool to room temperature. Weigh the bottle with sand along with stopper to the nearest of 0.001 g.

Step 3. The stopper is removed and with the help of a pipette 4 ml of liquid admixture is distributed evenly over the dry sand.

Step 4. Afterwards, the stopper is placed back and the entire setup is placed inside the electric oven for a period of 17 ± ¼ hours at 105 ± 2°C.

Step 5. Remove the setup from the oven and place it in a desiccator for cooling to room temperature. Now the setup is weighed to the nearest of 0.001 g.

Step 6. The setup along with the stopper is placed back inside the oven and dried for a period of one hour and weighed again. The setup is placed in the oven and dried again until the sample reaches a constant mass.

Step 7. In case of a non-liquid admixture, place about 3 g of non-liquid admixture in the glass bottle, place the stopper and place it in an electric oven at 17 ± ¼ hours at 105 ± 2°C. Similar measurements are adopted and recorded in Table 2.2.

Observations and Calculations

Initial mass and final mass after weighing is recorded in Table 2.2.

TABLE 2.2
Determination of Dry Material Content or Solid Content

	Sample 1	Sample 2	Sample 3
Weight of bottle and sand (W_1)			
Weight of bottle, sand and sample (W_2)			
Weight of sample ($W_2 - W_1$)			
Weight of bottle, sand and dried sample W_3			
Weight of dried sample ($W_3 - W_1$)			
Percentage of dried sample $\frac{W_3 - W_1}{W_2 - W_1} \times 100$			

Report

Dry material content or solid content ....

Student Remark

....

....

....

....

$$\frac{\textbf{\textit{Marks obtained}}}{\textbf{\textit{Total marks}}} = -$$

Instructor signature

Instructor Remark

....

....

....

....

2.6.3 Determination of Chloride Content (Volumetric Method)

During the synthesis of admixtures, several inorganic reagents are used. The presence of more than permissible content of chloride in admixtures may result in the start of corrosion in reinforced cement concrete.

Test Summary

In this test method, a pre-determined mass of admixture is mixed with the prepared standardised solution. By the volumetric method, the titration is carried out to determine the chloride content of admixture (IS 6925 2018).

Use and Significance

- Usage of admixture with chloride content more than the prescribed limit may result in corrosion of RCC. Consequently, this method is used to determine the amount of chloride content in the admixture.

Required Apparatus and Reagents

- Distilled water
- Nitric acid (1:2.5) – 6 N
- Sodium or potassium chloride solution (standard) – 0.1 N
- Silver nitrate solution – 0.1 N
- Nitrobenzene
- Ferric alum indicator solution
- Ammonium thiocyanate solution – 0.1 N
- Weighing balance
- Pipette
- Conical flask
- Burette

Preparation and Standardisation

Step 1. 8.5 g of silver nitrate is dissolved in 500 ml of distilled water in a volumetric flask. The solution is standardised against 0.1 N sodium or potassium chloride solution using potassium chromate as an indicator. Adjust the normality to 0.1 exactly.

Step 2. Similarly, 8.5 g of ammonium thiocyanate is dissolved in 1000 ml of distilled water in the volumetric flask. Shake well and the sample is standardised against 0.1 N silver nitrate solution using ferric alum solution as indicator. Adjust the normality to 0.1 precisely.

Procedure

Step 1. The sufficient quantity of admixture is accurately weighed, which has approximately minimum chloride content of 0.1 g. Hot water is added to the plasticiser and stirred well to attain the complete dissolution.

Step 2. The solution of 150 ml is prepared. If the presence of undissolved residue is detected, the sample is filtered. The filtered solution is added and shaken with water to make the solution to 250 ml.

Step 3. 50 ml of solution is pipetted out in a conical flask containing 5 ml of 6 N nitric acid. 10 to 15 ml of 0.1 N silver nitrate solution from the burette is added. Add 2 to 3 ml of nitrobenzene and 1 ml of ferric alum indicator and shake vigorously to coagulate the precipitate.

Step 4. Titration is carried out with 0.1 N ammonium thiocyanate solution until a permanent faint reddish-brown colour appears (excess silver nitrate is neutralised). Titration is repeated with another 50 ml portion.

Step 5. From the volume of silver nitrate solution added, subtract the volume of the thiocyanate solution required. Take the average of two determinations. Calculate the percentage of chloride in the sample.

1 ml of 0.1 N $AgNO_3 = 0.003546$ g of Cl^-.

Observations and Calculations

The volume of thiocyanate and silver nitrate solution is recorded in Table 2.3.

TABLE 2.3
Determination of Chloride Content

	Sample 1	Sample 2	Sample 3
Volume of thiocyanate solution (V_1)			
Volume of silver nitrate solution (V_2)			
$V_2 - V_1$			
Amount of Cl^- in (g)			
Percentage of Cl^- in sample			

Report

Chloride content ....

Student Remark

....
....
....
....

$$\frac{\textit{Marks obtained}}{\textit{Total marks}} = -$$

Instructor signature

Instructor Remark

....
....
....
....

2.6.4 Determination of PCE Charges by Acid-Base Titration

Charges carried by PCE on their backbone is one among the essential characteristics of PCEs that decides the affinity of PCE to adsorb on the surface and disperse cementitious materials (Mantellato 2017).

Test Summary

- This test method is used to identify backbone charges carried by PCE based superplasticisers by a titrimetric method.
- In this method, the base is used as a titrant and acid is used to protonate, un-protonated carboxylic groups in PCE based superplasticisers.
- The point at which the stoichiometric ratio is satisfied between the analyte and the titrant is considered to be an equivalent point.
- The equivalent point is signalised in terms of "volume". The indication of an equivalent point is captured in terms of an indicator.

Apparatus Required

- Burette
- Volumetric flask
- Conical flask
- Weighing balance
- pH meter

Procedure

Step 1. Potassium hydroxide (KOH) of concentration 0.4275 ± 0.004 M is prepared and used as a titrant. KOH is standardised by potassium hydrogen phthalate (KHPt).

Step 2. The prepared KOH is filled in the burette until zero mark.

Step 3. In a glass beaker, 1.5 g of polymer is taken and mixed with hydrochloric (HCl) acid (fuming ≥ 37%) drop by drop until the pH value drops to 1.6. This is done basically to protonate all the carboxylic groups available in the PCEs.

Step 4. This titration evolves with two equivalent points (EQPs). The first EQP corresponds to the titration of excess HCl, and the second EQP corresponds to the titration of carboxylic groups.

Step 5. To capture both the EQPs, a polymer solution is divided into two equivalent parts after acidification.

Step 6. Titration is done in one part to identify the EQP corresponding to acid. Titration is also done in another part to identify the EQP corresponding to acid and polymer. This point usually happens around a pH of 9.

Observations and Calculations

Required volume of titrant is noted down in Table 2.4.

$$\text{Number of carboxylic groups} = \frac{n}{m_{polymer}} = \frac{n_{EQP2}}{m_{polymer2}} - \frac{n_{EPQ1}}{m_{polymer1}}$$

Where, $m_{polymer2}$, $m_{polymer1}$ = mass of polymer in g.

TABLE 2.4
Charge Quantification to Identify EQPs

Titrant	Analyte	1.5 g of polymer with HCl (fuming ≥ 37%)
KOH 0.4275 ± 0.004 M	Volume of titrant	
	Initial (mL)	Final (mL)
EQP1		
EQP2		

$m_{polymer1} = m_{polymer2}$ = if the mass of polymer solution is split into two equivalent parts in g.

n = moles expressed in mmoles

Report

The number of charges can be expressed as .... mmol carboxylate/g of polymer.

Student Remark

....
....
....
....

$$\frac{\textbf{\textit{Marks obtained}}}{\textbf{\textit{Total marks}}} = -$$

Instructor signature

Instructor Remark

....
....
....
....

2.6.5 Determination of Optimum Dosage of Superplasticisers by Marsh Cone Test

Marsh cone test is used to understand the fluidity of cementitious pastes with respect to different types and dosage of superplasticisers. This test can be used to determine the optimum dosage of superplasticiser (Roussel and Le Roy 2005).

Use and Significance

- This test is used to identify relative fluidity of different cementitious pastes by measuring the time taken by a fixed quantity of paste to pass through the Marsh cone.
- Data obtained from this test can also be used to identify the compatibility of different superplasticisers with cement.

Limitations

- Optimised superplasticiser dosage for the cement paste cannot be used directly in the corresponding concrete even though the similar paste is being used.
- The usage of optimised dosage at the site that undergoes higher shearing needs to be taken care off.

Required Apparatus

- Marsh cone
- Stopwatch
- Graduated container

Precautions to Be Observed before Performing the Test

- The test needs to be conducted in an entirely controlled environment, since relative changes in environmental condition can bring about initial erroneous results.
- The procedure adopted, including the mixing duration and the operating person, need to be uniform for different trials.
- In case if the used superplasticiser is liquid in nature, water correction needs to be carried out by determining the solid content.

Procedure

Step 1. Cement is thoroughly mixed with 70% of water with the help of a Hobart mixer at a lower speed level.

Step 2. Afterwards, 30% of remaining water is taken in a beaker and the measured superplasticiser is adequately mixed with the water.

Step 3. This 30% of water is poured into the mixer and additional mixing of paste is adopted for 120 seconds.

Step 4. The Hobart mixer is stopped and the paste which sticks to the walls of the bowl is scrapped quickly. Afterwards, the paste needs to be mixed instantaneously for another 120 seconds to achieve a proper paste for testing. This method of paste preparation needs to be followed for different dosages of admixtures.

Step 5. About 1000 ml of the prepared paste is poured into the truncated long neck of Marsh cone by closing the orifice with an index finger

Step 6. A Marsh cone is shown in Figure 2.1.

Step 7. Through the small orifice flow of time taken for the flow of the poured paste is noted.

Step 8. The time taken for the paste to flow corresponds to the fluidity of cement paste and indicates superplasticiser's ability to disperse effectively.

Step 9. Tests are done at particular replacements of mineral admixtures by volume or weight at constant water to binder ratio for superplasticisers at different dosages.

Step 10. Flow time measurements are made correspondingly for each mix after mixing.

Step 11. Optimum dosage is observed through the dosage at which flow time observed significantly dropped by an increase in relative fluidity attributed due to maximum adsorption of superplasticisers by the cement paste. Further alteration in the flow time cannot be obtained after saturation dosage of superplasticiser.

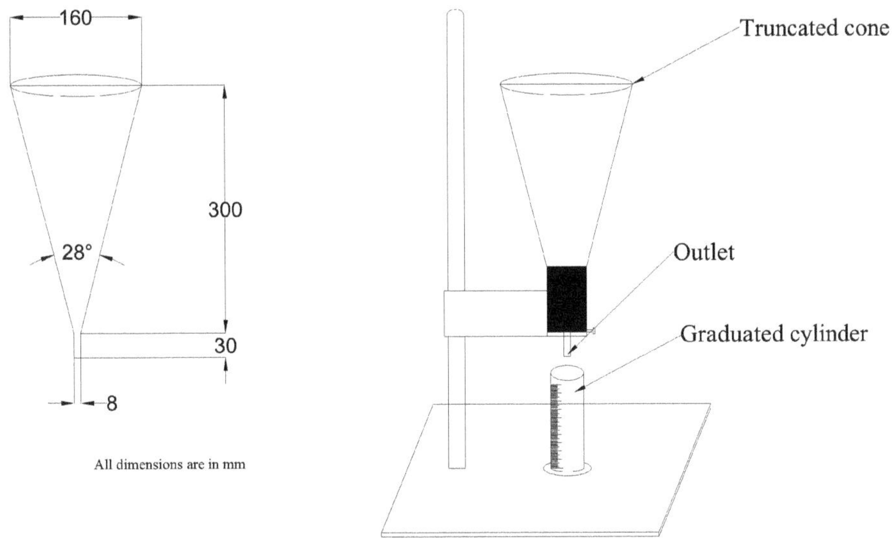

FIGURE 2.1 Schematic view of Marsh cone.

Step 12. A semi-log graph is plotted between the percentage of superplasticiser added in the x-axis and flow time in the y-axis (log). After the optimum dosage, there is no rise in the graph. It means that further addition of superplasticiser beyond the optimum dosage does not help to enhance the fluidity of the paste.

Step 13. It is interesting to note that the compatibility of superplasticiser and cement can be determined with the help of flow times measured at 5 minutes and 60 minutes. Just after the mixing of paste, flow is measured at 5 minutes. Once the test for 5 minutes is completed, then the paste is again collected in the bowl and closed with a plastic sheet to avoid the evaporation of water from the paste. After 60 minutes, in a similar way to the semi-log graph of 5 minutes, another semi-log graph is drawn for 60 minutes flow time. Coincidence at the common point indicates the compatibility between superplasticiser and binder.

Observations and Calculations

Cementitious materials used ....

Water to binder ratio (w/b) ....

Observed values are noted down in Table 2.5.

TABLE 2.5
Superplasticiser Dosage and the Corresponding Flow Time

Percentage of superplasticiser (by weight of cementitious material)	
Flow time (s)	

Report

Optimum dosage.............................

Note: Saturation dosage is considered to be the dosage at which a plot is drawn between flow time (s) (y-axis) and the percentage of superplasticiser added (x-axis) approaches a minimum and starts to remain at the same level (no significant decrease in flow time thereafter).

Student Remark

....
....
....
....

$$\frac{\textit{Marks obtained}}{\textit{Total marks}} = -$$

Instructor signature

Instructor Remark

....
....
....
....

2.6.6 Determination of Optimised Dosage of Superplasticiser by Mini-Slump Test

Similar to Marsh cone test, mini-slump test is also used to understand the fluidity of cementitious pastes. This test is also used to obtain optimised superplasticiser dosage (Kantro 1980).

Use and Significance

- This test is used to identify relative fluidity of different cementitious pastes by measuring the spread after pouring through mini-slump.
- This test can be used as a tool to optimise the dosage of superplasticiser.
- Data obtained from this test can be used to identify the effectiveness of different superplasticisers.
- The superplasticiser dosage at which the paste starts to bleed can be determined from this test.

Limitations

- Optimised superplasticiser dosage for the cement paste cannot be used directly in the corresponding concrete even though the similar paste is being used.
- Usage of optimised dosage in processing that undergoes higher shearing needs to be taken care of.

Required Apparatus

- Mini-slump
- Non-absorbent square board or a glass plate (minimum size 400×400 mm) with centre markings or graduations as required
- Vernier calliper
- Wooden mallet

Precautions to Be Observed before Performing the Test

- The test needs to be conducted in a perfect temperature- and humidity-controlled environment, since relative changes in environmental condition can bring about primary erroneous results.
- The procedure adopted for different trials need to be uniform.
- If the used superplasticiser is liquid in nature, then water correction needs to be carried out by determining the solid content.

Procedure

Step 1. The mini-slump cone is greased inside and placed on top of the non-absorbent square board at the centre.

Step 2. Paste prepared according to the requirement is poured into the mini-slump cone (approximately 40 ml).

Step 3. A mini-slump cone is shown in Figure 2.2.

Step 4. To ensure that no air bubbles are present after pouring, the mini-slump is tapped gently on its sides by a wooden mallet.

Step 5. After tapping, the mini-slump cone is lifted upwards. Ensure that there is no agitation to avoid the disturbance on the sample.

Step 6. Wait for a minute. Measure the diameter of spread obtained in two perpendicular directions. It is highly recommended to measure the diameter of the spread in more direction to attain accurate measurement.

Step 7. The spread of the paste corresponds to the fluidity of cement paste and indicates the superplasticiser's ability to disperse effectively.

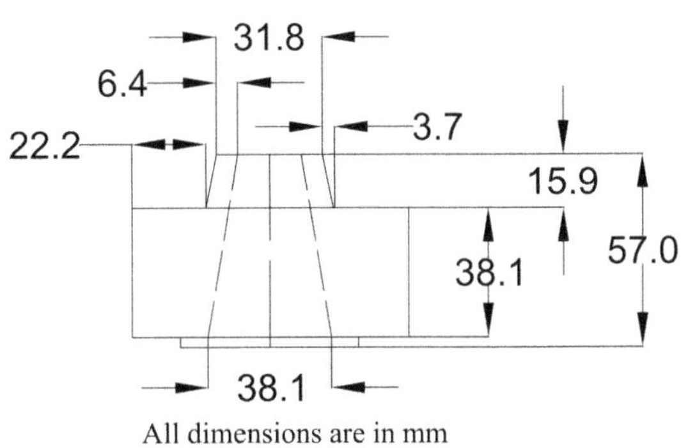

FIGURE 2.2 Schematic view of mini-slump.

Step 8. Tests are done at particular replacements of mineral admixtures by volume or weight at constant water to binder ratio for superplasticisers at different dosages.

Step 9. Spread measurements are made correspondingly for each mix.

Step 10. The optimum dosage corresponds to the dosage after which the change in the spread is minimal. Moreover, bleeding may have occurred at the higher dosages, which is more than the optimum level of dosage.

Step 11. A semi-log graph is plotted between the percentage of superplasticiser added in the x-axis and spread in the y-axis (log).

Observations and Calculations

Cementitious materials used ....

Water to binder ratio (w/b) ....

Observed values are noted down in Table 2.6.

TABLE 2.6

Superplasticiser Dosage and the Corresponding Average Spread

Percentage of superplasticiser (by weight of cementitious material)	
Average spread (mm)	

Report

Optimum dosage.............................

Student Remark

....

....

....

....

$$\frac{\textit{Marks obtained}}{\textit{Total marks}} = -$$

Instructor signature

Instructor Remark

....

....

....

....

2.6.7 Determination of Optimised Dosage of Viscosity-Modifying Admixtures

Viscosity-modifying admixtures (VMA) are water-soluble organic polymers of high molecular weight or inorganic materials such as colloidal silica or fine carbonate fillers. VMA's can be used as a stabiliser to reduce segregation and bleeding in highly flowable concrete such as self-compacting concrete (SCC). This is one among the various test methods to optimise VMA's (Nanthagopalan and Santhanam 2010).

Use and Significance

- VMA's are generally used to control the segregation and bleeding of highly flowable concrete.
- Optimisation of VMA is highly essential, since over-dosage of VMA will result in a loss in workability, higher retardation, increase in coarse air bubbles and reduced placing rate.
- Under-dosage of VMA will result in segregation and bleeding, especially wash out in underwater concreting.
- The optimum dosage of VMA will be capable of suspending the aggregates in the paste medium without segregation in both static and dynamic processes.

Limitation

- This test method is suitable for obtaining an optimised dosage for the static process.

Required Apparatus

- Marble test apparatus (schematic diagram as shown in Figure 2.3)
- Beaker of 1000 ml capacity
- A glass marble (specific gravity 2.51; diameter 18 mm) tied to a nylon wire
- A stand to hold the nylon wire along with the glass marble

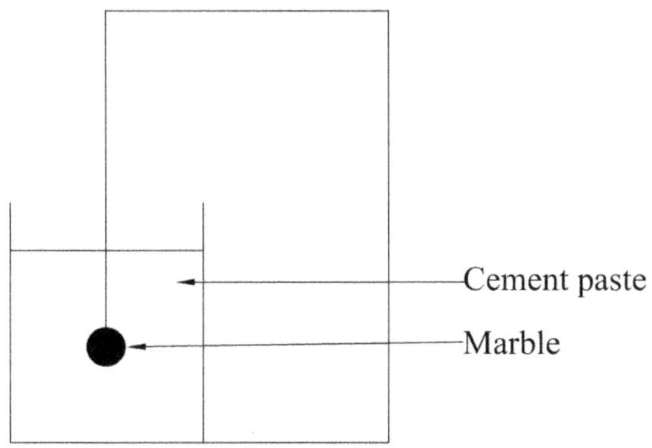

FIGURE 2.3 Schematic view of marble test.

Note:

- The size of the glass marble should be near to the maximum size of the aggregates used in the concrete.
- The surface of the glass marble must be wiped with a wet cloth each time before conducting the experiment.
- Fully stretched nylon wire with the glass marble corresponds to 82 mm maximum depth of penetration in the current setup.

Procedure

Step 1. 800 ml of cementitious paste at required water to binder ratio and necessary or the optimum dosage of superplasticiser as obtained from sections 2.6.5 and 2.6.6 is taken in the beaker.

Step 2. The beaker is placed on the base of the stand. In the meantime, the nylon wire along with the glass marble is held gently on the surface of the paste in the beaker and released.

Step 3. Based on the internal resistance of the paste (viscosity of the paste), the glass marble starts to move down in the paste.

Step 4. After a time-lapse of 4 minutes, when there is no further movement of the glass marble, the height of immersion of the marble is measured.

Step 5. The downward movement of the glass marble can be obtained by measuring the height of the paste adhering to the nylon wire from the bottom of the glass marble.

Step 6. Steps from 1 to 5 are repeated for different dosages of VMA's.

Step 7. At a low dosage of VMA, the viscosity of the paste will be low. Therefore the depth of penetration of marble will be higher.

Step 8. On the other hand, at a high dosage of VMA, the viscosity of the paste will be relatively high, thus resulting in less depth of penetration.

Step 9. Therefore, the optimum dosage can be considered as the dosage above, which the paste starts to resist the downward movement of the glass marble.

Step 10. A plot is obtained between percentage dosages of VMA (% by weight of binder) in x-axis and depth of penetration in the y-axis.

Observations and Calculations

Cementitious materials used ....

Water to binder ratio (w/b) ....

Observed values are noted down in Table 2.7.

TABLE 2.7
VMA Dosage and the Corresponding Depth of Penetration

Percentage of VMA added (by weight of cementitious material)	
Depth of penetration (mm)	

Report

Optimum dosage..............................

Student Remark

....
....
....
....

$$\frac{\textit{Marks obtained}}{\textit{Total marks}} = -$$

Instructor signature

Instructor Remark

....
....
....
....

2.7 PRACTICE QUESTIONS AND ANSWERS FROM COMPETITIVE EXAMS

1. The working mechanism of sulphonated naphthalene formaldehyde (SNF) is based on
   a) Electrostatic repulsion
   b) Steric hindrance
   c) Both a and b
   d) None of the above

2. Name two different mechanisms through which superplasticisers act.
   a) Electrostatic repulsion and Steric hindrance
   b) Hydrophilic action
   c) Electrostatic attraction and Steric hindrance
   d) Van der Waals attraction

3. With respect to cement and concrete what kind of environment makes superplasticisers active?
   a) High pH and high alkalinity
   b) Low pH and high alkalinity
   c) High pH and high acidity
   d) Low pH and low alkalinity

4. What is the reason behind identifying solid content of a superplasticiser?
   a) To maintain cement/superplasticiser ratio constant
   b) To maintain solid volume fraction constant
   c) To maintain water/binder ratio constant
   d) None of the above

5. What is the purpose behind the optimisation of superplasticiser dosage?
   a) If used below optimised dosage required performance may not be achieved
   b) If used above optimised dosage bleeding may occur
   c) Both a and b
   d) None of the above

6. What is the purpose behind determining ash content of chemical admixtures?
   a) Ash content will increase the water demand
   b) Ash content will reduce the workability
   c) Ash content beyond the specified limit will interfere in the process of hydration
   d) All the above

7. In general, usage of lignosulfonates not only increases fluidity but also
   a) Accelerates hydration
   b) Retards hydration
   c) Does not influence hydration
   d) Increases early age strength

8. Viscosity-modifying admixtures (VMA's) are generally used to
   a) Increase viscosity
   b) Increase thixotropy
   c) Both a and b
   d) Decrease thixotropy

9. What is the mechanism behind the action of a viscosity-modifying admixture (VMA)?
   a) Electrostatic repulsion
   b) Van der Waals attraction
   c) Steric hindrance
   d) Hydrophilicity

10. What are the shrinkages that can be reduced by shrinkage-reducing admixtures?
   a) Autogenous shrinkage
   b) Drying shrinkage
   c) Plastic shrinkage
   d) All the above

11. Superplasticisers on adsorption with cement particles .... pore solution.
    a) Reduces surface tension of
    b) Increases surface tension of
    c) Causes no changes to
    d) First reduces and then increases the surface tension of

12. Action of shrinkage-reducing admixtures will be on
    a) Water-cement surface interface
    b) Air-cement surface interface
    c) Water-air interface
    d) Cement-cement interface

13. In general, shrinkage-reducing admixtures are …………….. because of their ....
    a) Less soluble; hydrophilic nature
    b) More soluble; hydrophilic nature
    c) Less soluble; hydrophobic nature
    d) More soluble; hydrophobic nature

14. Which among the following corrosion inhibitors cannot be used with fresh concrete?
    a) Calcium nitrite
    b) Mono-fluoro-phosphate
    c) Sodium benzonate
    d) Dimethyl ethanolamine

15. Which among the following is a retarder?
    a) Calcium sulphate
    b) Chromium oxide
    c) Tributyl acetate
    d) Sodium penta chloro phenate

[AE 2015]

16. Admixtures which are used to lower the permeability of concrete are known as
    a) Accelerators
    b) Water-repellent admixtures
    c) Air-entraining agents
    d) Bonding admixtures

[AE 2015]

17. Admixtures which are used to lower the permeability of concrete are known as
    a) Accelerators
    b) Water-repellent admixtures

c) Air-entraining agents
d) Bonding admixtures

[AE 2018]

18. Sulphonated melamine formaldehyde is for .... climate.
    a) Hot
    b) Cold
    c) Rainy
    d) None of the above

19. According to Forsen set-controlling chemicals are classified into .... types.
    a) 4
    b) 3
    c) 5
    d) 2

1.	a	2.	a	3.	a	4.	c	5.	c	6.	d	7.	b	8.	c	9.	d	10.	d
11.	a	12.	c	13.	c	14.	b	15.	a	16.	b	17.	b	18.	b	19.	c		

REFERENCES

ACI 116R. (2000). *Cement and concrete terminology.* American Concrete Institute. Michigan, US.

Assistant Engineer (AE). 2020. Tamil Nadu Public Service Commission (TNPSC), Tamil Nadu, India. http://www.tnpsc.gov.in/previous-questions.html (accessed September 18, 2020).

Engineering Service Exam (ESE). 2020. Union Public Service Commission, New Delhi, India. https://www.upsc.gov.in/examinations/previous-question-papers (accessed September 18, 2020).

Graduate Aptitude Test in Engineering (GATE). 2020. GATE Office, Chennai, India. http://gate.iitm.ac.in/gate2019/previousqp18.php (accessed September 18, 2020).

IS 6925. (2018). *Methods of test for determination of water soluble chlorides in concrete admixtures.* Bureau of Indian Standards. New Delhi.

IS 9103. (2018). *Chemical admixtures – specifications.* Bureau of Indian Standards. New Delhi.

Junior Engineer (JE), Staff Selection Commission (SSC), New Delhi, India. https://ssc.nic.in/Portal/SchemeExamination (accessed September 18, 2020).

Kantro, D. L. (1980). Influence of water-reducing admixtures on properties of cement paste – a miniature slump test. *Cement, Concrete and Aggregates, 2*: 95.

Mantellato, Sara. (2017). *Flow loss in superplasticized cement pastes.* ETHZurich. Zurich.

Nanthagopalan, Prakash, and Santhanam, Manu. (2010). A new empirical test method for the optimisation of viscosity modifying agent dosage in self-compacting concrete. *Materials and Structures, 43*, 203–12.

Roussel, N., and Le Roy, R. (2005). The marsh cone: A test or a rheological apparatus? *Cement and Concrete Research, 35*, 823–30.

3 Aggregates

3.1 INTRODUCTION

Aggregates form 70–80% of the total volume of concrete, along with binders, i.e. cementitious materials. Aggregates can be either obtained naturally or produced by different crushing methods. The necessity for understanding the properties of aggregates arises since any changes brought about by the inherent properties of the aggregates can affect the integrity of the structure, which can be fatal in nature due to lack of durability and required strength. This necessitates the essentiality to know about the aggregates' origins, their production methods, the influence of different aggregate types and their characteristics in fresh and hardened properties of concrete. In general, aggregates can be classified into different heads, as shown in Figure 3.1.

3.2 CLASSIFICATION BASED ON GEOLOGICAL ORIGIN

Classification of aggregates based on their geological origin is of major importance because, based on the properties of the parent rock, the chemical, physical and surface characteristics of the aggregates differ. Consequently, identifying the suitable properties of parent rock can help in the selection of better aggregates that can increase the durability and strength of the constructed structures. Aggregates of natural origin can be of three different types based on their parent rock's geological origin. They are igneous, sedimentary and metamorphic.

Rocks are formed by the solidification of magma and lava, as magma and lava get exposed to variation in temperature and pressure that result in crystallization. Igneous rocks are of above-mentioned types, where the change in phase takes place from liquid to solid due to the exothermic process, where heat is released. The surface of the earth is mainly made up of igneous rock. The classification of igneous rocks is based on their composition and texture.

The name "sedimentary rocks" itself clearly states its nature of origin: through the collective deposition of minerals and organic materials. Generally, they are formed underwater. The utilization of sedimentary rocks for construction processes requires scrupulous selection skills. Subsequently, their nature varies from soft to hard, porous to dense and heavy to light. Further, consolidation degree, cementation type, layer thickness and contamination are critical factors for the selection of sedimentary rocks.

Any rock can be transformed into metamorphic rocks. Meta means "change" and morph means "form"; consequently, "metamorphic" means "change of form". It refers to the change in the form of their mineralogical composition by altering them from equilibrium to unstable condition by the application of external pressure and temperature. Finally, restoration of equilibrium, i.e. changes from unstable form to

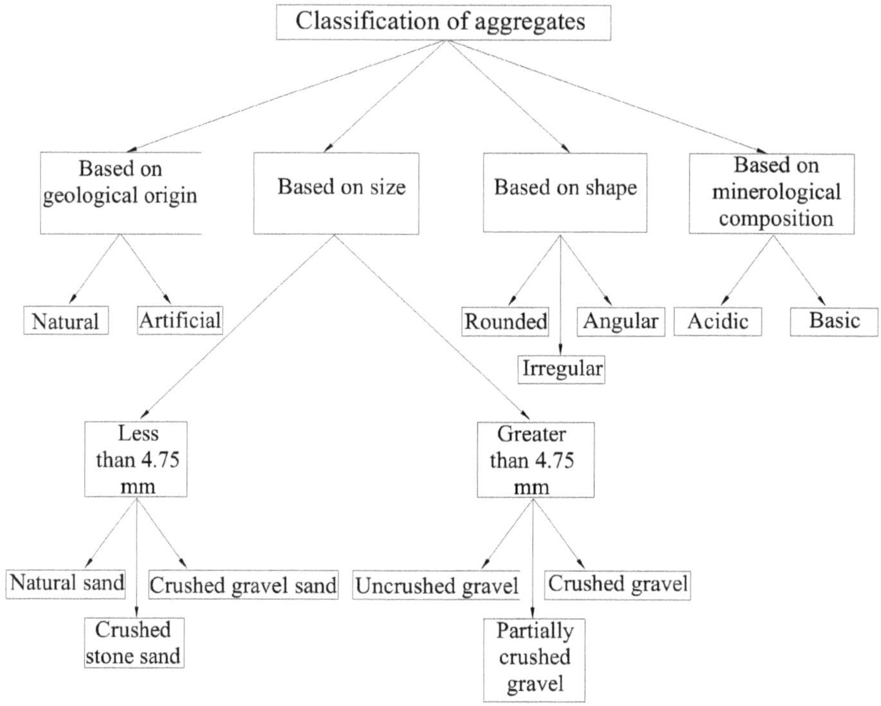

FIGURE 3.1 Classification of aggregates under different heads.

the stable form, results in the formation of transformed rock. Generally occurring metamorphic rocks are slate, schist, gneiss and marble. Some of the artificial aggregates are coarse blast furnace slag, ceramic waste, etc.

3.3 CLASSIFICATION BASED ON THE SIZE

Based on the size, aggregates can be classified into coarse, fine, all-in, filler and single-sized aggregates. Aggregates that pass through a 4.75 mm IS sieve are termed as fine aggregates. Fine aggregates can be further classified into three types depending upon their origin.

1. *Natural sand*: Those aggregates that are dis-integrated naturally from rocks and deposited by means of wind, water and glacier action.
2. *Crushed stone sand*: Hard stones are crushed to produce this kind of fine aggregates.
3. *Crushed gravel sand*: Natural gravels are crushed to produce crushed gravel sand.

Based on their size, fine aggregates can be classified into coarse sand, medium sand and fine sand. IS code has classified fine aggregates into different zones based on their gradation, primarily based on their fines content, i.e. the percentage of particles passing through a 150 μm sieve and the percentage of particles passing different sieves from 4.75 mm to 150 μm. Fine content generally varies from 0% to 15%. Another set of aggregates are coarse aggregates, where the particles that are retained in a 4.75 mm sieve according to IS specifications.

Based on the source, coarse aggregates can be of different types:

1. *Uncrushed gravel or stone*: These aggregates are obtained from the natural disintegration of rocks.
2. *Crushed gravel or stone*: These aggregates result from the crushing of gravel or hard stone.
3. *Partially crushed gravel or stone*: When the above two products are mixed together, these types of aggregates are formed.

Coarse aggregates are generally represented in terms of their nominal sizes, i.e. aggregates that pass a particular sieve size, for example, when most of the aggregates that pass 12.5 mm sieve, then the aggregates are termed as nominal size of 12.5 mm. The nominal size of aggregates is generally denoted by sieve sizes, for example, 40 mm, 20 mm, 12.5 mm, etc.

"All-in aggregates" is the technical term generally used to represent the aggregates that constitute both fine and coarse aggregates. For example, all-in aggregates of nominal size 12.5 mm means that it constitutes aggregates where most of the aggregates pass through a 12.5 mm sieve, but it includes other aggregates also.

3.4 CLASSIFICATION BASED ON THE SHAPE

Table 3.1 lists the classification of aggregates based on their shape but also their influence on the fresh and hardened properties of concrete. This type of aggregate is mostly formed due to attrition and crushing.

The desired shape and size of aggregates can be obtained by varying the production method during the process of quarrying and also adopting different crushers—for

TABLE 3.1
Classification of Aggregates Based on the Shape and Their Consequences in Concrete Properties

Shape of aggregate	Percentage of voids	Workability	Compressive strength
Rounded – Smooth surface	32–33	Higher workability	Less
Irregular or partly rounded	35–37	Less when compared to rounded	High compared to rounded because of increased bond strength
Angular	38–45	Lesser workability	10–20% higher due to strong aggregate-mortar bond

example, jaw crusher, vertical shaft crushers, etc. Difference in size and shape is predominantly due to the action of crushers that can vary their operations by imparting different forces: impact, attrition, shearing and compression.

3.5 CLASSIFICATION BASED ON MINERALOGICAL COMPOSITION

Based on the mineralogical classification, aggregates can be acidic or basic. Aggregates that have a higher content of silica (more than 65%) are termed as acidic, and those aggregates that have a silica content less than 55% are termed as basic in nature.

3.6 TESTING OF AGGREGATES

Aggregates are major constituents of the concrete. Consequently, properties of the aggregates added in concrete for construction are of utmost importance. For example, considering their shape, the usage of aggregates with rounded shape is more preferred in the case of normal construction. But the usage of similar types of aggregates in flexible pavement may lead to rutting related problems. In the same way, other properties of aggregates such as their chemical and mineralogical properties can have a severe effect on hardened properties of concrete in later stages. Coarse aggregates that have a high amount of reactive silica content may result in alkali-aggregate reaction and result in immature deterioration of concrete before its expected service life. Therefore, the usage of aggregates in a particular construction activity requires proper selection. Therefore, there is a need to understand the properties of aggregates to avoid undesirable consequences. Some of the essential properties that need to be taken care of are listed below:

1. Gradation and size
2. Particle shape angularity and surface texture
3. Porosity and absorption
4. Moisture sensitivity of aggregates
5. Toughness and abrasion resistance
6. Durability and soundness
7. Expansive characteristics
8. Polish and frictional characteristics
9. Density and specific gravity of aggregates
10. Cleanliness and deleterious materials
11. Surface area of aggregates
12. Thermal properties of aggregates
13. Reactive properties of aggregates

The need for proper testing of aggregates is to identify their properties and to decide on their suitability for use as a potential constituent of the concrete. Different types of testing procedures have been developed based on their usage in concrete structures. Some of the potential properties that are highly relevant to aggregate usage in concrete for construction are listed below.

3.6.1 Sieve Analysis or Gradation Test

In general, gradation of any granular material can be identified by sieve analysis, also called a gradation test. The sieve analysis can be used for a wide range of powdered materials, clay, sands, feldspar, crushed rocks, soils, etc. Gradation of sand is considered to be one among the critical parameter that governs its performance. Especially, construction of any structure constitutes aggregates with varied composition, diverse sizes and different proportions. Consequently, it is important to pack different constituents more appropriately in order to establish a densely arranged packing with minimum voids. To achieve a dense arrangement, the defined proportion of different-sized aggregates needs to be packed thoroughly. This can be achieved by choosing appropriate proportions. Proper proportion can be chosen by sieve analysis or gradation test.

3.6.2 Fineness Modulus (FM)

Through a gradation test, fineness modulus can be determined. In due course of the gradation test, aggregates are subjected to pass through sieves of different sizes. Sieve sizes vary from 150 μm to 4.75 mm for fine aggregates. Mostly, the size of the next higher sized sieve is obtained by approximately doubling the size of lower sized sieves. For example, the next sizes to 150 μm are 300 μm, 600 μm, 1.18 mm, 2.36 mm and 4.75 mm.

In the same way, for coarser sized aggregates, it extends from 4.75 mm to 38.1 mm above which increases in 2:1 ratio. When the aggregates pass through different sieve sets depending upon the crushing method or source of aggregates, a particular proportion of aggregates will be retained in each sieve. Fineness modulus is given by the sum of this proportion obtained in each sieve divided by 100. Since fineness modulus is a collective calculation of different-sized aggregates passing through different sieve sizes, similarly, fineness modulus can be obtained for different particle size distribution. In practice, aggregates with lower fineness modulus are said to be fine aggregates and higher values represent coarser aggregates. In general, FM of fine aggregates ranges from 2.0 to 4.0 and FM of coarse aggregates ranges from 6.5 to 8.0 for aggregates within 38.1 mm; apparently, the combination of fine and coarse aggregates yields intermediate values of FM.

Similar to particle size distribution, one among the critical parameter that needs to be quantified is their shape, angularity and surface texture. Parameters such as flakiness index, elongation index, un-compacted void content and the number of fractured faces can be used to understand shape, angularity and surface texture characteristics.

3.6.3 Flakiness Index

An aggregate is said to be flaky if its least dimension (thickness) is 0.6 times (3/5th) lesser than that of its mean dimension. When a similar investigation is done for a set of aggregates, then the index shows the percentage by weight of particles. For particles less than 6.3 mm, this test is not applicable.

3.6.4 Elongation Index

An aggregate is said to be elongated if its greatest dimension (length) is 1.8 times (9/5th) greater than that of its mean dimension. When a similar investigation is done for a set of aggregates, then the index shows the percentage by weight of particles. For particles less than 6.3 mm, this test is not applicable. The measure of void content can give a comprehensive view of the workability of the mixture when a similar aggregate is used in concrete. Further, a quick indication of angularity, sphericity and surface texture can be measured for known grading of aggregates by comparing with different grading of aggregates by means of measuring the **un-compacted void content**.

In order to identify the presence of deleterious material as well as cleanliness of the material, different tests are performed. Some of the tests are the sand equivalent test, liquid limit/plastic limit and methylene blue test.

3.6.5 Methylene Blue Test

Based on the absorption of methylene blue dye, the presence of clay-like material in a water-base drilling fluid can be determined.

To measure strength, toughness and abrasion of aggregates, different tests are available. Commonly used test methods for measuring strength, toughness and abrasion are the Los Angeles test and impact value test.

3.6.6 Los Angeles Test

This test is generally used to abrade the aggregates by loading it with some pre-fixed quantity of steel charges, also called as steel balls of specified weight. They are mixed with aggregates and rotated in a drum for specified revolutions in order to impart, impact as well as abrade on the aggregates. Abrasion value is determined by measuring the percentage wear of the aggregates due to the impact of steel charges. The value obtained is also known as the Los Angeles abrasion value.

3.6.7 Impact Value Test

Toughness is an intrinsic property of most of the construction materials, where the material possesses resistance against an impact. This characteristic value is determined by applying a sudden shock or impact on the aggregates. This value differs from the value obtained from a gradually loaded compressive load.

3.6.8 Aggregate Soundness

The soundness of aggregates can be determined by repeated wetting and drying of aggregates in sodium sulphate or magnesium sulphate solution. Whenever immersed, aggregate's water-soluble pores form salt crystals, resulting in higher internal pressure that results in the breaking of aggregates. Percentage loss of material is identified after sieving, which is usually performed after repeated submerging and drying.

Moreover, water absorption and specific gravity test are performed to identify the percentage of water to be added at the time of concrete mix proportioning. Since aggregates that are not in surface saturated condition tend to absorb water, they influence the actual water available during the process of mix proportioning.

3.7 TESTING PROCEDURES

3.7.1 Sieve Analysis of Fine and Coarse Aggregates

Packing of aggregates plays a vital role in enhancing both fresh and hardened properties of concrete. Moreover, the packing of aggregate skeleton is directly related to the size of aggregates. Therefore, it is essential to estimate the percentage of particles passing or retained in different sieves of standard sizes. This test method is used to determine the particle size distribution of fine and coarse aggregates.

Test Summary

A known mass of dry aggregates is passed through a set of progressive smaller opening sieves to determine the particle size distribution of the sample (IS 2386 Part 1, 2016a).

Use and Significance

- Packing of aggregates determines the porosity of the concrete. Packing in turn is determined by available percentage aggregates of all sizes (i.e. particle size distribution).
- To have an accurate estimation of particles finer than 75 µm, a separate washing technique needs to be adopted.
- Based on the particle size distribution obtained from this test, compliance of the size distribution is checked for the required specification based on the application. A control in aggregate production and products containing aggregates can be emphasized.

Required Apparatus

- Balance (0.1 g accuracy for fine aggregate and 0.5 g accuracy for coarse aggregate)
- Sieves
- A mechanical sieve shaker
- An oven (capable of maintaining $100 \pm 5°C$)

Sampling

Step 1. A sampling of aggregates is an essential practice to determine the correct particle size distribution of aggregates. Thorough mixing needs to be carried out before obtaining a sample for particle size distribution. It is preferred to use anyone of the well-defined sampling techniques used in regular practices.

Step 2. For fine aggregates, minimum of 300 g of dry sample is required.

Step 3. For coarse aggregate, the minimum sample size should be as prescribed in Table 3.2.

Procedure

Step 1. The sample is dried at a constant temperature of 100 ± 5°C. In general, for coarse aggregates, there is no need to dry, since the particle size distribution will be less affected by moisture content except:

- If the nominal maximum size of aggregate is smaller than 12.5 mm.
- If the appreciable amount of materials finer than 4.75 mm is present.
- If the lightweight aggregate is used, where the aggregates can absorb a higher amount of moisture.

Step 2. Select the required sieves and nest the sieves in their decreasing order of sizes from top to bottom and place the sample in the top sieve.

Step 3. Continuous manual shaking or mechanical shaking is applied to a prescribed timing (determined by a standard sample prior to conducting the test for the prescribed period).

Step 4. Quantity of the sample on each sieve should be limited in such a way that for particles that are smaller than 4.75 mm in size, the quantity retained should not exceed 7 kg/m^2 of sieving surface area. If the particle is more than 4.75 mm, the quantity retained should not exceed 2.5 × sieve opening size in mm × effective sieving area (mm^2). In any case, if overloading of an individual sieve is possible, then it should be prevented by necessary actions, e.g.

TABLE 3.2
Minimum Test Sample Required for the Testing Particle Size Distribution of Coarse Aggregates

Nominal maximum size (square opening) (mm)	Minimum test sample size (kg)
9.5	1
12.5	2
19	5
25	10
37.5	15
50	20
63	35
75	60
90	100
100	150
125	300

by adding an additional sieve with intermediate size between two sieves or splitting the sample and sieving it more times.

Step 5. The total percentage passing and retained in each sieve is recorded in Table 3.3 and calculated to the nearest of 0.1% of the total mass of the initial dry sample taken for the test.

Step 6. Fineness modulus is calculated by adding the total percentage of the material in the sample that is coarser than each of the following sieves and dividing it by 100. (150 μm, 300 μm, 600 μm, 1.18 mm, 2.36 mm, 4.75 mm, 10 mm, 12.5 mm, 16 mm, and 20 mm and larger).

Observations and Calculations

Total weight of aggregate ....

Observed weights are noted down in Table 3.3.

TABLE 3.3
Determination of Particle Size Distribution of Fine and Coarse Aggregate

IS sieve size (mm or μm)	Weight of aggregate retained (gram)				Percentage of total weight retained	Cumulative percentage of total weight retained	Percentage passing
	Trial 1	Trial 2	Trial 3	Average			
20 mm							
16 mm							
12.5 mm							
10 mm							
4.75 mm							
2.36 mm							
1.18 mm							
600 μm							
300 μm							
150 μm							
75 μm							
Pan							

Report

Depending upon the usage of material the final report should include the following:

- Total percentage of material passing each sieve
- Percentage of material retained on each sieve
- Fineness modulus reported to the nearest of 0.01
- Material passing 75 μm reported to the nearest of 0.1%

Student Remark

....
....
....
....

$$\frac{\textit{Marks obtained}}{\textit{Total marks}} = -$$

Instructor signature

Instructor Remark

....
....
....
....

3.7.2 Sieve Analysis for the Determination of Particles Finer Than 75 mm by Washing

This test method is used to determine the materials that are finer than 75 μm by washing. This method can remove clay as well as other aggregate particles by washing.

Test Summary

The sample to be determined with materials finer than 75 μm is washed in plain water or water containing wetting agents. Wash water that contains dissolved and suspended materials is then passed through 75 μm sieve. Loss in mass due to the washing is calculated and reported as a percentage finer than 75 μm (IS 2386 Part 1, 2016a).

Use and Significance

- Wet sieving can be used as an indication in order to assess the degradation of the aggregate.

Required Apparatus

- Balance (0.1 g precision)
- Sieves (1.18 mm sieve and 75 μm sieve)
- Container for vigorous agitation of wet aggregates sample
- An oven (can maintain $110 \pm 5°C$) and wetting agent

Sampling

The sampling of aggregate is an essential practice in order to determine the correct particle size distribution of aggregates. Thorough mixing needs to be carried out before obtaining a sample for particle size distribution. Any one of the standard sampling techniques needs to be followed. The minimum amount of sample should conform to Table 3.4.

TABLE 3.4
Nominal Maximum Size and Minimum Mass

Nominal maximum size	Minimum mass (g)
4.75 mm or smaller	300
Greater than 4.75–9.5 mm	1,000
Greater than 9.5–19 mm	2,500
Greater than 19 mm	5,000

Procedure

Step 1. Washing of the sample can be performed either by plain water or by a wetting agent. Unless mentioned, the wetting agent should not be used for washing.

Step 2. The test sample is dried in an oven at $110 \pm 5°C$ until it reaches a constant mass. Mass of the dry specimen is determined to the nearest of 0.1% of the mass of the test sample.

Step 3. The dried sample is placed in the container, and a sufficient amount of water is added to cover the sample. The sample is agitated vigorously, to bring the fine materials to suspension. The wash water that contains suspended and dissolved particles are poured in sieve sets with a coarser sieve at the top. As much as possible, the decantation of coarse aggregates of the sample is avoided.

Step 4. Water is added again, and the same procedure is repeated until the clean wash water is obtained.

Step 5. Samples retained in different sieve sets during washing operation are collected and dried in an oven at $100 \pm 5°C$ until a constant mass is obtained. The mass of the sample nearest to 0.1% of the original sample is measured.

Step 6. If the wetting agent needs to be used follow the same procedure instead of adding plain water during agitation, the wetting agent is added additionally, and a similar process is continued for the first time. For successive operations, only plain water is used. Follow successive steps similar to that of using only plain water.

Step 7. Materials passing 75 µm sieve is determined by

$$A = \left[\frac{B - C}{B}\right] \times 100$$

Where

A – Percentage of materials finer than 75 µm sieve by washing
B – Dry mass of sample (g)
C – Dry mass of the sample after washing (g)

Report

Following information reported

- If the particle passing 75 µm is less than 10%, then it should be reported to the nearest of 0.1%. If more than 10% report to the nearest whole number.
- Particles passing 75 µm ....

Usage of wetting agent (if any) ....

Student Remark

....
....
....
....

$$\frac{\textit{Marks obtained}}{\textit{Total marks}} = -$$

Instructor signature

Instructor Remark

....
....
....
....

3.7.3 Flakiness Index

Coarse aggregates with length-to-width ratio higher than a specified value are called elongated aggregate. Similarly, aggregates with length-to-thickness ratio greater than a specified value are called both elongated and flat particles. Likewise, the ratio of width to the thickness more than the specified value is called flat particles.

Test Summary

The relative shape characteristics of aggregates can be determined from this test procedure (IS 2386 Part 1, 2016b).

Use and Significance

- The shape of the particle affects packing and the workability of concrete during construction.
- This test can be used as one among the test to check the compliance of coarse aggregates towards a specific shape.

Required Apparatus

- Set of sieves (depends upon the size of aggregates under test)

- Thickness gauge confirming to IS: 2386 (Part 1)
- Weighing balance

Procedure

Step 1. Measure the weight of a minimum of 200 pieces of coarse aggregate samples collected from a lot.

Step 2. Sieves are arranged from descending order based on the requirement with the pan at the bottom and cover at the top.

Step 3. Sieves are arranged in the following order with cover at the top followed by other sieve sizes in the decreasing order (Cover, 20 mm, 16 mm, 12.5 mm, 10 mm, 6.3 mm, and pan).

Step 4. Measured coarse aggregates are thoroughly sieved, either manually or mechanically. Aggregates passing through and retained on two consecutive sieves are taken. Afterwards these aggregates are passed through the corresponding gauges marked in thickness, as shown in Figure 3.2. It is based on the two consecutive sieve sizes. The width of a slot is 0.6 times the average size of the two consecutive sieves. If the sample is passing through 20 mm size and retailed on 16 mm size, then $[0.6 \times (20+16)/2]$ is considered as the width of a slot.

Step 5. The particles passing through the corresponding gauge are weighed, and the determined mass is recorded. In the same way, the particles retained through the corresponding gauge are weighed and the determined mass is recorded in Table 3.5.

Observations and Calculations

Step 1. Total weight of the sample, m_1 (g) =

Step 2. Nominal size of aggregate, mm =

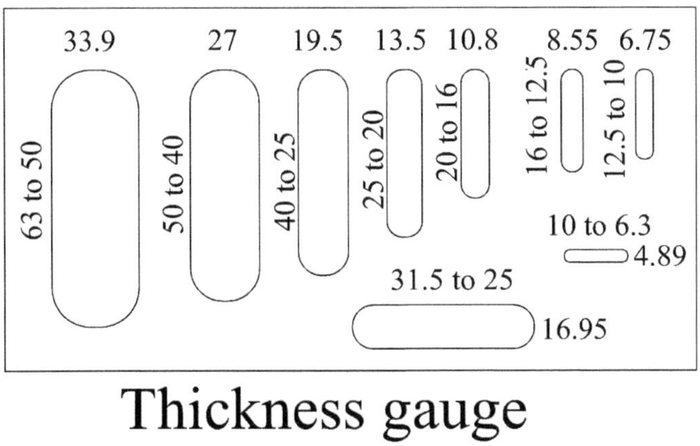

FIGURE 3.2 Thickness gauge (all dimensions are in mm).

TABLE 3.5
Flakiness Index of Aggregates

Size of aggregate			Weight of aggregate passing through thickness gauge	
Passing through IS sieve	Retained on IS sieve	Corresponding thickness gauge size	1	2
mm	mm	mm		
20	16	10.80		
16	12.5	8.55		
12.5	10	6.75		
10	6.3	4.89		
		Total weight, m_2 (g)		
		Flakiness index $\frac{m_2}{m_1} \times 100$		
		Average flakiness index in (%)		

Note: Flakiness index of coarse aggregates upper limit is 35–40%. An excess more than the above value is considered to be undesirable (IS 383, 2016).

Flakiness index =

- $$\frac{\text{Total weight of material passing through various thickness gauge}}{\text{Total weight of the sample taken}} \times 100$$

Report

Flakiness index is ....

Student Remark

....
....
....
....

$$\frac{\textit{Marks obtained}}{\textit{Total marks}} = -$$

Instructor signature

Instructor Remark

....
....
....
....

3.7.4 Elongation Index

Coarse aggregates with length-to-width ratio greater than a specified value are called elongated aggregates. Similarly, aggregates with length-to-thickness ratio greater than a specified value are called both elongated and flat particles. Likewise, aggregates with a ratio of width to the thickness more than the specified value are called flat particles.

Test Summary

Relative shape characteristics of aggregates can be determined from this test procedure (IS 2386 Part 1, 2016c).

Use and Significance

- The shape of the particle affects packing and the workability of concrete during construction.
- This test can be used as one among the test to check the compliance of coarse aggregates towards a specific shape.

Required Apparatus

- Set of sieves (depends upon the size of aggregates under test)
- Length gauge showing length slot confirming to IS: 2386 (Part 1)
- Weighing balance

Procedure

Step 1. Measure the weight of a minimum of 200 pieces of coarse aggregate samples collected from a lot.

Step 2. Sieves are arranged from descending order based on the requirement with the pan at the bottom and cover at the top.

Step 3. Sieves are arranged in the following order with cover at the top followed by other sieve sizes in the decreasing order (Cover, 20 mm, 16 mm, 12.5 mm, 10 mm, 6.3 mm, and pan).

Step 4. Measured aggregates are thoroughly sieved, either manually or mechanically. Aggregates passing through and retained on two consecutive sieves are taken and passed through the corresponding gauge marked in length gauge as shown in Figure 3.3. It is based on the two consecutive sieve sizes. The slot is 1.8 times of the average size of the two consecutive sieves. If the sample passing through 20 mm size and retailed on 16 mm size, then $[1.8 \times (20 + 16)/2]$ is considered as the length of the slot.

Step 5. The particles passing through the corresponding gauge are weighed, and the determined mass is recorded. Similarly, the particles retained through the corresponding gauge are weighed and the determined mass is recorded in Table 3.6.

Observations and Calculations

Total weight of the sample, m_1 (g) =

Nominal size of aggregate, mm =

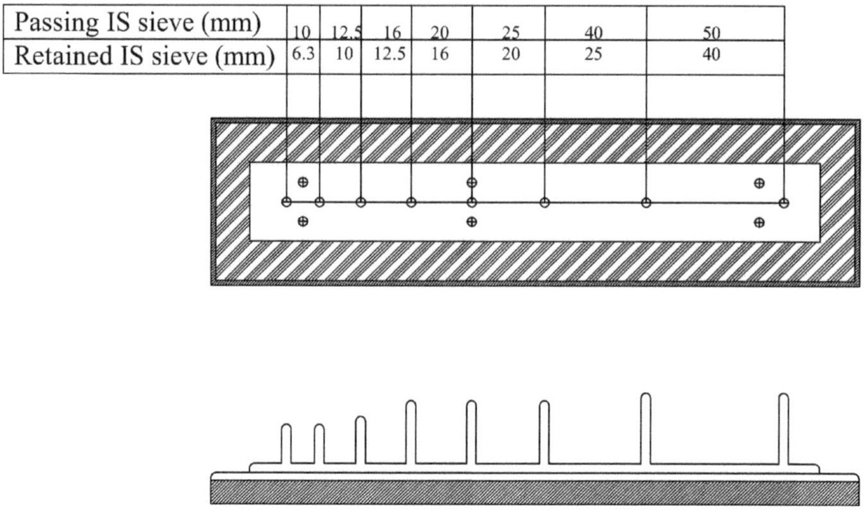

FIGURE 3.3 Length gauge.

TABLE 3.6
Elongation Index of Aggregates

Size of aggregate			Weight of aggregate retained on length gauge	
Passing through IS sieve	**Retained on IS sieve**	**Corresponding length gauge size**	**1**	**2**
mm	**mm**	**mm**		
20	16	32.4		
16	12.5	25.6		
12.5	10	20.6		
10	6.3	14.7		
		Total weight, m_3 (g)		
		Elongation index $\frac{m_3}{m_1} \times 100$		
		Average Elongation index in (%)		

Note: Elongation index of coarse aggregates upper limit is 35–40%. An excess more than the above value is considered to be undesirable (IS 383, 2016). As per BIS, the elongation index is not applicable for aggregates smaller than 6.3 mm (IS 2386 Part 1, 2016c).

Elongation index =

- $$\frac{\text{Total weight of material retained through various length gauge}}{\text{Total weight of the sample taken}} \times 100$$

Report

Elongation index is.................

Student Remark

....
....
....
....

$$\frac{\textit{Marks obtained}}{\textit{Total marks}} = -$$

Instructor signature

Instructor Remark

....
....
....
....

3.7.5 Aggregate Crushing Value

Coarse aggregates used in construction works, mainly in pavements, undergo higher crushing during vehicular movement. The crushing of aggregates results in detachment and deterioration of aggregates to smaller sizes. The decrease in the size of aggregates during continuous vehicular movement results in instability and undulation in pavements.

Test Summary

Specified compressive load is applied to aggregates in the prescribed rate and disintegration of aggregates due to the applied crushing load is calculated (IS 2386 Part 4, 2016a).

Use and Significance

- Poor quality of aggregates in terms of crushing may result in undulation and disorientation of flexible pavements as well as cracking in the rigid pavement.
- Higher crushing value means the resistance of the aggregate towards crushing is less.

Required Apparatus

- An open-ended steel cylinder of diameter 150 mm with a plunger and base plate
- A cylindrical metal measure of 115 mm diameter and 180 mm height that can withstand rough usage

- A steel taping rod of 16 mm diameter and 450–600 mm long with a rounded edge on its one end
- A weighing balance (accurate to 1 g)
- IS sieves of 12.5 mm, 10 mm and 2.36 mm
- A compression testing machine capable of applying 40 tonnes of load at a uniform rate of 4 tonnes per minute
- An oven capable of maintaining 100–110°C

Procedure

Step 1. Coarse aggregates are taken for testing using any standard sampling technique. The sample for testing is prepared according to the following requirements, as shown in Table 3.7.

Step 2. The sample to be tested should be in surface dry condition or dried by heating for 4 hours at 100–110°C. The sample should be at room temperature before testing. Aggregates passing through 12.5 mm sieve and retained on 10 mm sieve are taken for the test.

Step 3. A cylindrical metal measure of 115 mm diameter and 180 mm height, as shown in Figure 3.4, is filled with the required size of aggregates in three equal layers with each layer tamped 25 times uniformly by the rounded edge of the steel tamping rod.

Step 4. After tamping the third layer for 25 blows, the top layer is levelled by tamping rod considering the tamping rod as a straight edge.

Step 5. A sample from the cylinder is removed, weighed and refilled in the open-ended steel cylinder of diameter 150 mm or 75 mm based on the requirement. Consider the mass of the sample to be A.

Step 6. Refilling of aggregate is done in three equal layers with each layer tamped 25 times uniformly by the rounded edge of the steel

TABLE 3.7
Aggregate Crushing Strength Details for the Non-standard Size of Aggregates

Nominal sizes (IS Sieves)		Diameter of the cylinder to be used	Size of IS sieves for separating fines
Passing through	Retained on		
mm	mm	mm	mm
25	20	150	4.75
20	12.5	150	3.35
10	6.3	150 or 75	1.70
6.3	4.75	150 or 75	1.18
4.75	3.35	150 or 75	850 μm
3.35	2.36	150 or 75	600 μm

Note: Natural aggregates of 6.5 kg is required for 150 mm diameter cylinder or 1 kg is required for 75 mm cylinder

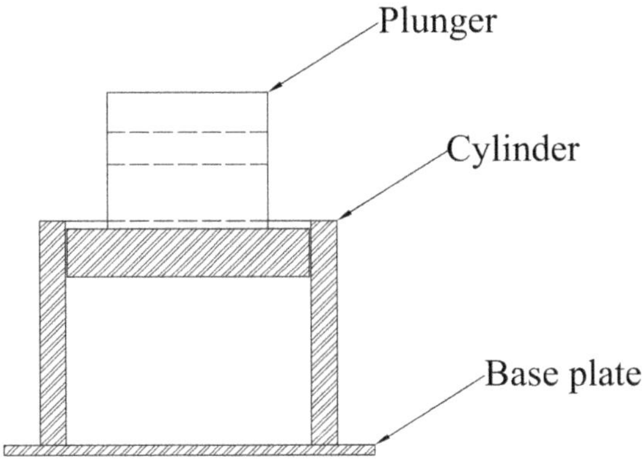

FIGURE 3.4 Cylindrical metal measure.

tamping rod. This step is followed by placing the plunger on the top of the specimen.

Step 7. Complete setup with a steel cylinder, specimen and plunger is placed on the compression testing machine. The load is applied at a rate of 4 tonnes per minute for 10 minutes until the maximum specified load of 40 tonnes.

Step 8. As soon as the loading is completed, aggregates are sieved through 2.36 mm IS sieve. The material passed through the 2.36 mm sieve is collected and weighed. The measured weight is denoted as B. This mass represents the amount of aggregate crushed to powder due to less resistance to the crushing load.

Caution: Care should be taken to avoid loss of fines while the above procedure is followed.

Note: In order to determine the crushing value of aggregates of size below 10 mm, a steel tamping rod of 8 mm diameter and 300 mm diameter is used. Compression load needs to be applied at a rate of 4 tonnes per minute, and the complete load needs to be applied within 10 minutes.

Observations and Calculations

Aggregate crushing value is given by $\frac{B}{A} \times 100$

Where,

A – the initial mass of the sample

B – the mass of the sample passing through 2.36 mm sieve

Report

Aggregate crushing value is ....

Specifications and Limits

Aggregate crushing value should not exceed 30% for concrete wearing surfaces such as runways, pavement and roads. Similarly, the aggregate crushing value should not exceed 45% for aggregate used for concrete other than wearing surfaces according to IS 383 (IS 383, 2016).

Student Remark

....
....
....
....

$$\frac{\textbf{\textit{Marks obtained}}}{\textbf{\textit{Total marks}}} = -$$

Instructor signature

Instructor Remark

....
....
....
....

3.7.6 Aggregate Impact Value

Vehicular movement on the road causes a higher amount of impact on the aggregates. A sudden impact on aggregates can cause different problems when compared to aggregate during crushing. Impact load varies from crushing in such a way that a uniform load is applied during crushing.

Test Summary

Impact load is applied to aggregates in a prescribed rate, and the disintegration of aggregates due to the applied impact load is calculated (IS 2386 Part 4, 2016b).

Use and Significance

- Poor quality of aggregates in terms of crushing may result in undulation and disorientation of flexible pavements as well as rigid pavements.
- Higher impact value means the resistance of aggregate towards impact is very less.

Required Apparatus

- Impact testing apparatus consists of a metal base weighing 22–30 kg and an open-ended cylindrical metal cylinder with an inner diameter of 102 mm and depth of 50 mm
- A cylindrical metal hammer weighing 13.5–14 kg with a diameter of 100 mm and a length of 50 mm attached to vertical guides by means of a support
- Preferably a counter needs to be attached along with the apparatus in order to count the number of blows

- Hammer on releasing should move freely on the vertical guides and drop on the sample placed in the cylindrical metal cylinder from a height of 380 ± 5 mm
- IS sieves of 12.5, 10 and 2.36 mm, tamping rod of 10 mm diameter and 230 mm long with rounded edge
- A metal measure of 75 mm diameter and 50 mm depth
- Balance of 0.1 gram accuracy
- Oven capable of maintaining 100–110°C

Test Sample Preparation

Step 1. The sample for testing is prepared according to the requirement, as shown in Table 3.8.

Step 2. The sample to be tested should be in surface dry condition or dried by heating for four hours at 100–110°C. The sample should be at room temperature before testing. Aggregates passing through 12.5 mm sieve and retained on 10 mm sieve are taken for the test.

Step 3. A cylindrical metal measure of 75 mm diameter and 50 mm height is filled with the required size of aggregates in three equal layers with each layer tamped 25 times uniformly by the rounded edge of the steel tamping rod.

Step 4. After tamping the third layer for 25 blows, the top surface is levelled by a tamping rod considering the tamping rod as a straight edge.

Step 5. The sample from the cylinder is removed and weighed. The weight is considered as the initial mass of the aggregate sample A.

Step 6. The aggregate sample is transferred in the open-ended steel cylinder (diameter 102 mm and depth 50 mm) as a single layer and 25 blows are adopted using the tamping rod.

TABLE 3.8

Aggregate Impact Value Details for the Non-standard Size of Aggregates

Nominal sizes (IS Sieves)		Diameter of the cylinder to be used	Size of IS sieves for separating fines
Passing through	Retained on		
mm	mm	mm	mm
25	20	150	4.75
20	12.5	150	3.35
10	6.3	150 or 75	1.70
6.3	4.75	150 or 75	1.18
4.75	3.35	150 or 75	850 microns
3.35	2.36	150 or 75	600 microns

Note: Natural aggregates of 6.5 kg are required for 150 mm diameter cylinder or 1 kg is required for 75 mm cylinder

Procedure

Step 1. Prepared test sample in the open-ended steel cylinder (102 mm diameter and 50 mm depth) is placed at the bottom of the impact testing apparatus. The impact testing apparatus is placed on a plane surface, ensuring that the guides are in the vertical position.

Step 2. After placing the specimen in the position of impact testing apparatus, the hammer is lifted to a height of 380 ± 5 mm and dropped 15 times, as shown in Figure 3.5. The hammer is allowed to fall without obstruction under the influence of gravity. The time between successive drops shall not be less than 1 second.

Step 3. After 15 blows, aggregates are passed through the 2.36 mm IS sieve. The weight of the material passed as well as retained through the 2.36 mm IS sieve is measured. These masses are considered as B and C, respectively.

Caution: Care should be taken to avoid loss of fines, while the above procedure is followed. If the loss of fines is more than 1 gram (A is greater than B + C by 1 gram), then the test needs to be repeated.

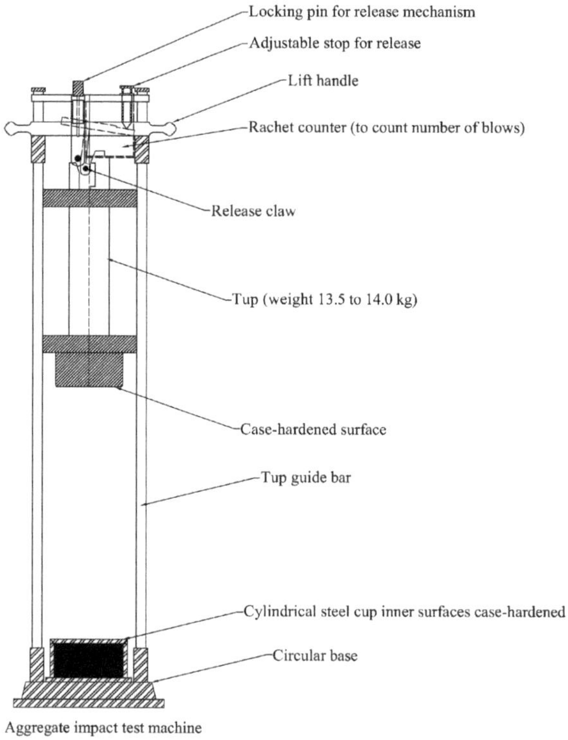

FIGURE 3.5 Aggregate impact test apparatus.

Observations and Calculations

Aggregate impact value is given by $\frac{B}{A} \times 100$

Where,

A – the initial mass of the sample

B – the mass of the sample passing through the corresponding IS sieve after testing

Report

Aggregate impact value is.................

Specifications and Limits

Aggregate impact value should not exceed 30% for concrete wearing surfaces such as runways, pavement and roads. Similarly, aggregate impact value should not exceed 45% for aggregate used for concrete other than wearing surfaces according to IS 383 (IS 383, 2016).

Student Remark

....

....

....

....

$$\frac{\textit{Marks obtained}}{\textit{Total marks}} = -$$

Instructor signature

Instructor Remark

....

....

....

....

3.7.7 Aggregate Abrasion Value (Los Angeles Test)

Aggregates used in the surface of the pavement will undergo wear and tear due to the friction between the vehicular tyres and the road during the due course of time. Moreover, this effect will be enhanced due to the presence of dust particles on the surface of the road.

Test Summary

Abrasion resistance of the aggregate is checked by abrading the aggregates by using charges in the Los Angeles test (IS 2386 Part 4, 2016c).

Use and Significance

- Poor quality of aggregates in terms of abrasion may result in quick abrasion of the surface of the pavement. By determining the abrasion value by this test method suitability of given aggregate for use in a particular road work can be assessed.
- Higher abrasive resistance means the aggregate can withstand higher resistance towards abrasion.

Required Apparatus

- Los Angeles apparatus of 700 mm internal diameter and 500 mm length that can rotate along its horizontal axis with a removable cover and a steel shelf provided radially projected throughout the length located at a distance of 1,250 mm away from the opening as per IS: 2386 (Part 4)
- Besides 12 abrasive charges of steel balls 48 mm diameter and weight 390–445 grams (the number of charges used depends upon the type of grade to be determined as shown in Table 3.9)
- IS sieve size of 1.7 mm and the other sieve size depends upon the type of grade selected
- Balance of 1 gram accuracy
- Oven capable of maintaining 100–110°C

Test Sample Preparation

Step 1. The sample for testing is prepared according to the requirement, as shown in Table 3.9. Let the mass of the sample taken be A.

Step 2. Sample to be tested should be in surface dry condition or dried by heating at 100–110°C until the sample attains substantially constant weight and conforming to the grading, as shown in Table 3.9.

Procedure

Step 1. Test samples and abrasive charges are placed in the apparatus and rotated at a speed of 20–33 revolutions per minute.

Step 2. For aggregates belonging to grades A, B, C and D, the apparatus is rotated for 500 revolutions. For other aggregate grades (E, F and G) the apparatus is rotated for 1,000 revolutions.

Step 3. Uniform peripheral speed is maintained by counterbalancing the apparatus.

Step 4. After the above-mentioned number of revolutions, the sample is taken out and sieved through 1.7 mm IS sieve. The weight of the sample passing through 1.7 mm sieve is noted; let the weight of the sample be B. Alternatively, samples coarser than 1.70 mm sieve are washed and placed in an oven at 105–110°C until the sample attains substantially constant weight. This weight is also considered for the calculation.

Step 5. Los Angeles apparatus is shown in Figures 3.6 and 3.7.

TABLE 3.9
Grading of Test Sample

Sieve size (Square mould)			Weight in a gram of test sample for grade						
Passing on	**Retained on**	**Grading**	**A**	**B**	**C**	**D**	**E**	**F**	**G**
(mm)	**(mm)**	**No of spheres**	**12**	**11**	**8**	**6**	**12**	**12**	**12**
		Weight of charge (g)	**5,000 ± 25**	**4,584 ± 25**	**3,330 ± 20**	**2,500 ± 15**	**5,000 ± 25**	**5,000 ± 25**	**5,000 ± 25**
80	63	–	–	–	–	–	2,500*	–	–
63	50	–	–	–	–	–	2,500*	–	–
50	40	–	–	–	–	–	5,000*	5,000*	–
40	25	–	1,250	–	–	–	–	5,000*	5,000*
25	20	–	1,250	–	–	–	–	–	5,000*
20	12.5	–	1,250	2,500	–	–	–	–	–
12.5	10	–	1,250	2,500	–	–	–	–	–
10	6.3	–	–	–	2,500	–	–	–	–
6.3	4.75	–	–	–	2,500	–	–	–	–
4.75	2.36	–	–	–	–	5,000	–	–	–

*Tolerance of ± 2 percentage is allowed

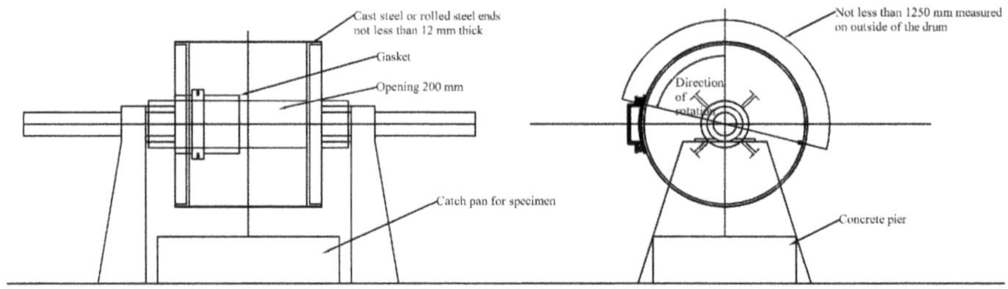

FIGURE 3.6 Los Angeles apparatus.

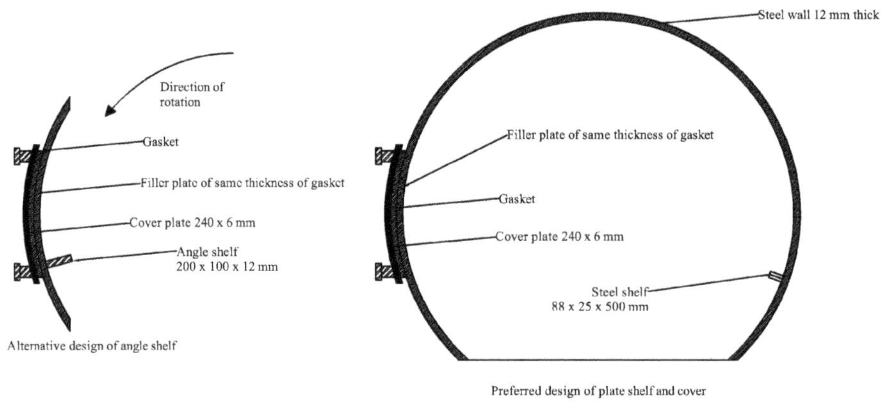

FIGURE 3.7 Los Angeles apparatus – plate shelf and cover.

Note: In order to understand that the experiment is conducted in a homogeneous sample, the test is conducted until 100 revolutions and the loss in mass is also accounted. While removing the samples during the above procedure, care should be taken to completely return back the samples into the apparatus without any loss.

Observations and Calculations

The initial weight of the sample, A =

Weight of the sample passing through 1.7 mm sieve, B =

$$\textbf{Percentage of wear } (\%) = \frac{B}{A} \times 100$$

Report

Percentage of wear is the difference between the original mass of the sample and the final mass of the sample expressed as a percentage of the original mass of the sample.

Specifications and Limits

Aggregate abrasion value should not exceed 30% for concrete wearing surfaces such as runways, pavement and roads. Similarly, aggregate abrasion value should not exceed 50% for aggregate used for concrete other than wearing surfaces according to IS 383 (IS 383, 2016).

Student Remark

....
....
....
....
....

$$\frac{\textbf{\textit{Marks obtained}}}{\textbf{\textit{Total marks}}} = -$$

Instructor signature

Instructor Remark

....
....
....
....
....

3.7.8 Stripping Value of Aggregates

The presence of water on the surface of the aggregate causes adhesion problems when these aggregates are used for the construction of pavement. This can be avoided or reduced by drying the aggregates or increasing the heating temperature during mixing. Excessive water during mixing can result in the stripping of bitumen from the surface of the aggregate. Therefore, it is essential to assess the effect of the presence of water and the subsequent loss of adhesion between bitumen and aggregate.

Test Summary

This test method is used to understand the effect of the presence of water on the adhesion of bitumen on the surface of aggregates (IS 6241, 2013).

Use and Significance

- Under controlled laboratory conditions, immersing a mixture of bitumen and aggregate underwater can be used to understand the loss of adhesion between bitumen and aggregate.

Required Apparatus

- Glass beaker (heat resistant) 500 ml capacity
- IS sieves 20 mm and 12.5 mm
- Mixer
- Balance (1 g accuracy)
- Water bath with thermostat

Sampling and Testing Procedure

Step 1. A sample of 200 grams of aggregate passing 20 mm IS sieve and retained in 12.5 mm IS sieve is taken and heated at 150°C.

Step 2. Bitumen of 5% of the weight of aggregate is taken and heated at 160°C. Bitumen and the aggregates are mixed until the aggregates are completely coated.

Step 3. Now the aggregates are transferred into 500 ml beaker and allowed to cool at room temperature for 2 hours.

Step 4. Distilled water is added until immersing the aggregates in a 500 ml beaker.

Step 5. Taking care of the level of water in the beaker to be maintained at least half of the volume of beaker, the beaker is transferred to a water bath and the temperature of the water bath is maintained at 40°C for 24 hours.

Step 6. Beaker is removed after 24 hours from the water bath and allowed to cool at room temperature.

Step 7. By visual observation uncovered area of the aggregate is estimated.

Observations and Calculations

Observations are noted down in Table 3.10.

TABLE 3.10
Observation Table for Stripping Value of Aggregates

Type of aggregate		**Type of binder**	
Percentage of binder used			
Total weight of aggregate			
Total weight of the binder			
Temperature of the water bath			
Samples	**1**	**2**	**3**
Stripping (%)			
Average			

Report

- Stripping value is expressed as a percentage of the uncovered areas of the aggregates observed visually to that of the total area of aggregates in each test.
- The average value of the three tests is considered to be stripping value of aggregates.
- Stripping value...........................

Note: IRC recommends 5% as maximum allowable stripping value of aggregate for surface dressing ((IRC 110, 2005) Clause: 4.2. (Table 2)).

For bituminous surfacing layers, the maximum allowable stripping value is recommended as 15% ((IRC SP 20, 2002) Clause: 4.10.6 (Table 4.14)).

Student Remark

....
....
....
....

$$\frac{\textit{Marks obtained}}{\textit{Total marks}} = -$$

Instructor signature

Instructor Remark

....
....
....
....

3.7.9 Relative Density and Absorption of Fine Aggregates

This test is used to determine the relative density (dimensionless quantity) at oven-dry, saturated surface dry condition. This test method does not apply to specified lightweight aggregates.

Test Summary

Aggregates for which specific gravity and absorption need to be determined are placed in water for 24 ± 4 hours. This procedure primarily fills the voids within the aggregates. After immersing for the recommended amount of time, aggregates are taken out and surface dried and mass is determined. By using the gravimetric method, the volume of the aggregate is determined. As a final point, the mass of the sample is determined again after oven drying (IS 2386 Part 3, 2016a).

Use and Significance

- Relative density or specific gravity is a ratio of the mass of aggregate to that of the mass of an equivalent volume of water displaced by the aggregate. This value is used to distinguish between the density of the aggregate and the bulk density of the aggregate, which includes the volume of voids between the particles of aggregates.
- Relative density can be used to estimate the volume of aggregates occupied in cement concrete, concrete bitumen and other mixtures.

Required Apparatus

- Balance (at least 0.1 g accuracy)
- Oven capable of maintaining $110 \pm 5°C$

- Mould and a metal tamper for surface moisture test
- Mould is a truncated cone with a bottom inner diameter of 90 ± 3 mm, a top inner diameter of 40 ± 3 mm with a total height of 75 ± 3 mm and a thickness not less than 0.8 mm
- Metal tamper should be of mass 340 ± 15 g and a flat tamping face with a diameter of 25 ± 3 mm

Test Specimen Preparation

Step 1. A test sample of approximately 500 g is placed in a suitable pan or vessel. The sample is dried to a constant weight at a temperature of 110 ± 5°C.

Step 2. Allow the sample to cool (until the sample reaches a temperature of 50°C approximately to handle without difficulty).

Step 3. Immerse the sample in water and allow it to stand for 24 ± 4 hours.

Step 4. If the specific gravity of the sample from the field needs to be used as such, then the above treatment is not required. In its place, specific gravity is reported in terms of its natural condition. Generally, in the field during concrete construction, aggregates are sprinkled with water and made to saturated surface dry condition or else water correction needs to be carried out by way of determining the water absorption of the aggregate.

Note: Water absorption and relative density of aggregates at saturated surface dry condition will be higher for the aggregates without the above treatments when compared to the aggregates with treatments.

Step 5. After immersing the sample for 24 ± 4 hours, the water is poured out without losing the fines. Afterwards, the sample is spread on a non-absorbent flat surface exposed to gently flowing warm air. If needed, mechanical means of tumbling and stirring are done to ensure a homogenous drying condition.

Step 6. In order to confirm that the sample has reached saturated surface dry condition, a flow test is conducted with the truncated cone (mould) and a metal tamper.

Step 7. The truncated cone is placed with the larger diameter on a flat surface. The sample is filled until a heap is formed and overflow from the truncated cone. The metal tamper is placed at a height of 5 cm above the sample and dropped throughout the surface of the sample for 25 times. For subsequent drops of metal tamper, the height is adjusted, according to the surface of the sample. The metal tamper must be under free fall.

Step 8. Free samples around the surface of the truncated cone are removed. The truncated cone is lifted without disturbing the sample. On lifting the cone vertically, a slight slump of the sample can be observed that indicates that the sample has attained a saturated surface dry condition.

Step 9. Although this is a general method to check the attainment of SSD by a sample, for certain samples that contain a high amount of angular aggregates and a higher amount of fines the above procedure does not work. In such cases, a handful of fine aggregate is dropped through a cone at the height of 100–150 mm. If the presence of fines is higher, more amounts of airborne fines can be observed. For such samples, a slight slump on one side during the above test indicates that the sample has reached a saturated surface dry condition.

Procedure

Gravimetric method (based on mass):

Step 1. The pycnometer is filled partially with water. Then the pycnometer is filled with saturated surface dry fine aggregate of 500 ± 10 g. Additional water is filled to the calibration mark. Then the pycnometer is agitated either manually or mechanically.

Step 2. Agitation is performed to avoid air bubbles in the sample. Manual agitation is done by rotating or inverting the pycnometer.

Step 3. While rotating the pycnometer on its side, the hole in the apex of the conical top is covered by a finger.

Step 4. Mechanical agitation is performed by a vibrator that sets individual particles in motion without degradation of the sample.

Step 5. After removing the air bubbles, froth formation is avoided by dipping a paper towel in the pycnometer where the froth is observed during the elimination of air bubbles. Occasionally, the froths formed are dispersed by using isopropanol.

Step 6. Once the air bubbles and froth formation are completely eliminated, the pycnometer with sample and water is weighed. Before that, the pycnometer with the sample and the water are brought to a temperature of 23 ± 2°C using a water bath.

Step 7. The sample from the pycnometer is removed and dried in an oven at 110 ± 5°C until the sample attains a constant mass. After cooling the sample at room temperature for 23 ± 2°C, the mass of the sample is determined.

Step 8. The mass of the pycnometer filled with water to its calibrated capacity is determined at 23 ± 2°C.

Notation for Calculation

For the calculation of water absorption and relative density of the sample, the following notations are given and observed readings were recorded in Table 3.11.

- Mass of oven-dry specimen in grams is denoted as A.
- Mass of pycnometer and water-filled to the mark in grams is denoted as B.
- Mass of pycnometer with water and sample up to the mark in grams is denoted as C.

- Mass of saturated surface dry sample in grams used in the gravimetric method is denoted as D.

TABLE 3.11
Determination of Specific Gravity and Water Absorption of Fine Aggregates

Details	Sample 1	Sample 2	Sample 3
Mass of oven-dry specimen in grams (A)			
Mass of pycnometer and water-filled to the mark in grams (B)			
Mass of pycnometer with water and sample up to the mark in grams (C)			
Mass of saturated surface dry sample in grams used in the gravimetric method (D)			
Based on the gravimetric method (relative density or specific gravity (SSD)) is $\frac{A}{\left(D-(C-B)\right)}$			
Based on the gravimetric method (apparent specific gravity) is $\frac{A}{\left(A-(C-B)\right)}$			
Absorption is calculated in percentage as $\frac{(D-A)}{A}\times 100$			

Observations and Calculations

Observations are noted down in Table 3.11.

Note: Huge variation in bulk relative density and absorption can be observed for fine aggregates due to the presence of aggregates finer than 75 μm.

Report

The final report is presented in the following form

- Relative density or specific gravity (SSD) or apparent relative density or specific gravity to a precision of 0.01...............................
- Absorption is reported to the nearest of 0.1%..............................

Student Remark

....
....
....
....

$$\frac{\textit{Marks obtained}}{\textit{Total marks}} = -$$

Instructor signature

Instructor Remark

....
....
....
....

3.7.10 Relative Density and Absorption of Coarse Aggregates

This test is used to determine the relative density (dimensionless quantity) at oven-dry, saturated surface dry condition. This test method does not apply to specified lightweight aggregates.

Test Summary

Aggregates for which specific gravity and absorption need to be determined are placed in water for 24 ± 4 hours. This procedure fundamentally fills the voids within the aggregates. After immersing for the prescribed amount of time, aggregates are taken out and surface dried and mass is determined. By using the gravimetric method, the volume of the aggregate is determined. Finally, the mass of the sample is determined again after oven drying (IS 2386 Part 3, 2016a).

Use and Significance

- Relative density or specific gravity is a ratio of the mass of aggregate and the mass of an equivalent volume of water displaced by the aggregate. This value is used to distinguish between the density of the aggregate and the bulk density of the aggregate, which includes the volume of voids between the particles of aggregates.
- Relative density can be used to estimate the volume of aggregates occupied in cement concrete, concrete bitumen and other mixtures.

Required Apparatus

- Balance (at least 0.5 g accuracy)
- Pycnometer or a watertight container
- Oven capable of maintaining 110 ± 5°C
- Airtight container
- Dry soft absorbent clothes

Test Specimen Preparation

Step 1. Test specimen of approximately 2,000 grams, sieved and retained in 4.75 mm sieve, is placed in a suitable pan or vessel and cleaned properly to remove the dust and fines.

Step 2. The specimen is placed in an oven and maintained at a temperature of 110 ± 10°C until the specimen reaches a constant mass.

Step 3. The sample is immersed in water and allowed to stand for 24 ± 4 hours.

Step 4. If the specific gravity of the sample from the field needs to be used as such, then the above treatment is not required. Instead, specific gravity is reported in terms of its natural condition.

Step 5. In general, in the field, during concrete construction, aggregates are sprinkled with water and brought to a saturated, surface dry condition or else water correction needs to be carried out by determining the water absorption of the aggregate.

Note: Water absorption and relative density of aggregates at a saturated surface dry condition will be higher for the aggregates without the above treatments when compared to the aggregates with the above treatments. Especially for an aggregate size larger than 75 mm, water may not penetrate to the centre of the aggregates during the prescribed soaking period.

Procedure

Step 1. After immersing the sample for 24 ± 4 hours, the water is poured out.

Step 2. Now, the sample is spread on a dry soft absorbent cloth exposed to gently flowing warm air until the sample is just surface dried. Care should be taken not to evaporate the water in voids of the aggregates.

Step 3. Determine the mass of the saturated surface dry aggregates. Record the mass to the nearest of 0.5 grams. This mass is denoted as A.

Step 4. Saturated surface dry aggregates are placed in the pycnometer or watertight container and filled with distilled water partially.

Step 5. Agitation is performed in order to avoid air bubbles in the sample. Manual agitation is done by rolling or inverting the pycnometer.

Step 6. Once the air bubbles and froth formation is completely eliminated, the pycnometer or the container is filled with distilled water completely. Now, the pycnometer or the container along with the sample and distilled water is weighed. This weight is denoted as B. Before that, the pycnometer with the sample and the distilled water is brought to a temperature of 23 ± 2°C using a water bath.

Step 7. Sample from the pycnometer or the container is removed and dried in an oven at 110 ± 5°C until the sample attains a constant mass. After cooling the sample in the airtight container, the mass of the sample is determined and denoted as D.

Step 8. Mass of the pycnometer or the container filled with distilled water completely is determined at 23 ± 2°C and denoted as C.

Step 9. Note down all the measurements in Table 3.12.

Notation for Calculation

For the calculation of water absorption and relative density of the sample, the following notations are given and followed.

- Mass of saturated surface dry specimen in grams is denoted as A.
- Mass of pycnometer or container along with saturated surface dry sample and distilled water in grams is denoted as B.
- Mass of the pycnometer or the container filled with distilled water completely is denoted as C.
- Mass of oven-dry sample after attaining a constant mass in grams is denoted as D.

TABLE 3.12
Determination of Specific Gravity and Water Absorption of Coarse Aggregate

Details	Sample 1	Sample 2	Sample 3
Mass of saturated surface dry specimen in grams (A)			
Mass of pycnometer or container along with saturated surface dry sample and distilled water in grams (B)			
Mass of the pycnometer or the container filled with distilled water completely in grams (C)			
Mass of oven-dry sample after attaining a constant mass in grams (D)			
Relative density or specific gravity is $\frac{D}{(A-(B-C))}$			
The apparent specific gravity of the specimen is $\frac{D}{(D-(B-C))}$			
Water absorption of coarse aggregate is given by $\frac{A-D}{D}\times 100$			

Observations and Calculations

Observations are noted down in Table 3.12.

The relative density or specific gravity of the specimen is calculated based on the condition of the sample. The relative density or specific gravity of the naturally obtained specimen is given by the following form.

- Relative density or specific gravity is $\frac{D}{(A-(B-C))}$
- The apparent specific gravity of the specimen is $\frac{D}{(D-(B-C))}$
- Water absorption of coarse aggregate is given by $\frac{A-D}{D}\times 100$

Report

Final results are reported in the following form:

- Relative density or specific gravity and apparent relative density or specific gravity to a precision of 0.01.....................
- Absorption is reported to the nearest of 0.1%.........................

Student Remark

....
....
....
....

$$\frac{\textit{Marks obtained}}{\textit{Total marks}} = -$$

Instructor signature

Instructor Remark

....
....
....
....

3.7.11 Bulking of Fine Aggregates

A concrete mix design is made as either a nominal mix or design mix. The nominal mix has a certain defined proportion (that has acceptable strength) based on which the mixtures are proportioned. In the case of mix design, each constituent of the concrete is tested in the laboratory. Based on the test properties of the constituents, the mix is designed. In the case of a simple mix design, certain constituents that have a significant variation in properties can result in unexpected variation in the proportioned mix. One of the properties is the bulking of fine aggregates. Notably, in the case of volume-based mixtures, bulking of fine aggregate is a significant problem.

Test Summary

Bulking of fine aggregate can be determined by adding a known quantity of water to the sample under consideration (IS 2386 Part 3, 2016b).

Use and Significance

- This test can be used to identify the abrupt loss of workability as well as unexpected expansion after the setting of concrete due to the absorption of water from the mix or from outside.

Required Apparatus

- A steel rod of 6 mm diameter
- A steel ruler
- Measuring cylinder of 250 ml capacity

Procedure

Step 1. Fill the measuring cylinder until 200 ml mark with the sample under consideration.

Step 2. Once, the aggregate is filled until the mark, level it by using the steel ruler without shacking the measuring cylinder.

Step 3. The sample of the aggregate is taken in the measuring cylinder until the mark is poured into another container.

Step 4. The measuring cylinder is filled with water (water taken is sufficient enough to submerge the sample completely), and the sample is transferred into the measuring cylinder. The sample is thoroughly mixed by the steel rod.

Step 5. The sample is allowed to settle down. As soon as the sample has settled down, the graduation in the measuring cylinder is measured.

Step 6. Let the new mark on the graduation cylinder be X. Note down in Table 3.13.

Caution: Take at least an average of three tests for a given aggregate sample.

Note: In general, bulking of fine aggregates with more fines will be more when compared to aggregates with fewer fines. This is due to the fact that the presence of higher fines in the aggregates results in more absorption of water and bulking. In general, the presence of 5–6% of fines will result in an approximate increase in the bulking of sand from 20% to 40%.

Observations and Calculations

Observations are noted down in Table 3.13.

TABLE 3.13
Determination of Bulking of Fine Aggregate

Details	Sample 1	Sample 2	Sample 3
Volume of loose sand (ml)	200	200	200
Volume of saturated sand (X) (ml)			
Percentage bulking $\frac{200-X}{X}\times 100$			
Average bulking of sand (%)			

Note: If the initial volume of the sample is Y then, percentage bulking is given by $\frac{Y-X}{X}\times 100$

Report

Bulking of fine aggregate is given by $\frac{200-X}{X}\times 100$

Bulking of fine aggregate is ....

Student Remark

....
....
....
....

$$\frac{\textit{Marks obtained}}{\textit{Total marks}} = -$$

Instructor signature

Instructor Remark

....
....
....
....

3.7.12 Bulk Density and Voids in Aggregates

Coarse aggregates and fine aggregates are vital constituents of concrete. In general, proportioning of concrete is done in mass as well as volume and it is quite easy to convert volumetric proportion to weight proportion and vice versa by understanding the bulk density of aggregates.

Test Summary

The bulk density of aggregates can be calculated by filling the aggregates in a known volume of a container. Based on the requirement of packing, bulk density can be calculated for different packing (IS 2386 Part 3, 2016c).

Use and Significance

- Bulk density is used in order to understand the proportioning of different constituents during concrete mix design.
- Volume to weight proportioning and vice versa can be done by understanding the bulk density of the constituents.
- Moreover, the presence of different surface moisture content in the aggregate may result in bulking, specifically in fine aggregates. In those cases, the values obtained by performing this experiment need to be taken care or need to be revised.

Required Apparatus

- Balances (0.1 g accuracy for fine aggregate and 100 g accuracy for coarse aggregate)

- Tamping rod
- Straight edge
- Cylindrical metal vessel of different capacity (3, 15 and 30 litres)
- For fine aggregates (particle size less than 4.75 mm), a cylindrical vessel of 3 litres is used
- For coarse aggregates (particle size more than 4.75 mm and less than 40 mm), a cylindrical vessel of 15 litres is used
- For coarse aggregates of size more than 40 mm, a cylindrical vessel of 20 litres is used

Procedure

Step 1. Based on the size of aggregate to be determined for bulk density and voids, a consistent cylindrical vessel is chosen. The mass of the cylindrical vessel (chosen for the experiment) is determined and denoted as A.

Step 2. By means of a scoop, the cylindrical vessel is filled until it overflows. The filling can be done either in a loose way or by rodding.

Step 3. While filling the aggregates, the height of the measure in the vessel should not exceed 5 cm from the top of the subsequent measure.

Step 4. Once the aggregates are filled until overflowing, the surface of the aggregate is shelved and levelled by means of a straight edge. The loosely filled aggregates, along with the container, are weighed, and the mass is denoted as B.

Step 5. In order to determine the bulk density of aggregates after rodding, the prescribed cylindrical vessel is filled with aggregates to its one third. Tamping is done by giving 25 blows uniformly through the sample with the rounded edge of the taming rod.

Step 6. In a similar way, two subsequent layers are filled and 25 blows are given by the rounded edge of the tamping rod. During filling the third layer, aggregates are filled to overflowing level and tamping is done.

Step 7. After filling and tamping the subsequent layers using a straight edge, the aggregates are shelved and levelled. Now, determine the mass of the cylindrical vessel with aggregates. This mass is denoted as C.

Step 8. Observed readings are noted down in Table 3.14.

Cautions: Care should be taken to avoid the segregation of aggregates in the vessel. Aggregates to be used should be in dry condition. However, in bulk density test aggregates with natural or given moisture content can be used, provided the condition of the aggregate is mentioned.

Observations and Calculations

Observations are noted down in Table 3.14.

TABLE 3.14
Determination of Bulk Density and Voids

Details	Sample 1	Sample 2	Sample 3
The capacity of the cylindrical vessel in L (V)			
Weight of cylindrical vessel (A)			
Weight of cylindrical vessel + Loosely filled aggregate (B)			
Weight of cylindrical vessel + compacted aggregate (C)			
Loose bulk weight, $R = \frac{B-A}{V}$			
Average			
Rodded bulk weight $\frac{C-A}{V}$			
Average			
Percentage voids in the aggregate $\frac{G_S - R}{G_S} \times 100$			
Average			

Step 1. Net lose weight of the aggregate is determined as (B-A)
Step 2. Loose bulk density (R) can be calculated as (B-A)/volume of the cylindrical vessel
Step 3. Net (rodded) weight of the aggregate is determined by (C-A)
Step 4. Bulk density after rodding can be calculated as (C-A)/volume of the cylindrical vessel
Step 5. Percentage voids in the aggregates can be calculated by $\frac{G_S - R}{G_S} \times 100$

- G_S is the specific gravity of the aggregate
- R is the loose bulk density of the aggregate

Report

Depending upon the usage of material, the final report should include the following,

- Loose bulk density.....................
- Bulk density after rodding.....................
- Percentage voids in the aggregate.....................

Student Remark

....
....
....
....

$$\frac{\textbf{\textit{Marks obtained}}}{\textbf{\textit{Total marks}}} = -$$

Instructor signature

Instructor Remark

....
....
....
....

3.8 PRACTICE QUESTIONS AND ANSWERS FROM COMPETITIVE EXAMS

1. Why is particle size analysis of coarse and fine aggregate important?
   a) It helps to understand the packing density.
   b) It is very useful in estimating the compressive strength of concrete.
   c) Workability of the concrete can be determined by the careful selection of aggregates.
   d) None of the above.

2. For 19 mm nominal maximum size of aggregate, the minimum test sample size required is
   a) 5 kg
   b) 10 kg
   c) 15 kg
   d) 20 kg

3. What is the flakiness index?
   a) Length-to-width ratio higher than the specified value
   b) Length-to-thickness ratio greater than the specified value
   c) Width to thickness ratio greater than the specified value
   d) Both b and c

4. What is the elongation index?
   a) Length-to-width ratio higher than the specified value
   b) Length-to-thickness ratio higher than the specified value
   c) Width to thickness ratio higher than the specified value
   d) Both a and b

5. Aggregate crushing value should not exceed .... for concrete wearing surfaces.
   a) 15%
   b) 25%
   c) 30%
   d) 35%

6. Aggregate impact value should not exceed .... for aggregates used for concrete other than wearing surfaces according to IS: 383.
   a) 30%
   b) 35%
   c) 40%
   d) 45%

7. Aggregates used in the surface of the pavement will undergo wear and tear due to the friction. What test is conducted to assess the wear and tear undergone by the aggregates?
   a) Impact value test
   b) Crushing value test
   c) Los Angeles test
   d) None of the above

8. To determine the water absorption of aggregates, why are aggregates placed underwater for a prescribed period of 24 ± 4 hours? Can it be reduced?
   a) To fill the voids in the aggregates; no
   b) To fill the voids in the aggregates; yes
   c) To fill the voids in the aggregates; except lightweight aggregates
   d) None of the above

9. In general, the presence of 5–6% of fines will result in an approximate increase in the bulking of sand from
   a) 10–30%
   b) 20–40%
   c) 15–35%
   d) 25–45%

10. Consider the following statements:
    1. A high aggregate impact value indicates strong aggregates.
    2. A low crushing value indicates a high crushing strength of aggregates
    3. Aggregates having elongation index values higher than 15% are generally considered suitable for pavement construction.
    4. Flakiness index of aggregates should not be lesser than 25% for use in road construction.

Which of the above statements are correct?
   a) 2 and 3 only
   b) 2 and 4 only
   c) 1 and 3 only
   d) 1 and 4 only

[IES 2017]

11. In elongation and flakiness test after sieving process, the aggregates passing through a 20 mm sieve and retained on a 16 mm sieve are used. A suitable slot is selected in the test apparatus for flakiness test, as specified in

the Indian standard. Similarly, a suitable slot is also selected in the length gauge for elongation test. The ratio of the selected slot length in elongation test to the selected width of the slot in the flakiness test is

a) 2
b) 3
c) 4
d) 5

12. Good quality sand is never obtained from
    a) River
    b) Nala
    c) Sea
    d) Gravel powder

[SSC JE 2008]

13. Grading of sand causes great variation in
    a) Workability of concrete
    b) Strength of concrete
    c) Durability of concrete
    d) All of the above

[SSC JE 2009]

14. An aggregate is said to be flaky if its least dimension is less than
    a) $\frac{2}{3}$ mean dimension
    b) $\frac{3}{4}$ mean dimension
    c) $\frac{3}{5}$ mean dimension
    d) $\frac{5}{8}$ mean dimension

[SSC JE 2011]

15. Bulking is
    a) Increase in the volume of sand due to moisture which keeps sand particles apart
    b) Increase in the density of sand due to impurities like clay, organic matter
    c) Ramming of sand so that it occupies minimum volume
    d) Compacting of sand

[SSC JE 2014]

16. The percentage of the aggregate of fineness modulus 2.6 to be combined with a coarse aggregate of fineness modulus 6.8 for obtaining the aggregate of fineness modulus 5.4 is
    a) 60%

b) 30%
c) 40%
d) 50%

[SSC JE 2014]

17. An aggregate is known as cyclopean aggregate if its size is more than
    a) 4.75 mm
    b) 30 mm
    c) 60 mm
    d) 75 mm

[SSC JE 2014]

18. If fineness modulus of a sand is 2.5, it is graded as
    a) Very fine sand
    b) Fine sand
    c) Medium sand
    d) Coarse sand

[SSC JE 2014]

19. The fineness modulus of an aggregate is roughly proportional to
    a) Average size of particles in the aggregate
    b) Grading of the aggregate
    c) Specific gravity of the aggregate
    d) Shape of the aggregate

[SSC JE 2015]

20. The aggregate impact value of the aggregate used in
    a) Building concrete is less than 45
    b) Road pavement concrete is less than 30
    c) Runway concrete is less than 30
    d) All the options are correct

[SSC JE 2017]

21. If aggregates completely pass through a sieve of size 75 mm, the particular aggregate will be flaky if its minimum dimension is less than
    a) 20.5 mm
    b) 30.5 mm
    c) 40.5 mm
    d) 50.5 mm

[SSC JE 2017]

22. The aggregate is called fine aggregate if it is completely retained on
    a) 0.15 mm sieve
    b) 0.30 mm sieve
    c) 4.75 mm sieve
    d) None of these

[SSC JE 2017]

23. If 20 kg of coarse aggregate is sieved through 80 mm, 40 mm, 20 mm, 10 mm, 4.75 mm, 2.36 mm, 1.18 mm, 600 micron, 300 micron and 150 micron standard sieves and the weights retained are 0 kg, 2 kg, 8 kg, 6 kg and 4 kg, respectively, the fineness modulus of the aggregate lies in the range of
    a) 6.85–7.10
    b) 7.20–7.45
    c) 7.50–7.75
    d) None of these

[SSC JE 2017]

24. If aggregates completely pass through a sieve of size 75 mm and are retained on a sieve of size 60 mm, the aggregates will be known as elongated aggregate if its length is not less than
    a) 81.5 mm
    b) 91.5 mm
    c) 101.5 mm
    d) 121.5 mm

[SSC JE 2018]

25. Spot the odd statement
    a) Rounded aggregate
    b) Irregular or partly rounded aggregate
    c) Angular flaky aggregate
    d) Single size aggregate

[SSC JE 2018]

26. What should be done to ensure constant moisture content in aggregates?
    a) Area of each aggregate pile should be large.
    b) Height of each aggregate pile should not exceed 1.50 m.
    c) The aggregate pile should be left for 24 hours before aggregates are used.
    d) All of these.

[SSC JE 2018]

27. The maximum amount of dust which may be permitted in aggregates is
    a) 5% of the total aggregate for low workability with a coarse grading
    b) 10% of the total aggregate for low workability with a fine grading
    c) 20% of the total aggregate for high workability with fine grading
    d) All options are correct

[SSC JE 2018]

28. An aggregate which passes through 25 mm IS sieve and is retained on 20 mm sieve is said to be flaky if its least dimension is less than
    a) 22.5 mm
    b) 18.5 mm

c) 16.5 mm
d) 13.5 mm

[SSC JE 2018]

29. The percentage of the aggregate of FM 2.6 to be combined with a coarse aggregate of FM 6.8 for obtaining the aggregates of FM 5.4 is
    a) 0.3
    b) 0.4
    c) 0.5
    d) 0.6

[SSC JE 2018]

30. The strength of concrete made with angular aggregate and rounded aggregate is practically the same at the water/cement ratio is
    a) 0.4
    b) 0.48
    c) 0.55
    d) 0.65

[SSC JE 2018]

31. The function of aggregates in concrete is to serve as
    a) Binding material
    b) Filler
    c) Catalyst
    d) All of these

[AE 2008]

32. Sea sand used in structures causes
    a) Dampness
    b) Efflorescence
    c) Disintegration
    d) All of these

[AE 2008]

33. The rock generally used for roofing is
    a) Granite
    b) Basalt
    c) Slate
    d) Pumice

[AE 2008]

34. The hardest rock is
    a) Marble
    b) Diamond
    c) Talc
    d) Quartz

[AE 2008]

35. The solidification of molten magma within the earth's crust results in the formation of
    a) Sedimentary rock
    b) Metamorphic rock
    c) Basalts and traps
    d) Granite

[AE 2010]

36. If the fineness modulus of sand is 3, then the sand is graded as
    a) Very fine sand
    b) Fine sand
    c) Medium sand
    d) Coarse sand

[AE 2010]

37. Chemically, marble is known as
    a) Metamorphic rock
    b) Argillaceous rock
    c) Calcareous rock
    d) Siliceous rock

[AE 2013]

38. A good building stone is one which does not absorb more than .... of its weight of water after one days immersion
    a) 5%
    b) 10%
    c) 15%
    d) 20%

[AE 2013]

39. The type of stone masonry commonly adopted in the construction of residential building is
    a) Uncoursed rubble masonry
    b) Coursed rubble masonry
    c) Random rubble masonry
    d) Dry rubble masonry

[AE 2013]

40. A type of flooring made with special aggregate marble chips mixed with white and coloured cement is called
    a) Mosaic flooring
    b) Terrazzo flooring
    c) Asphalt flooring
    d) None of these

[AE 2013]

41. The exposed edges of stones are bevelled for a depth of 2.5 cm in a
    a) Ashlar rough tooled masonry
    b) Ashlar rock faced masonry
    c) Ashlar changed masonry
    d) Ashlar block in course

[AE 2014]

42. Mohr's scale for stones is used to determine
    a) Toughness
    b) Hardness
    c) Durability
    d) Specific gravity

[AE 2014]

43. The specific gravity of most of the stones lie between
    a) 1.8 and 2.2
    b) 2.5 and 3.0
    c) 3.0 and 3.5
    d) 3.5 and 4.5

[AE 2014]

44. The volume of standard gauge box used for volume batching of concrete is
    a) 25 litres
    b) 35 litres
    c) 42 litres
    d) 50 litres

[AE 2014]

45. Which of the following shape of aggregates gives the maximum strength in concrete?
    a) Rounded aggregate
    b) Elongated aggregate
    c) Flaky aggregate
    d) Angular aggregate

[AE 2015]

46. Which type of the following mix will be produced if the fineness modulus of an aggregate is high?
    a) Lean concrete
    b) Stiff concrete
    c) Harsh concrete
    d) Workable concrete

[AE 2015]

47. If water content, specific gravity and void ratio of given soil are 11.11%, 2.7 and 0.5, respectively, the degree of saturation is equal to
    a) 40%
    b) 50%
    c) 60%
    d) 70%

[AE 2015]

48. Pick the incorrect principle in stone masonry construction
    a) Stone should be properly dressed to requirements.
    b) Header and bond stones should be of dumb-bell shapes.
    c) Stones to be used should be hard, tough and durable.
    d) Construction work of stone masonry should be raised uniformly.

[AE 2016]

49. Type of stone masonry without any mortar is
    a) Random rubble
    b) Polygonal rubble
    c) Flint rubble
    d) Dry rubble

[AE 2016]

50. The resistance of aggregate to failure by an impact is called
    a) Hardness
    b) Tenacity
    c) Toughness
    d) Soundness

[AE 2016]

51. In case of volume batching a smaller mass of sand occupying the fixed volume of the measuring box is due to which of the following effect?
    a) Bulking
    b) Hardening
    c) Softening
    d) Texture modification

[AE 2016]

52. The common source for flaky aggregates is
    a) Sea shore gravel bed
    b) Sandstone bed
    c) Laminated rock bed
    d) Talus

[AE 2016]

53. The fire resistance characteristics of concrete depend upon the
    a) Coefficient of thermal expansion of coarse aggregate
    b) Thermal expansion of sand
    c) Degree of compaction
    d) W/C ratio

[AE 2018]

54. For the given W/C ratio, higher the maximum size of aggregate, the strength of concrete
    a) Increases
    b) Decreases
    c) Remains the same
    d) W/C ratio does not have any effect on the strength of concrete

[AE 2018]

55. Which is the most dangerous for decaying the stones?
    a) Efflorescence
    b) Water absorption
    c) Incorrect bedding
    d) Vegetation growth

[AE 2018]

56. Select the incorrect joint considered in stone masonry.
    a) Butt joint
    b) Rebated joint
    c) Grooved joint
    d) Contraction joint

[AE 2018]

57. The masonry constructed using more than one material to improve the appearance and durability is called
    a) Stone masonry
    b) Brick masonry
    c) Rubble masonry
    d) Composite masonry

[AE 2018]

58. Which one of the following shape of aggregates gives the maximum strength in concrete?
    a) Round
    b) Elongated
    c) Flaky
    d) Angular

[AE 2018]

59. Which type of following mix will be produced if the fineness modulus of an aggregate is high?
    a) Lean concrete
    b) Stiff concrete
    c) Harsh concrete
    d) Workable concrete

[AE 2018]

1.	a	2.	a	3.	d	4.	d	5.	c	6.	d	7.	c	8.	b	9.	b	10.	a
11.	b	12.	b	13.	d	14.	c	15.	a	16.	d	17.	d	18.	b	19.	a	20.	d
21.	c	22.	a	23.	b	24.	d	25.	c	26.	d	27.	d	28.	d	29.	c	30.	d
31.	b	32.	d	33.	c	34.	b	35.	d	36.	d	37.	c	38.	a	39.	b	40.	b
41.	c	42.	b	43.	b	44.	b	45.	d	46.	c	47.	c	48.	b	49.	d	50.	c
51.	a	52.	c	53.	a	54.	a	55.	c	56.	d	57.	d	58.	d	59.	c		

REFERENCES

Assistant Engineer (AE). 2020. Tamil Nadu Public Service Commission (TNPSC), Tamil Nadu, India. http://www.tnpsc.gov.in/previous-questions.html (accessed September 18, 2020).

Engineering Service Exam (ESE). 2020. Union Public Service Commission, New Delhi, India. https://www.upsc.gov.in/examinations/previous-question-papers (accessed September 18, 2020).

Graduate Aptitude Test in Engineering (GATE). 2020. GATE Office, Chennai, India. http://gate.iitm.ac.in/gate2019/previousqp18.php (accessed September 18, 2020).

IRC 110. (2005). *Standard specifications and code of practice for design and construction of surface dressing.* Bureau of Indian Standards. New Delhi.

IRC SP 20. (2002). *Rural roads manual.* Indian Road Congress. New Delhi.

IS 2386 Part 1. (2016a). *Methods of test for aggregates for concrete: Particle size and shape.* Bureau of Indian Standards. New Delhi.

IS 2386 Part 1. (2016b). *Methods of test for aggregates for concrete: Particle size and shape.* Bureau of Indian Standards. New Delhi.

IS 2386 Part 1. (2016c). *Methods of test for aggregates for concrete: Particle size and shape.* Bureau of Indian Standards. New Delhi.

IS 2386 Part 3. (2016a). *Methods of test for aggregates for concrete: Specific gravity, density, voids, absorption and bulking.* Bureau of Indian Standards. New Delhi.

IS 2386 Part 3. (2016b). *Methods of test for aggregates for concrete: Specific gravity, density, voids, absorption and bulking.* Bureau of Indian Standards. New Delhi.

IS 2386 Part 3. (2016c). *Methods of test for aggregates for concrete: Specific gravity, density, voids, absorption and bulking.* Bureau of Indian Standards. New Delhi.

IS 2386 Part 4. (2016a). *Methods of test for aggregates for concrete: Mechanical properties.* Bureau of Indian Standards. New Delhi.

IS 2386 Part 4. (2016b). *Methods of test for aggregates for concrete: Mechanical properties.* Bureau of Indian Standards. New Delhi.

IS 2386 Part 4. (2016c). *Methods of test for aggregates for concrete: Mechanical properties.* Bureau of Indian Standards. New Delhi.

IS 383. (2016). *Coarse and fine aggregates for concrete – specification.* Bureau of Indian Standards. New Delhi.

IS 6241. (2013). *Methods of test for determination of stripping value of road aggregates.* Bureau of Indian Standards. New Delhi.

Junior Engineer (JE), Staff Selection Commission (SSC), New Delhi, India. https://ssc.nic.in/Portal/SchemeExamination (accessed September 18, 2020).

4 Concrete

4.1 FRESH PROPERTIES OF CONCRETE

The construction industry uses different construction materials; one among the widely used construction materials is concrete. It consists of different proportions of mixtures such as water, well-graded coarse and fine aggregates, a binder and chemical admixtures. Among the binders used, cement is the most prominent one because of its qualities such as better moulding ability, early age hardening, controlled setting and flexibility towards the usage of different admixtures. As concrete is a constituent of different materials, altogether concrete needs to be appropriately mixed in its fresh state in order to obtain a better-hardened state property. Consequently, the properties of concrete need to be checked in its fresh state as well as hardened state.

4.1.1 Workability

Workability is an intrinsic property of any material in its fresh state, where it can be moulded and made into other forms. This property is especially applicable for concrete and mortar in their freshly mixed state. IS: 6461 (Part-7) defines workability as the ease with which the concrete can be placed, mixed, compacted and finished homogeneously. Workability is a complex measure of parameters such as consistency and homogeneity.

4.1.2 Workability Tests

Different empirical tests are available in order to test the workability of the concrete. Some of them are slump test, flow test, compaction factor test and Vee-Bee consistency test. Most commonly used is the slump test because of its easy test procedure and the practical utility that it provides when measured in sites practically. A short description of each of the above tests is given below.

4.1.2.1 Slump Test

The slump test is performed in a truncated cone, where subsidence of concrete is measured by filling the truncated cone in three layers with concrete and with regular tamping on each layer. When the truncated cone is lifted vertically from its horizontal surface, subsidence of concrete takes place. This subsidence is an indication of the workability of the concrete. The slump test is limited to concrete with a maximum aggregate size of 38 mm.

4.1.2.2 Vee-Bee Test

This test is applicable for concrete that is stiff in nature and the workability of concrete is very low. In this test, concrete undergoes testing methods that are similar to the testing that it undergoes in the real fields.

4.1.2.3 Compaction Factor Test

This test is performed more prominently in laboratory conditions. This test is suitable for concrete that is varying from medium workability to low workability.

4.1.2.4 Flow Test

This test is primarily used for concrete that is highly workable in nature, although concrete that can be used in the slump test can also be used. Subsequently, the slump test and flow test are performed in the same cone. This test is generally used for self-compacting concrete (SCC), where their slump is more than 625 mm.

Although several tests are available to identify the workability of concrete, when applied to field, only specific tests are more useful. One among them which is used still in the field is the slump test. Although, usage of this test gives erroneous results in some instances, especially during the prediction of pumping pressure requirements. In order to improve pumping prediction ability and requirements, a new way of quantifying concrete workability is required. This led to the study of the deformation of matter, i.e. concrete as a fluid material, whenever deformed by applying specific stress, it starts to move. This study is mainly called as rheology is a new evolutionary study that is revolutionizing the concrete fresh state properties. Some of the crucial factors that affect the workability of concrete are its constituent materials, their mix proportion and environmental conditions.

Whenever a study is made on the stability of the concrete, two crucial parameters that need to be defined are segregation and bleeding. Concrete in its fresh state tends to separate out its individual constituents which may result in affecting its homogeneity. In a similar sense, bleeding can be defined as the rising of water or separation of water from the fresh concrete mix in the initial stages of mixing. The above two processes result in the reduction of concrete quality.

4.2 HARDENED PROPERTIES OF CONCRETE

Concrete is considered to be a two-phase system which constitutes of cement mortar and aggregates. Generally, aggregates are considered to be inert in nature, although this may change. For high strength concrete, cement mortar phase is considered to give more strength when compared to aggregates. Cement binders are hydraulic in nature, and their properties keep on evolving with respect to time. Hardened concrete properties that are considered to be very important are its strength, durability and permeability.

Water-to-cement ratio and compaction directly decide the strength of the concrete. Since the amount of water decides the porosity of the concrete, it, in turn, determines its strength.

4.2.1 Compressive Strength Test (IS: 516)

The compressive strength test is done in cube specimens of 150 mm. After a proper sampling of concrete constituents, fresh concrete can be moulded into different shapes according to requirements. In general, concrete cubes of 150 mm are considered to be a standard specimen for compression testing. After casting, cubes are stored under wet straw or gunny bags for the first 24 hours. Then, the cubes are de-moulded and placed in a room of temperature 24–30°C for 28 days, which includes the initial one day. On the prescribed days, 7th and 28th-day concrete is tested in the compression testing machine (CTM) at a prescribed loading rate of 14 N/mm^2 under saturated (wiped dry) condition. The load at which the specimen fails is considered to be compressive strength and the average of three specimens is reported as the final compressive strength.

4.2.2 Tensile Strength Test (IS: 5816)

The tensile test is done on a cylindrical concrete specimen of 15 cm in diameter and 30 cm in height. Tensile strength of concrete is estimated indirectly by applying a compressive load on the longitudinal axis of the cylinder. For the period of loading of concrete in its longitudinal axis, cylindrical specimens experience tension in the lateral axis and hence split into two halves along the diameter.

4.2.3 Flexural Strength

Flexural strength is essential for the structural elements under tension. The flexural test can be used to estimate the modulus of rupture or flexural tensile strength at failure. Failure in a concrete member is determined when it starts to crack under failure. The standard specimen used for flexural testing is a beam of 150 × 150 × 700 mm. In general, the beam is the structural member that mostly undergoes flexure in practice.

4.3 DURABILITY OF CONCRETE

In general, all the structural materials undergo deterioration process. This deterioration process is directly proportional to the member durability. Durability may be defined as the ability of hydraulic cement concrete to resist the action of the weather, chemical, abrasion and other processes that ultimately bring about some kind of deterioration. Concrete is said to be durable if it can retain its original form, quality and serviceability under its exposed conditions. Eventually, while durability is a factor of a concrete matrix, the level of durability directly depends on the environment to which the concrete is exposed. Some of the common problems in concrete related to durability are discussed in the following sections.

4.3.1 Corrosion of Steel in Reinforced Concrete

Corrosion of reinforcements in RCC structures results in the formation of rust. This rust formed occupies more volume than its predecessor (steel), resulting in stresses that cause cracks. Usually, the stresses that cause these cracks are expansive in nature. An excessive water-to-cement ratio, insufficient cover concrete, excessive moisture and chlorides are mostly responsible for the corrosion.

4.3.2 Sulphate Attack

In many cases, durability problems are generally due to an increase in volume that leads to cracking. Sulphate salts of sodium, magnesium, calcium and ammonium are harmful in nature; they cause expansion of concrete and subsequently lead to cracking. When external sulphate ions ingress into concrete, they react with calcium hydroxide and the concrete is converted into gypsum. Afterwards, this gypsum further reacts with the available C_3A and forms ettringite. The ettringite is expansive in nature and leads to cracking in concrete.

4.3.3 Alkali Aggregate Reaction

Similar to other durability problems, this reaction is between a high concentration of alkalis and reactive silica present in the aggregates that leads to the formation of alkali-silica gels. This gel is formed around the aggregate and it absorbs moisture and causes an increase in volume. Therefore, it creates expansive stresses around the aggregate. Successively, expansion causes cracking and structural failures.

4.3.4 Freezing and Thawing Damage

Cracks are formed due to internal pressure, which is mainly due to freezing and thawing. At a lower temperature, the water present inside the concrete freezes and leads to stresses. Hence concrete cracks.

4.3.5 Carbonation

Carbonation is mainly due to the reaction of atmospheric CO_2, which results in the conversion of the basic nature of concrete to acidic. CO_2 diffuses into the concrete and reacts with calcium hydroxide and is converted into calcium carbonate and subsequently carbonic acid. Due to this reaction, the pH of the concrete is dropped. Steel reinforcements require a basic environment. Once the pH drops, corrosion of steel reinforcement starts resulting in an increase in the volume of steel which in turn results in expansion and cracking.

4.4 SPECIAL CONCRETES

Cement concrete is considered to be versatile in nature, even though some shortcomings such as low tensile strength, high permeability to aggressive agents, the consequence of corrosion, chemical attack and low durability require some adaptation to be done in concrete by addition of some special additives. Moreover, special requirements need specific modifications in concrete. Commonly used special concretes are discussed as below

Lightweight concrete – Lightweight concrete is useful in places where the reduction of dead weight is necessary in order to reduce the cost of construction by reducing the foundation design and other structural members. Lightweight concrete is prepared by reducing the weight of the coarse aggregates. In general, the bulk density of concrete is 2400 kg/m^3. Lightweight concrete's bulk density varies typically from 500 kg/m^3 to 1,800 kg/m^3. Ultra-lightweight concrete's bulk density is around 1,000 kg/m^3. One of the essential properties of lightweight concrete is the improvement in its thermal property.

Heavyweight concrete – Certain structures like nuclear power plants require high-density concrete for shielding purpose and hence heavyweight concrete is used. It is concrete having a density greater than 2,600 kg/m^3 (EN: 206-2013). It is made up of aggregates of higher density, such as magnetite and hematite. It is costly and prone to bleeding.

Underwater concrete – Proper control is required as an anti-washout admixture is used.

Ultra high strength concrete – Ultra high strength concrete can be obtained by reducing the water-to-cement ratio (by adding high-range water reducers) by means of high binder content and highly reactive supplementary cementitious materials such as silica fume.

Ultra-durable concrete – Permeability is one of the main causes for durability issues. Therefore, durability can be increased by reducing pores in concrete by deciding on the proper mix proportion of concrete. This can be done by selecting a suitable water-to-cement ratio, SCM's and superplasticizers. To improve durability, sometimes stainless steel bars, as well as coated bars, are also used.

Self-compacting concrete – SCC is mainly used in places where handling of the material is tedious and manpower used needs to be reduced. They are generally prepared by adding high-range water reducers, viscosity modifiers and fillers in appropriate dosages.

Roller compacted concrete – Concrete that needs to be compacted is similar to other concretes, proportioned in such a way that it is stiff and has zero slump value. It is commonly mixed in a continuous flow system.

Shotcrete – Shotcrete is generally used in repair and rehabilitation works in order to protect the exposed concrete from external attack. It is a mixture of cement and fine aggregates (mortar) or sometimes with small-size coarse aggregates forcefully deposited by the jetting process. They are most widely used in tunnel lining, canal lining and reservoir lining, but also in the lining of overhead vertical and horizontal surfaces' uniquely curved and folded sections.

Geopolymer concrete – It is a type of concrete made by the activation of a precursor such as slag or rice husk ash by sodium- and potassium-based alkaline activators. It is highly durable and possesses high strength and do not use any cement.

Pervious concrete – They are specially used for water recharging. Their applications are in car parking, flooring and pavements in order to allow water to penetrate through the surface. Fines content is nil in this type of concrete.

Polished concrete – Especially used for decoration and industrial flooring.

Fibre reinforced concrete – Flexural toughness of concrete can be enhanced by the usage of glass, steel and other fibres. Fibres bridge the cracks, resulting in significant resistance against the crack formation and also increase post-peak behaviour.

4.5 TESTING PROCEDURES

4.5.1 Determination of Density of Fresh Concrete

This test can be used on the site to determine the fresh density of the concrete. It can also be used for calculating the volume of concrete per batch (yield per bag of concrete).

Test Summary

Fresh density and yield of concrete can be measured by use of a container of fixed volume (IS 1199 Part 3, 2018).

Use and Significance

- Determining the fresh density of concrete is of utmost practical importance in the field. This gives an indication of the design of formwork to be used during concreting.
- The yield of concrete is also used as an essential parameter by the ready-mix concrete (RMC) plants to understand and stock the required quantity of ingredients such as coarse aggregate, fine aggregate and cementitious materials.
- Fresh density can also be used as a quality check in the field.

Limitations

- This test method does not apply to aerated concrete or very stiff concrete (which cannot be compacted using the normal vibration method).

Required Apparatus

- Watertight, sufficiently rigid container with a smooth internal face (the smallest dimension of the internal diameter and the height of the container shall be at least four times the maximum size of the aggregate used. Moreover, it shall not be less than 150 mm and also the volume of the container shall not be less than 5 litres; the ratio of diameter (d) to that of height (h) of the container should be $1.25 \geq \frac{d}{h} \geq 0.50$)

- Weighing balance
- An internal vibrator (minimum frequency of 120 Hz), a vibrating table (minimum frequency of 50 Hz), compacting rod with rounded ends made up of steel of diameter 16 ± 1 mm and a length of 600 ± 5 mm, compaction bar (square or rounded cross-section with mass greater than 1.8 kg for hand compaction)
- A straight-edge scrapper made of steel with edge size more than 100 mm of the internal size of the container used
- Square mouthed shovel
- Remixing tray (non-absorbent)
- Steel float
- Mallet
- A trowel 100 mm wide
- A circular or a square plate (non-absorbent) with a minimum thickness of 5 mm and 25 mm extension on both the sides

Procedure

Step 1. The weight of the container is measured by weighing balance and is denoted as m_1.

Step 2. Concrete is filled in a minimum of two layers in the container. The concrete must be compacted immediately after placing the concrete with neither segregation nor laitance.

Step 3. During hand compaction, each layer is compacted by applying 25 strokes evenly distributed throughout the surface of the container.

Step 4. Ensure that the compaction rod does not forcibly strike the bottom or sides of the container. Moreover, care must be taken to ensure that compaction does not disturb the previous layers.

Step 5. In order to ensure that the concrete is free from entrapped air, after each compaction the container is stricken gently by a wooden mallet around the sides until no large air bubbles are found on the surface of the container.

Step 6. In case of other compaction methods such as an internal vibrator or vibrating table, vibration is given until a minimum time period necessary to achieve full compaction. Over-vibration is avoided to reduce the loss of entrained air.

Step 7. While using a needle vibrator, the vibrator must be held vertically and the needle must not touch the bottom or sides of the container.

Step 8. After compacting the final layer, the top layer is smooth levelled with a steel float. The surface is skimmed and rimmed with a straight edge. The outside of the container is cleaned.

Step 9. The container with the concrete is weighed and the weight denoted as m_2. The volume of the container is determined according to the following steps.

Step 10. The container is weighed along with the circular or square plate of 5 mm thickness.

Step 11. The container is placed in a horizontal position and filled with water maintained at 27 ± 2°C. Water is added until it overflows, and the glass plate is slid over the container to expel air bubbles and the excess water from the container.

Step 12. The container is weighed along with the water and the glass plate. The total volume of the container is calculated by dividing the mass of the water by 996.5 kg/m³ (density of water at 27 ± 2 °C). Volume is expressed in m³.

Note: Container needs to be calibrated at each use.

Observation and Calculations

Weight of the container, m_1 (kg) =

Weight of the container + concrete, m_2 (kg) =

The volume of the container, V (m³) =

Fresh density of the concrete, $\rho = \dfrac{m_2 - m_1}{V}$

Fresh density to be expressed to the nearest of 10 kg/m³.

Additionally, the volume of concrete per batch and yield per bag of cement can be calculated.

Volume of concrete per batch is given by $V_T = \dfrac{M_T}{\rho}$

ρ – density of the fully compacted concrete (kg/m³)

M_T – the sum of masses of all the constituents of the concrete as batched in kg

Yield per bag of cement bag (50 kg) is given by $Y = \dfrac{V}{N}$

V – the volume of concrete produced per bag in m³

N – number of 50 kg cement bags used per batch

Report

Fresh density of concrete ....

Volume of concrete per batch ....

Yield per bag of cement (50 kg) ....

Ambient temperature ....

Student Remark

....

....

....

....

$$\frac{\textbf{\textit{Marks obtained}}}{\textbf{\textit{Total marks}}} = -$$

Instructor signature

Instructor Remark

....
....
....
....

4.5.2 Determination of Setting Time of Concrete by Penetration Resistance

This test is used to determine initial and final setting times of concrete using penetration resistance apparatus.

Test Summary

This test is used to measure the time taken by a specific concrete to set, i.e. the initial setting time beyond which the concrete cannot be worked, and it represents stiffening of concrete. This test is applicable for concrete with slump more than 0 mm (IS 1199 Part 7, 2018).

Use and Significance

- Setting time can be used as an indication for different purposes.
- Initial setting time of concrete indicates that the concrete cannot be used if not cast, i.e. the ability of the concrete to cast in a formwork into a particular shape or size. It will be complicated to use the concrete for any other purpose.
- Final setting time of concrete indicates stiffened concrete. The stress required by concrete to withstand on its own. i.e. the concrete has reached its green strength. However, strength gaining will continue beyond the final setting time.
- This test method can be used in both controlled laboratory conditions as well as field conditions.

Required Apparatus

- Penetration resistance apparatus (a spring reaction type graduated from 50 N to 600 N in increments of 10 N or a hydraulic reaction type apparatus graduated from 700 N to 900 N in intervals of 10 N or less)
- A different set of removable needles with following areas: 645, 323, 161, 65, 32 and 16 mm^2. Each needle has a peripheral ring at a distance of 25 mm above the bearing face in the shaft of the needle. The shaft length of 16 mm^2 needles should not be more than 90 mm to minimize bleeding
- Pipette for drawing out free water from the surface of the test sample
- Tamping rod of circular cross-section with rounded ends made up of steel of diameter 16 $\pm$ 1 mm and a length of 600 $\pm$ 5 mm

- A thermometer with an accuracy of 0.5°C along with graduation from 0°C to 100°C is used
- 4.75 mm IS sieve

Preparation of Test Specimen

- A representative sample of concrete from the mix proportion is taken and sieved through a 4.75 mm IS sieve in a non-absorptive surface.
- The sieved mortar is thoroughly mixed in the non-absorptive surface.
- The required volume of the homogenized sample is filled in the container, at least 140 mm deep in a single layer.
- The sample is compacted in the container to eliminate air pockets and the surface of the sample in the container is levelled. (Compaction is done by shaking the container To and fro or by tapping the sides of the container.)
- In most of the cases, the mortar samples will be fluid enough in nature to fill the container. If the mortar is stiff in nature, then the mortar is filed in 50 mm layers with compaction by means of a tamping rod for every 6.5 cm^2. Tamping is distributed uniformly throughout the surface (35 tamps for 150 × 150 mm surface area).
- The sides of the container are tamped in order to remove the air pockets and to level the surface of the specimen.
- The prepared specimens are stored at 27 ± 2°C. The specimens are covered with damp burlap or a water-impermeable cover during the duration of the test.
- In general, three samples are prepared from each concrete mix.

Procedure

Step 1. Bleed water is removed using the pipette before each penetration measurement. Bleed water is collected by tilting the container at 10 degrees from the horizontal by means of a block placed beneath one bottom edge of the container before 2 minutes of each measurement.

Step 2. The area of the needle is chosen based on the hardened state of the mortar. At the initial stage, the mortar will be in the less hardened state; therefore, the force required by the needle to penetrate will be less. Therefore, the needle with the highest area is chosen, followed by the remaining needles based on the force required to penetrate the mortar.

Step 3. Penetration resistance apparatus is held vertically, and force is applied within a time period of 10 s until the needles penetrate to a depth of 25 mm.

Step 4. Care should be taken to avoid the effect of previous penetration on the subsequent penetrations. This is done by maintaining a minimum separation distance of 25 mm from the surface of the container and also keeping a distance of 13 mm from previous penetrations.

Step 5. During each penetration, the time elapsed is recorded (the time at which the water is added to the cement until the time of penetration) along with the corresponding force in Table 4.1.

Step 6. As a recommendation, the initial penetration is done after a time period of 2–3 hours for normal concrete at room temperature and 1–2 hours for special concrete or at high temperatures. Subsequent readings are noted down every 30 minutes. On the other hand, for low temperature or retarded concrete, the initial test is done after a time period of 4–6 hours. Subsequent readings are taken at 1-hour intervals. In all the above conditions based on the rate of change in the resistance of the concrete, the time period of testing can be reduced or increased.

Step 7. In each rate of hardening test, six penetrations should be made. The time interval between each penetration resistance is chosen in such a way that the obtained curve is satisfactory. The test is continued until a penetration resistance of 27.5 MPa is reached.

Step 8. Penetration resistance can be obtained by dividing the force required for a penetration of 25 mm depth by the corresponding cross-sectional area of the needle.

Step 9. A plot is made with the penetration resistance of the concrete (MPa) on the y-axis and the elapsed time in hours on the x-axis.

Note: In general, each needle can be used repeatedly until penetration to a depth of 25 mm is impossible using that particular needle.

Observations and Calculations

Observations are noted down in Table 4.1.

TABLE 4.1
Penetration Resistance and the Corresponding Time Elapsed

Time (hours)	The corresponding area of the needle used (mm^2)	Force (N)	Penetration resistance (MPa)

In the plot between time elapsed (hours) and penetration resistance (N/mm^2) two lines are dropped from y-axis to x-axis at a penetration resistance of 3.5 MPa and 27.5 MPa, respectively. Time obtained for the two drop-down lines corresponds to initial setting time (3.5 MPa) and final setting time (27.5 MPa).

Report

Ambient temperature ....
Initial setting time ....
Final setting time..................
Data on the concrete mix (Grade/mix proportion/maximum nominal size of aggregates) ....
Consistency of concrete as determined in slump test ....

Student Remark

....
....
....
....

$$\frac{\textit{Marks obtained}}{\textit{Total marks}} = -$$

Instructor signature

Instructor Remark

....
....
....
....

4.5.3 Slump Cone Test

This test is used in the site for assessing the workability of the concrete, i.e. the ease with which the concrete can be placed.

Test Summary

This test is used to measure the workability of concrete qualitatively by the use of a truncated cone (IS 1199 Part 2, 2018a).

Use and Significance

- This test method is used as a qualitative means to understand the consistency of concrete.

Required Apparatus

- Slump cone test consists of a truncated cone of 300 mm height with a bottom diameter of 200 mm and a top diameter of 100 mm with a base plate
- A steel tamping rod with 16 mm diameter and 60 cm height is used for compacting the concrete in different layers
- A trowel for levelling top layer of the concrete

Note: This test is extensively used in the site for understanding the workability of concrete. This test is conducted at the commencement of concreting operation or change in the workability of the concrete during the process of concreting.

Cautions to be Observed before Performing the Test

- The internal surface of the cone should be cleaned and should be free of moisture and any hardened concrete should be removed before the commencement of the test.
- The cone should be placed in a smooth, horizontal, rigid and non-absorbent surface.
- The apparatus is lubricated before performing the test.

Procedure

Step 1. Concrete is filled in three layers. The thickness of each layer is approximately equivalent to 1/3rd of the height of the truncated cone, as shown in Figure 4.1.

Step 2. Each layer is tamped 25 times evenly using the rounded end of the tamping rod such that the strokes are evenly distributed through the cross-section.

Step 3. After the top layer is tamped, the concrete is levelled by a trowel so that the mould is completely filled.

Step 4. The mould is then removed by raising it gradually. The concrete will subside; this subsidence is called a slump.

Step 5. During the performance of the above operation, no sort of disturbances, such as vibration or jerks, should be there and the test

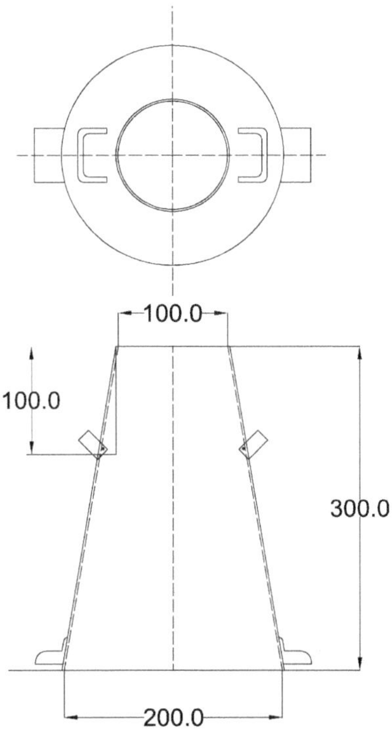

FIGURE 4.1 Slump cone (all dimensions are in mm).

needs to be completed within a period of 2 minutes after the completion of final tamping.

Step 6. The difference in level between the height of the mould and that of the highest point of the height of subsided concrete is called a slump of the concrete.

Step 7. The measured slump can be a true slump or a shear slump or a collapsible slump.

Step 8. True slump is the desired form of the slump. If the mix is harsh, then the ability of the concrete to retain its subsidence is less due to its particle packing, resulting in shearing or collapse. In such cases, the concrete mix proportion is redesigned in order to obtain the required slump.

Observations and Calculations

The initial height of slump (mm), $h_1 = 300$
The final height of slump (mm), $h_2 =$
Slump of concrete h_2-h_1 =

Report

Type of slump ....
Slump of the concrete is ....

Student Remark

....
....
....
....

$$\frac{\textbf{\textit{Marks obtained}}}{\textbf{\textit{Total marks}}} = -$$

Instructor signature

Instructor Remark

....
....
....
....

4.5.4 Compaction Factor Test

This method is one of the tests used widely to determine the workability of the concrete. But this test is not used widely in the field when compared to the slump cone test.

Test Summary

This test is used to measure the workability of concrete qualitatively by using a set of hoppers and a cylinder (IS 1199 Part 2, 2018b).

Use and Significance

- Two different concretes with very low workability that give the same slump value can be tested in the compaction factor test.
- This test is not suitable for concrete that has a compaction factor value less than 0.7.

Required Apparatus

- Compacting factor apparatus that consists of two hoppers with a cylindrical mould placed at the bottom
- Needle vibrator

Precautions to Be Observed before Performing the Test

- This test is highly user-dependent compared to the slump test, since the compaction done by a tamping rod depends upon the force and uniformity of spread. The above two factors may vary from person to person.
- Usage of compacting equipment such as needle vibrator or tamping rod needs to be the same as that of the site in order to ensure the usage of this test result.
- Apparatus is lubricated before performing the test.

Procedure

Step 1. Dry materials of the required proportion are weighed and mixed thoroughly in order to obtain a uniform mixture.

Step 2. The proportion of different ingredients must be sufficient enough to fill the first hopper, as shown in Figure 4.2.

Step 3. The empty weight of the bottom cylinder is noted. Let the empty weight of the cylinder be W_1.

Step 4. The desired weight of water is added and mixed thoroughly to obtain concrete with uniform consistency.

Step 5. After thorough mixing, the concrete is transferred to the top hopper. Concrete is filled in the top hopper until the concrete reaches hopper's plinth level.

Step 6. Top hopper's trap door is opened and the concrete is allowed to fall into the second hopper. Afterwards, the second hopper's trap door is opened so that the concrete fills in the bottom cylinder.

Step 7. Once the cylinder is full, the concrete is levelled to the topmost surface of the cylinder and the weight of the cylinder with concrete is noted. Note that the bottom-most cylinder is filled with concrete that is partially compacted. Let the weight of the cylinder with partially compacted concrete be W_2.

Step 8. Weight of the partially compacted concrete can be calculated using W_2-W_1.

Step 9. The same cylinder is refilled with concrete in three layers, each layer approximately 5 cm in height. Each layer is adequately tamped or vibrated in such a way that full compaction is achieved.

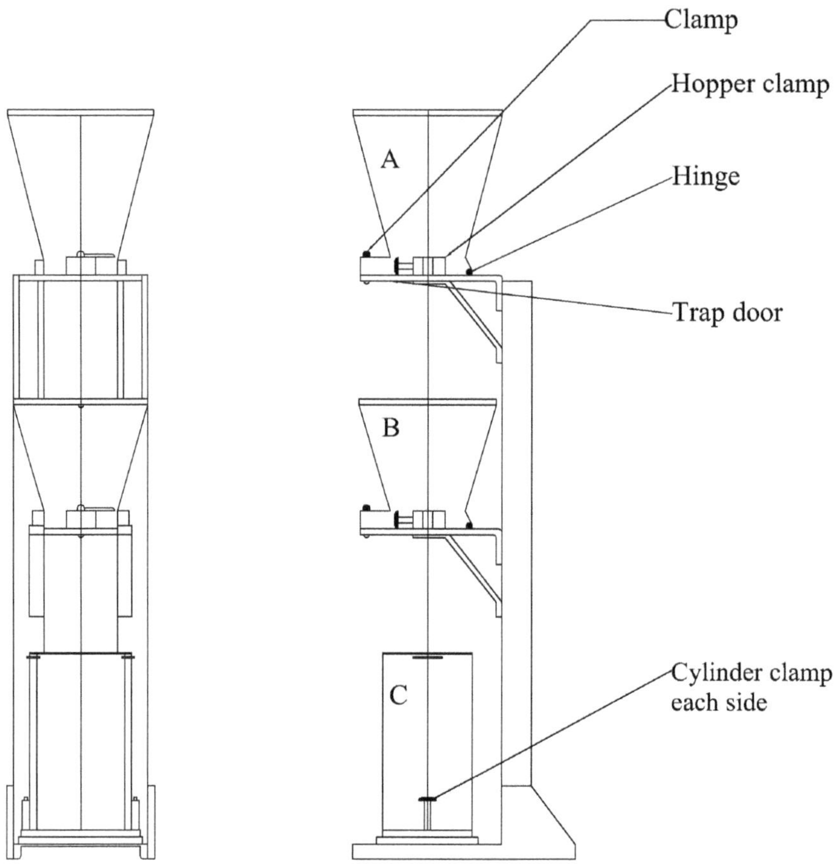

FIGURE 4.2 Compaction factor apparatus.

Choosing a tamping rod or vibrator used is of utmost importance. The rod that is used in the site needs to be used in order to get the same effect that we can achieve in the field.

Step 10. After filling the cylinder in three layers with compaction, the topmost layer is levelled with excess concrete if required and the weight of the cylinder with compacted concrete is noted. Let the weight of the cylinder with compacted concrete be W_3.

Step 11. The weight of compacted concrete can be calculated using W_3-W_1.

Step 12. Compaction factor can be calculated by taking the ratio of the weight of partially compacted concrete to that of compacted concrete. The compaction factor test is more useful in the case of concrete with low and medium slump values. The compaction factor ratio varies between 0.80 and 0.90 for low and medium slump concretes.

Observations and Calculations

Details	Observed value
Proportion of concrete	
w/c ratio used	
Weight of empty cylinder, W_1	
Weight of cylinder filled with partially compacted concrete, W_2	
Weight of cylinder filled with fully compacted concrete, W_3	
Weight of partially compacted concrete, W_2-W_1	
Weight of fully compacted concrete, W_3-W_1	
Compacting factor, (W_2-W_1/W_3-W_1)	

Report

Compaction factor ....

Student Remark

....
....
....
....

$$\frac{\textbf{\textit{Marks obtained}}}{\textbf{\textit{Total marks}}} = -$$

Instructor signature

Instructor Remark

....
....
....
....

4.5.5 Vee-Bee Consistency Test

Vee-Bee consistency test is used to determine the workability of the concrete. This test is also not used in the field widely (IS 1199 Part 2, 2018c).

Use and Significance

- Two different concretes with the same slump value can be differentiated by this test, although two different concretes can have similar Vee-Bee consistency time as well.
- This test is used for testing concrete with very low slump value.

Required Apparatus

- Vee-Bee consistency test consists of a cylinder of 300 mm diameter and 200 mm height
- A cone frustum of 200 mm bottom diameter, 100 mm top diameter and 300 mm height with a collar
- Vibrating table
- Scoop
- Trowel
- Measuring jar
- Standard tamping rod
- A stopwatch
- Ruler

Precautions to Be Observed before Performing the Test

- Similar to other tests, this method is also operation sensitive.
- Apparatus is lubricated before performing the test.

Procedure

Step 1. Coarse aggregate, fine aggregate and cement of the required proportion are taken and mixed thoroughly with water.

Step 2. Cylinder, cone with its frustum and collar are correctly positioned and fastened in the vibrating table in order to avoid movement during testing.

Step 3. Concrete is filled inside the cone in three layers, with each layer tamped 25 times by standard tamping rod.

Step 4. After pouring the third layer excess concrete is removed, the surface is levelled and the cone is removed gradually.

Step 5. The vibrating table is switched on, and simultaneously a stopwatch is also started. The time taken for the concrete to subside and fill the cylindrical mould is noted down.

Step 6. A relative measure of this subsidence time for different concretes indicates the consistency of the concrete.

Observations and Calculations

Initial reading in the ruler before remoulding (mm), $h_{1} =$

Final reading in the ruler after remoulding (mm), $h_{2} =$

Slump h_2-h_1 (mm) =

Time take for the concrete to subside and fill the cylinder completely (s) =

Report

Consistency of concrete measured in Vee-Bee consistency time (s) ....

Specification

According to IS code, the description of workability and corresponding Vee-Bee time (s) is shown in Table 4.2.

TABLE 4.2
Specification of Workability and Corresponding Vee-Bee Time

Description of workability	Vee-Bee time (s)
Extreme dry	32–18
Very stiff	18–10
Stiff	10–5
Stiff Plastic	5–3
Plastic	3–0

Student Remark

....
....
....
....

$$\frac{\textit{Marks obtained}}{\textit{Total marks}} = -$$

Instructor signature

Instructor Remark

....
....
....
....

4.5.6 U-Box Test

This test is used to determine the filling and passing ability of self-compacting concrete (SCC) (EFNARC, 2002a).

Use and Significance

- Determination of passing and filling ability is of utmost importance in places where the degree of congestion of reinforcement increases.

Required Apparatus

- U-box test consists of U-shaped container with three 10 mm diameter rebars at the height of 190 mm from the bottom, the total height of the container being 680 mm
- The U-shaped box consists of two rooms with a partition at the centre with a total width of 280 mm
- A partition gate is located at the bottom of the cylinder that can be raised to allow SCC to flow

Precautions to Be Observed before Performing the Test

- Any sort of mechanical vibration needs to be avoided.

Procedure

Step 1. Apparatus is set on levelled terrain and the internal surface is lubricated with oil.

Step 2. Based on the practical applications, suitable bars can be used (10 mm or 13 mm), as shown in Figure 4.3.

Step 3. SCC is poured inside the apparatus without agitation or mechanical compaction, and the SCC is allowed to stand for one minute.

Step 4. The partition gate is lifted and the SCC is allowed to flow. After the concrete attains self-level on both sides of U tube, the height of the concrete is measured from the top surface of the concrete to the top of the container on both sides and the respective heights are noted as h_1 and h_2.

Step 5. The difference between h_1 and h_2 gives the filling height.

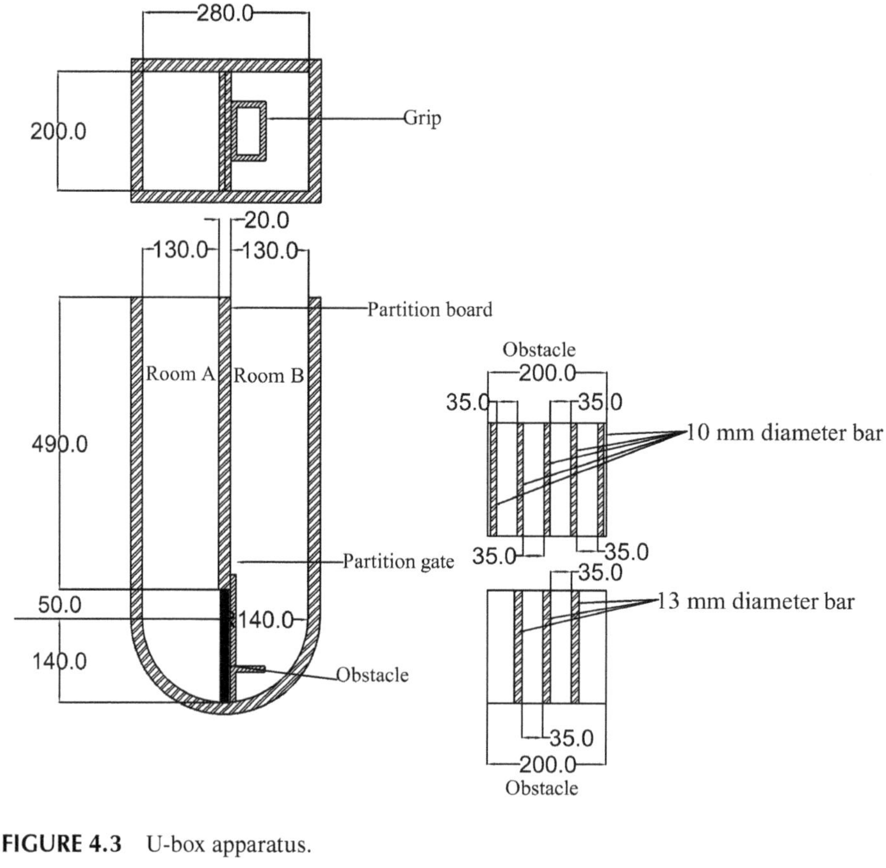

FIGURE 4.3 U-box apparatus.

Step 6. The time taken for the SCC to pass through the re-bars and rise in the other room is measured. This gives the passing time.

Observations and Calculations

Height of the concrete in one side after the concrete self-levels, h_1 (mm) =
Height of the concrete in another side after the concrete self-levels, h_2 (mm) =
Filling height, $h_1 \pm h_2 =$
Time taken for the concrete to pass through the rebar and self-level (s) =

Report

Filling ability of the concrete……………
Passing ability of the concrete………………

Student Remark

………………………………………………………………………………………
………………………………………………………………………………………
………………………………………………………………………………………
……………………………………………………………………………………

$$\frac{\textit{Marks obtained}}{\textit{Total marks}} = -$$

Instructor signature

Instructor Remark

………………………………………………………………………………………
………………………………………………………………………………………
………………………………………………………………………………………
……………………………………………………………………………………

4.5.7 J-Ring Test

J-ring test is used to determine the passing ability, slump flow and flow time taken for reaching 500 mm slump of self-compacting concrete (EFNARC, 2002b).

Use and Significance

Determination of passing and filling ability is of utmost importance in places where the degree of congestion of reinforcement increases.

Required Apparatus

- J-ring, base plate
- Slump cone
- Funnel
- Steel collar weight
- Stopwatch.

Caution

- The test needs to perform in a vibration-free ground.
- Apparatus used in the test are dampened with a wet cloth.
- Once the slump cone is filled with concrete, the cone needs to be lifted within 30 seconds.

Procedure

Step 1. The base plate is placed in a vibration-free ground with J-ring, slump cone and collar weight. J-ring is shown in Figure 4.4.

Step 2. The SCC that needs to be tested is poured in the slump cone through a funnel. After filling, excess concrete is levelled.

Step 3. The slump cone is lifted gradually within 2–3 seconds and simultaneously a stopwatch is started. The time taken for concrete to reach 500 mm graduation is noted down. The above recorded time is noted as t_{500} of the J-ring.

Step 4. Once the concrete stops flowing, the perpendicular graduations are noted down as sp_1 and sp_2. The average of the above values gives the slump flow through J-ring.

Step 5. The passing ability of concrete can be identified by measuring the thickness of spread concrete at different positions. Four readings

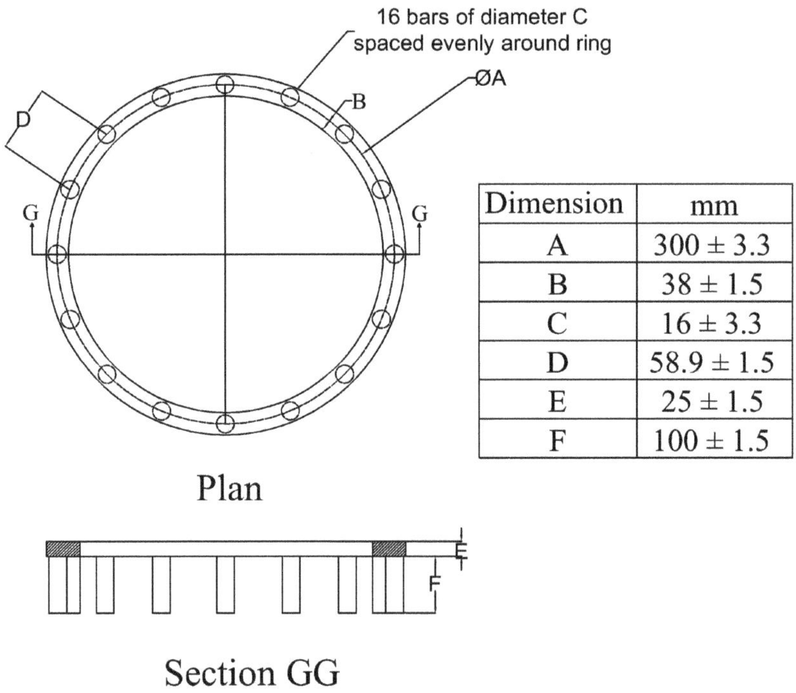

Dimension	mm
A	300 ± 3.3
B	38 ± 1.5
C	16 ± 3.3
D	58.9 ± 1.5
E	25 ± 1.5
F	100 ± 1.5

FIGURE 4.4 J-ring apparatus.

are taken at exterior portions near the J-ring, two readings at the end of the centre lines near the exterior portion of J-ring and two readings at perpendicular directions to the previous readings. Let us denote them as Δhx_1, Δhx_2, Δhy_1 and Δhy_2, respectively. One more reading is taken at the centre of the J-ring and denoted as Δh_0.

Step 6. Passing ability can be computed by $\frac{\Delta hx_1 + \Delta hx_2 + \Delta hx_3 + \Delta hx_4}{4} - \Delta h_0$

Step 7. To obtain time taken for concrete to reach 500 mm graduation without any obstruction, concrete is filled in slump cone without J-ring. Once the concrete is filled and levelled, the cone is lifted as before.

Step 8. The time taken for concrete to reach the 500 mm graduation is noted. To obtain the slump flow without obstruction after concrete stops flowing perpendicular spreads are noted down. The average of these spreads gives an average spread.

Observations and Calculations

Time taken to reach a flow of 500 mm, t_{500} (s) =

Slump flow through J-ring (mm) =

Passing ability, $\frac{\Delta hx_1 + \Delta hx_2 + \Delta hx_3 + \Delta hx_4}{4} - \Delta h_0$ (mm) =

Slump flow without J-ring (mm) =

Report

Time taken to reach a flow of 500 mm, t_{500} (s)...................

Slump flow through J-ring (mm) ....

Passing ability, $\frac{\Delta hx_1 + \Delta hx_2 + \Delta hx_3 + \Delta hx_4}{4} - \Delta h_0$ (mm) ....

Slump flow without J-ring (mm) ....

Student Remark

....

....

....

....

$$\frac{\textbf{\textit{Marks obtained}}}{\textbf{\textit{Total marks}}} = -$$

Instructor signature

Instructor Remark

....

....

....

....

4.5.8 V-Funnel Test

This test method is used to understand the filling ability of concrete with a maximum coarse aggregate size of 20 mm. The time taken by 12 litres of concrete to flow through the funnel is measured and reported (EFNARC, 2002c).

Use and Significance

- Filling ability of concrete is one among the essential factors that determine the ability of the concrete to reach the nooks and corners of the formwork.
- This test can also be used as a qualitative method to check the segregation of concrete. Shorter flow time means good flowability and vice versa.
- For self-compacting concrete (SCC), a period of 10 seconds is considered as appropriate.

Required Apparatus

- V-funnel
- A container of 15 litres capacity
- A Trowel
- Scoop
- A Stopwatch

Caution

- The test needs to perform in a vibration-free ground.
- The apparatus used in the test are dampened with a wet cloth.
- Once the V-funnel is filled with concrete, the trap door needs to be opened within 10 seconds.
- The test needs to be conducted with the good lightning condition to have clear visibility during emptying off the V-funnel to identify the endpoint of the test.

Procedure

Determination of flow time:

Step 1. Concrete of required quantity is mixed and approximately 15 litres of concrete is sampled to assess its filling ability using V-funnel test.

Step 2. The V-funnel is set on a firm ground and 12 litres of concrete is poured into the funnel without tamping or compacting. In the meantime, the trap door is kept closed. A container is placed underneath the funnel to collect the concrete. The line diagram of the V-funnel is shown in Figure 4.5.

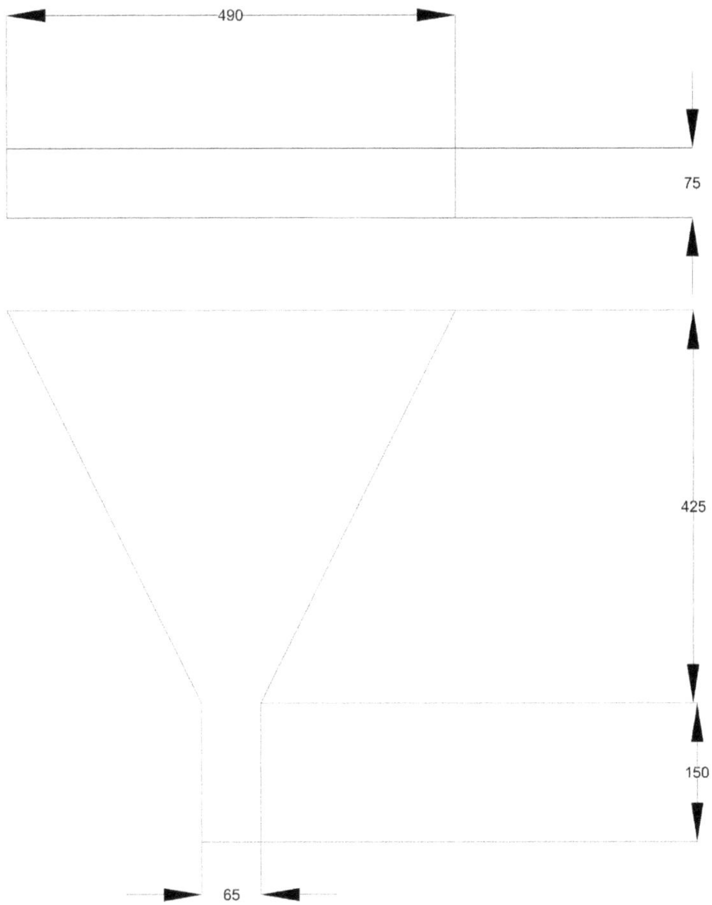

FIGURE 4.5 Line diagram of V-funnel test apparatus (all dimensions are in mm).

Step 3. Now, the trap door is opened and the concrete is allowed to pass through. The concrete is collected in the underneath container. Simultaneously, a stopwatch is started to note down the time taken by the concrete to pass through.

Step 4. During this procedure, the concrete should be only under the action of gravity.

Step 5. The complete test should be performed within a period of 5 minutes.

Determination of flow time after a rest period of 5 minutes:

Step 1. Once again, the trap door is closed, and the funnel is filled with the concrete collected in the container kept underneath the funnel.

Step 2. During the pouring of the concrete, no compaction energy should be provided.

Step 3. Before pouring the concrete, the funnel should not be cleaned or moist inside.

Step 4. The trap door is opened after 5 minutes of filling the concrete. Simultaneously, a stopwatch is started and the time taken by the concrete to pass through under the action of gravity is noted down.

Step 5. The time is noted down as flow time for the concrete after a rest period of 5 minutes.

Observations and Calculations

Flow time =

Flow time after 5 minutes of rest =

Report

Flow time (s)..................

Flow time after 5 minutes of rest (s) ....

Student Remark

....

....

....

....

$$\frac{\textit{Marks obtained}}{\textit{Total marks}} = -$$

Instructor signature

Instructor Remark

....

....

....

....

4.5.9 Compressive Strength

Compressive strength of concrete is one of the essential parameters of the concrete to be tested,since concrete is designed based on grades which imply the load-carrying capacity of the concrete. Further, most of the characteristics of the concrete are related to the compressive strength of the concrete qualitatively (IS 516, 2018).

Use and Significance

- Compressive strength is considered as an essential parameter that gives an idea about the rate at which the strength of the concrete gets developed macroscopically.

- Moreover, the design of any member in a concrete structure requires the required load-carrying capacity of the concrete.

Required Apparatus

- Moulds of (standard size 150 x 150 x 150 mm) (in case of cylinder 150 x 300 mm (diameter × height))
- Tamping rod of 16 mm in diameter and 600 mm in height
- Trowels

Notes

- Concrete, in general, is designed for certain specific grades, For example, M40 in which M denotes the mix and 40 is the corresponding strength in N/mm^2.
- In general, concrete is cast in both cylindrical specimens and cube specimens.
- Due to the difference in slenderness effect (the ratio between the length and the diameter of any shape) the strength of the cylinder will always be lesser than cubes with similar grades. This effect is manifested and explained by a phenomenon called shape effect.

Cautions

- The rate of loading needs to be taken care of since an increase in the rate of loading can give higher strength and vice versa than expected.
- In general, dry concrete gives higher strength when compared to wet concrete.
- After casting, concrete needs to be covered with hessian/jute cloths for 24 hours in order to avoid evaporation of moisture from the surface of the concrete.
- Mould should have a rigid connection with base necessarily, to prevent leakage of mortar during compaction.
- Site specimen should be stored between 22° C and 32° C.
- Tamping of concrete is to make the concrete constituents to be spread uniformly through the cross-section.
- In the case of specific site mixes, the type of tamping used in the site needs to be adopted to understand the actual effect of tamping on the compressive strength.

Procedure

Step 1. After arriving at the desired mix proportion according to IS: 10262, mixing of concrete is done manually or mechanically.

Step 2. The thoroughly mixed concrete is filled in the cube mould or cylindrical mould in three layers (approximately 1/3rd of the height of the cube or cylinder).

Step 3. Each layer is tamped using a tamping rod not less than 35 times for a cubical specimen of 150 mm and not less than 25 times for a cubical specimen of 100 mm. For a cylindrical specimen, the number of strokes should not be less than 30 per layer. Tamping should be spread uniformly through the surface of the concrete.

Step 4. When filling the consecutive layers (i.e. 2nd and 3rd layers), the tamping rod should pass through the subsequent layers.

Step 5. After the final surface is compacted, concrete is levelled by a trowel, and the mould is covered immediately with damp hessian cloth or glass or any non-absorptive material and left undisturbed for 24 hours.

Step 6. Cast concrete need to be marked with suitable identification numbers after slight hardening, in order to facilitate easy identification at different ages.

Step 7. After 24 hours, the mould is stripped off and the cast concrete is stored in water (or at 100% relative humidity) for further curing until the specified days of testing.

Step 8. The cast specimens should be taken out just prior to the testing, and surfaces need to be wiped off with a dry cotton cloth to remove the surface moisture.

Step 9. The curing water temperature should be maintained at 24–30° C.

Step 10. The cured specimens are tested in a compression testing machine after the required days according to the specifications.

Step 11. In general, strength at the different days of curing is obtained (3rd, 7th, and 28th day)

Step 12. Compressive strength can be obtained by dividing the obtained compressive load at the failure by the corresponding cross-sectional area of the specimen.

Step 13. The average strength obtained on the 28th day is called as a characteristic strength of the concrete. In general, a set of three specimens is cast to obtain the average strength of concrete.

Observations and Calculations

Details	Specimen 1	Specimen 2	Specimen 3
Length (mm)			
Breadth (mm)			
Diameter (mm)			
Height (mm)			
Compressive load at failure (kN)			
Compressive strength (N/mm^2)			

Report

Compressive strength of concrete……………

Student Remark

…………………………………………………………………………………………
…………………………………………………………………………………………
…………………………………………………………………………………………
…………………………………………………………………………………………

$$\frac{\textit{Marks obtained}}{\textit{Total marks}} = -$$

Instructor signature

Instructor Remark

…………………………………………………………………………………………
…………………………………………………………………………………………
…………………………………………………………………………………………
…………………………………………………………………………………………

4.5.10 Split Cylinder Test

This test is used to determine the split tensile strength of the concrete. Tensile strength estimated by this procedure will generally be more than the tensile strength estimated directly and less than the flexural strength (modulus of rupture) (IS 5816, 2018).

Test Summary

- This test consists of applying a diametrical compressive load along the length of the cylindrical specimen at a prescribed rate until failure of the specimen occurs. The plane containing the applied load induces tension, while the area immediately surrounding the load undergoes relatively higher compressive stress. It results in tensile failure rather than a compression failure of the specimen.
- The load applied in the cylinder is distributed along its length by thin, plywood bearing strips.
- Splitting tensile strength is obtained by dividing the maximum load sustained by suitable geometrical factors.

Use and Significance

- Tensile strength obtained by performing spilt tensile strength of cylindrical specimens is greater than the tensile strength obtained directly and less than the modulus of rupture obtained by flexural strength test.
- Split tensile strength of concrete is used to determine the development length of reinforcement.

Required Apparatus

- Testing machine (capable of applying a constant rate of loading continuous without shock at 0.7–1.4 MPa/min)
- Supplementary bearing bar or plate
- Bearing strips

Caution

- The specimen should be of required size, shape and must attain a specified curing condition. Specimen during testing should be kept moist by blanket covering and should be tested as soon as possible in its moist condition.
- Testing conducted on lightweight concrete at 28th day should be cured for 7 days in moist condition and kept in air-dry condition for 21 days at 23 ± 2°C and relative humidity of 50 ± 5%.

Procedure

Step 1. The alignment of a specimen is of utmost importance while performing split tensile strength. For proper alignment, marking of specimens is mandatory. Diametrical lines are marked on the top and bottom faces of the specimen, ensuring that the two lines are in the same axial plane. Alternatively, markings can also be carried out by aligning a jig.

Step 2. The dimensions of the cylindrical specimen are marked to the nearest of 0.25 mm and 2 mm for the diameter and length of the specimen, respectively.

Step 3. The positioning of the specimen is carried out by placing the specimen over a plywood strip aligning diametrically marked lines vertical and centred.

Step 4. After positioning, the load is applied continuously within a range of 0.7–1.4 MPa/min at a constant rate until the specimen fails by splitting. For a cylindrical specimen of 150 × 300 mm, the required total load range is from 50 kN/min to 100 kN/min, which corresponds to its splitting tensile stress.

Step 5. The maximum applied load at failure is recorded along with the failure type and concrete appearance after the failure.

Step 6. Splitting tensile strength of the concrete specimen is calculated as follows

$$T = \frac{2P}{\pi l D}$$

Where,

T – Tensile strength (MPa)

P – Maximum applied load (N)

l – Length (mm)

D – Diameter (mm)

Observations and Calculations

Details	Specimen 1	Specimen 2	Specimen 3
Age of the specimen			
Length (mm)			
Diameter (mm)			
Height (mm)			
Maximum applied load, *P* (kN)			
Tensile strength by splitting (nearest to 0.05 MPa) (MPa or N/mm^2)			
Fracture type			

Report

The following information are reported

- Identification number
- Length and diameter (mm)
- Maximum load (N)
- Tensile strength by splitting (nearest of 0.05 MPa)
- Specimen age
- History of curing
- Specimen defects
- Fracture type

Student Remark

....
....
....
....

$$\frac{\textbf{\textit{Marks obtained}}}{\textbf{\textit{Total marks}}} = -$$

Instructor signature

Instructor Remark

....
....
....
....

4.5.11 Oxygen Permeability Test

Oxygen permeability test is considered to be one among the most vital tests to understand the diffusion of oxygen into concrete (CoMSIRU, 2018a).

Use and Significance

- Ingress of oxygen in concrete can result in adverse effects in reducing the durability of concrete.

Required Apparatus

- Oxygen permeability test setup consists of an inlet and an outlet valve with a permeability cell capable of withstanding a pressure range between 0 kPa and 120 kPa; the cell needs to be checked regularly by impermeable specimen for airtightness. In general, a drop of pressure from 100 kPa to 0 kPa is required in 24 hours
- Oxygen cylinder with a pressure regulator
- Metal collar
- Rubber casket
- Concrete specimen (diameter 70 ± 2 mm, thickness 30 ± 2 mm)
- A water-cooled diamond-tipped core barrel with a nominal internal diameter of 70 mm with coring drill attachment
- A water-cooled movable bed diamond saw
- A holder to clamp cubes firmly
- Oven capable of maintaining temperature (50 ± 2°C)
- Vernier calliper
- Desiccator (RH controlled at 60%, desiccant: anhydrous silica gel)

Specimen Preparation

Step 1. The prepared specimen is either core cut from laboratory concrete specimen or from a real scale structure for which the oxygen permeability needs to be assessed.

Step 2. Preparation of test specimens from cast concrete cubes (150 × 150 × 150 mm) constitutes of core drilling the specimen with a core bit of dimension 70 ± 2 mm.

Step 3. Apply epoxy coating on the side surface of the core drilled specimen and allow it to cure as specified by the manufacturer of epoxy to attain better coating.

Step 4. After the epoxy coating is dried, mark 15 mm from one end and subsequent points are marked at 30 mm to slice 30 mm thick specimens. In the case of concrete, four specimens of 70 mm diameter and 30 mm thickness can be obtained by a diamond cutting machine. (Safety should be ensured before starting the cutting process.)

Step 5. During cutting, the machine can pull the concrete towards the blade; hence complete care should be taken.

Step 6. Specimens are kept at 50°C for 7 days in a temperature-controlled oven. After 7 days, specimens are taken out of the oven and cooled to room temperature for 2 hours. The cooled specimen is used for testing.

Caution

- This test should not be used for concrete that has aggregates of maximum nominal size 26.5 mm and more.
- The oven-drying method chosen (50°C for 7 days) instead of 105°C is to minimize the degree of micro-structure alteration in the concrete during test specimen preparation.
- Before commencing the calculation, the data obtained should be until the pressure reaches 50 kPa, and the time of reach should not exceed 6 hours.

Procedure

Step 1. The specimen is placed in the compressible collar with a rigid sleeve with the test face facing the bottom. The specimen should be placed in a position such a way that the outer face will be in the lip of the collar.

Step 2. The above arrangement is placed inside the permeability cell covering the hole, as shown in Figure 4.6. A solid ring is placed on the top of the collar (optional) ensuring that there is no gap between the collar and the sleeve. Finally, a cover plate is placed on the top of the solid ring.

Step 3. The cover plate is placed at the centre and tightened so that gas leakage can be avoided. Air inlets and outlets are opened in the permeability cell and the supply of oxygen in the tank is maintained between 100 kPa and 120 kPa and oxygen is allowed to flow through the cell for 5 seconds. This will expel other gases in the permeability cell other than oxygen.

Step 4. Ensuring that there are no leaks, the outlet valve in the permeability cell is closed. The inlet valve is closed after increasing the pressure to 100 ± 5 kPa in the permeability cell. Initial pressure (P_0) and time (t_0) are recorded to the nearest minute and 0.5 kPa, respectively.

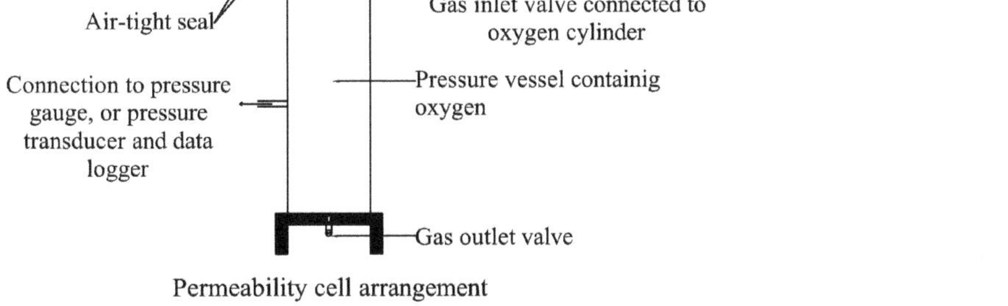

FIGURE 4.6 Oxygen permeability apparatus.

Step 5. The first reading is recorded at the 5th minute and subsequent readings are recorded at an interval of 5 ± 1 kPa drop in the pressure. A drop in pressure of more than 5 kPa indicates leakage. During such cases, the pressure in the chamber is released and the test is restarted after the specimen is tightly fit inside the collar.

Step 6. The test is stopped when either the pressure drops to 50 ± 2.5 kPa or after 6 hours ± 15 minutes, whichever occurs first. Minimum of 8 readings are recorded.

Step 7. A plot is obtained between $\ln \frac{P_0}{P_t}$ and time and a linear regression fit is obtained.

Where,

t – the time since the start of the test (to the nearest minute in seconds)

P_0 – the recorded initial pressure at the time t_0 to the nearest of 0.5 kPa

P_t – the pressure recorded at time t to the nearest of 0.5 kPa

Step 8. If the coefficient of correlation obtained is more than 0.99 then the values obtained in the test are used, if not, then retest is conducted on the same specimen. Again, if the obtained coefficient of correlation is less than 0.99, then the specimen is discarded, and a new specimen is obtained and tested again.

Step 9. To improve the coefficient of correlation, no data points should be dis-regarded. The slope of the linear regression line can be obtained by

$$Z = \frac{\Sigma\left[\ln\left(\frac{P_0}{P_t}\right)\right]^2}{\Sigma\left[\ln\left(\frac{P_0}{P_t}\right)t\right]}$$

Where,

Z – the slope of the linear regression

Step 10. Coefficient of correlation can be calculated using

$$r^2 = 1 - \frac{\Sigma\left[t_i - t_{p,i}\right]^2}{\Sigma t_i^2 - \left(\Sigma t_{p,i}^2\right)/n}$$

Where,

t_i – the time at any given pressure reading recorded to the nearest minute in seconds

$t_{p,i}$ – the predicted time at the same pressure reading (based on linear regression in seconds)

n – the number of data points being considered

Step 11. The value of $t_{p,i}$ can be calculated from

$$t_{p,i} = \frac{\ln \frac{P_0}{P_t}}{Z}$$

Step 12. Darcy's coefficient of permeability can be calculated by

$$k = \frac{\omega * V * g * d * Z}{R * A * T}$$

Where,

k – the coefficient of permeability of test specimen in (m/s)
ω – molecular mass of oxygen (0.032 kg/mol)
V – volume of permeability cell recorded to the nearest of 0.0001 m^3 or 0.01 L (which includes the volume of the opening in the top plate and rubber collar annulus below the sample)
g – the acceleration due to gravity (9.81 m/cm^2)
d – the average specimen thickness to the nearest 0.02 mm (in meters)
Z – the slope of the linear regression line forced towards (0, 0) in (s^{-1})
R – the gas constant (8.313 Nm/Kmol)
A – cross-sectional area of the specimen in m^2
T – the absolute temperature in Kelvin (K)

Step 13. For each specimen coefficient of permeability is calculated and the oxygen permeability index (OPI) is calculated as

$$OPI = -\log_{10} k$$

Observations and Calculations

Observed readings are noted down in Table 4.3.

TABLE 4.3
Observed Readings for Oxygen Permeability Test

	Specimen 1		Specimen 2		Specimen 3	
Details	**Pressure drop (kPa)**	**Time (minute)**	**Pressure drop (kPa)**	**Time (minute)**	**Pressure drop (kPa)**	**Time (minute)**
Initial pressure drop (P_0)						

Report

- Coefficient of permeability (*k*) of an individual specimen to three decimal places
- Oxygen permeability index of each specimen to the nearest of two decimals
- Average oxygen permeability test (OPT) of all the specimen
- Source of the specimen
- Location of the specimen (within the core or member)
- Type of concrete (binder type, w/c ratio)
- The age of concrete at the time of testing
- History of curing

Student Remark

....
....
....
....

$$\frac{\textbf{\textit{Marks obtained}}}{\textbf{\textit{Total marks}}} = -$$

Instructor signature

Instructor Remark

....
....
....
....

4.5.12 Sorptivity Test

This test is used to determine the water absorption rate by hydraulic cement concrete. Water absorption is determined by measuring the increase in specimen mass with respect to time resulting from the absorption of water when only one of its surfaces is exposed to water (CoMSIRU, 2018b).

Use and Significance

- Pore system in concrete determines the penetrability of aggressive agents from the environment, which in turn decides the performance of the concrete. Rise in capillarity due to absorption decides the rate of ingress of water and other liquids in unsaturated concrete.
- Different factors determine the absorption of water by concrete surface, some of them include: mixture proportion of concrete, presence of supplementary cementitious materials and chemical admixtures, physical and chemical characteristics of cement-based components and aggregates, air content, curing type and duration, age or hydration degree, micro-cracks, surface treatments

such as sealants and surface oil, method of placing that includes finishing and consolidation, moisture content of the concrete at the time of testing, etc.

- Exterior concrete which is mostly unsaturated and most susceptible to the attack of aggressive exterior agents such as penetration of water is tested. Due to the difference in saturation and other conditions, absorption of water is different for interior and exterior surfaces. Using this method, water absorption both in the interior and on the surface of the concrete can be measured. Water absorption in the interior surface can be determined by drilling a core in the existing structure and cutting the core transversely absorption at different depths can be estimated. Cores can be drilled both vertically as well as horizontally.
- This test should not be disordered with the other test procedure, where the complete specimen is oven-dried and immersed in water at 21°C and then boiled for 5 hours underwater. In this test, the total specimen is immersed in the water to measure the total water-permeable pore space.

Required Apparatus

- Pan (to carry the specimen exposed to water to be tested)
- Vacuum saturation facility
- Supporting devices such as pins and rods used to allow water access to the concrete surface
- Weighing balance (±0.01 g)
- Timer (±1 s)
- Cloth (absorbent paper towel)
- Water-cooled saw
- An environmental chamber or over with desiccator that can maintain a temperature of 50 ± 2°C and relative humidity of 80 ± 3%
- Polyethylene container with sealable lids large enough to carry one test specimen and calliper (0.02 mm)

Materials and Reagents

- Potassium Bromide (to maintain RH in a desiccator)
- Saturated calcium hydroxide solution prepared from tap water maintained at 23 ± 2°C (3 grams of $Ca(OH)_2$ in 1 litre of water)
- Sealants (epoxy paints, vinyl tapes, duct's tape or aluminium tap)
- Sheeting or plastic bag attached to the specimen in order to control the water evaporation from other portion of the specimen

Test Specimen

- The test specimen is a concrete disc of dimension 70 ± 2 mm and 30 ± 2 mm diameter and thickness, respectively. The specimen can be obtained either from moulded cylinders or cores drilled from a specified location of the structure under investigation.
- Cores obtained should have even area in the top and bottom surface, while the average error in the area should not exceed 1%.

Caution

- The specimen under test needs to be conditioned before the test by maintaining standard relative humidity so that the capillary pore system in the concrete is maintained at a consistent moisture content.
- The specimen surface exposed to the environment is immersed in the water. During initial contact of the concrete surface with water, water absorption in concrete takes place by capillary suction.
- The cores taken from the structure under investigation should be marked with caution. The surface under testing should not be marked, since the marking may change the rate of absorption of the specimen.
- While testing for samples taken from any structures repeated, testing should be done on cores drilled from similar locations at a similar depth, since concrete is not a homogeneous material. Further, the exterior surface and interior surface seldom have the same porosity. To ensure repeatability, measurements are done on more than one core taken from the same depth, which reduces the data scatter.

Conditioning of Specimens

Step 1. If new specimens are used, then they are placed in an oven at least for 7 days and not more than 8 days at 50 ± 2°C.

Step 2. After the drying period, the specimens are transferred immediately into a desiccator.

Step 3. The specimens are allowed to cool to a temperature of 23 ± 2°C within 4 hours. After cooling the test is started immediately within 30 minutes.

Step 4. The curved side of the specimen is sealed using a sealant.

Step 5. If the specimens are used previously in oxygen permeability cell, then the specimen is used as such without any pre-treatment, provided that the specimens are not exposed to the environment to absorb moisture.

Step 6. If the specimen from oxygen permeability cell cannot be used directly, then pre-treatment is done as per step number 3 overnight.

Step 7. Vacuum saturation facility is shown in Figure 4.7, and sorptivity test setup is shown in Figure 4.8.

Procedure

Step 1. Water sorptivity test is conducted in a temperature-controlled room at 23 ± 2 °C.

Step 2. To support the specimen pins and rods are placed in position. Alternatively, ten layers of absorbent paper towels can also be used.

Step 3. Prepared calcium hydroxide solution is placed in the tray. If paper towels are used as support, then the towels should be saturated by water, any air bubbles in the paper towel should be removed by smoothing the paper towel to the edges. Alternatively, if the

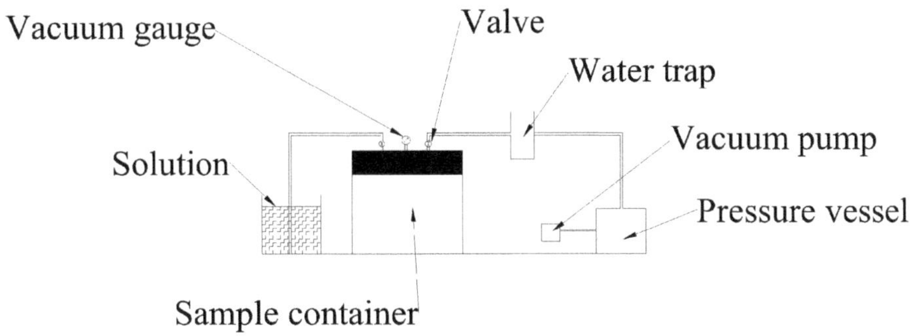

FIGURE 4.7 Vacuum saturation facility.

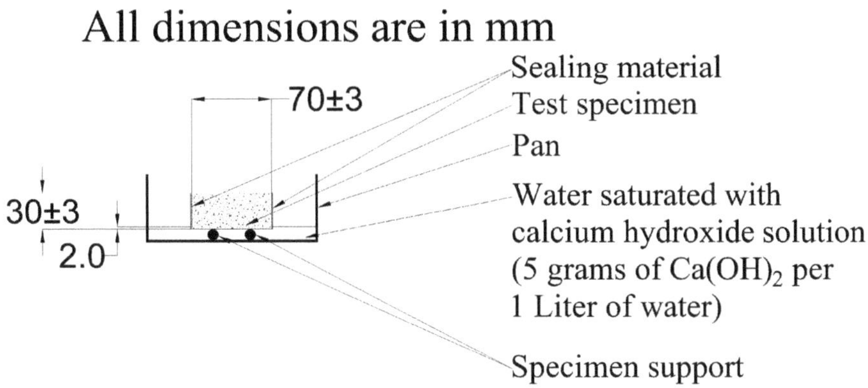

FIGURE 4.8 Sorptivity test setup (all dimensions are in mm).

rods are used for the support, then the calcium hydroxide solution should be filled above the top of the support.

Step 4. The final solution level should be just above the support and not more than 2 mm up the side of the specimen. Additional dampened paper towels are kept aside in order to remove excess water from the specimen.

Step 5. After the specimens are removed from the desiccator or oxygen permeability cell, they are weighed to a precision of 0.01 gram immediately. Let the dry weight be M_{so}.

Step 6. The diameter and thickness of the specimen are measured at different positions (0.02 mm precision).

Step 7. The test face of the specimen is placed immediately on the support at a time t_0.

Step 8. The specimens are weighed at different time intervals (3, 5, 7, 9, 12, 16, 20 and 25 minutes) after touching it on the damp paper towels. While weighing it, the test face of the specimen should look like a saturated surface dry condition. If any free water is available on the surface, it should be wiped off by dampening paper towel.

Step 9. The weight of the specimens is recorded within 10 seconds on taking it outside of the tray. Once the weight is noted down the specimens are placed back with the test face on the support. During this measurement, the stopwatch is not stopped.

Step 10. After weighing the specimens for a day, the specimens are placed in a vacuum saturation tank with the sealants around the perimeter of the specimens intact.

Step 11. The specimens are placed in the desiccator in such a way that maximum surface area is exposed, i.e. the specimens are standing upon their curved surface. The lid is placed over the desiccator with petroleum jelly as a sealant.

Step 12. Now the tank is evacuated until the pressure reaches a value between -75 kPa and -80 kPa. Specimens are placed in the same condition for 3 hours $\pm$ 15 minutes.

Caution: Pressure is not allowed to rise more than -75 kPa during the above time period.

Step 13. After 3 hours $\pm$ 15 minutes, the tank is isolated and saturated calcium hydroxide solution is allowed to flow inside the tank. The tank is filled until the solution reaches a height of 4 cm above the specimens.

Caution: During this period, air should not be allowed to flow inside the chamber.

Step 14. The vacuum is re-established and the pressure is reduced between -75 kPa and -80 kPa and the specimens are left for 1 hour $\pm$ 15 minutes. After 1 hour $\pm$ 15 minutes vacuum is released and the air is allowed to flow inside the tank. Specimens are allowed to stand in the solution for 18 $\pm$ 1 hours.

Step 15. After the prescribed period of 18 $\pm$ 1 hours, specimens are removed from the tank and wiped off with damp paper cloths so that the specimens reach saturated surface dry condition. Then they are weighed to the nearest of 0.01 g accuracy. This weight is recorded as vacuum saturated mass M_{sv} of the specimens.

Observations and Calculations

Step 1. Porosity (n) of each specimen can be determined by the following equation

$$n = \frac{M_{sv} - M_{s0}}{Ad\rho_w}$$

Where,

M_{sv} – vacuum saturated mass of the specimen to the nearest of 0.01 g

M_{s0} – the mass of the specimen at time t_0 (start of the test) to the nearest of 0.01 g

A – the cross-sectional area of the specimen to the nearest of 0.02 mm^2

d – the average thickness of the specimen to the nearest of 0.02 mm;

ρ_w – the density of water 10^{-3} g/mm^3

Step 2. A plot is made between the gain in mass (M_{wt}) and the square root of time $\sqrt{t}$.

$$M_{wt} = F\sqrt{t}$$

Where,

F – the slope of the best fit line from the plot between M_{wt} and $\sqrt{t}$

t – the time in hours after a specimen was first exposed to water on its lower face, to the nearest of 0.01 hours

Step 3. The following equation is used to determine the gain in mass

$$M_{wti} = M_{st} - M_{s0}$$

Where,

M_{st} – the mass of the specimen nearest to 0.01 g at any specific time 't'

M_{s0} – the mass of the specimen at time t_0 (start of the test) to the nearest of 0.01 g

***Note*:** Time taken to wipe and weigh the specimen shall be considered during the calculation. Zero time reading should not be included.

Step 4. A correlation coefficient is determined for the data by using the following equation

$$r^2 = \left[\frac{\Sigma\left(\sqrt{t_i} - T\right)\left(M_{wti} - \overline{M_{wt}}\right)}{\sqrt{\Sigma\left(\sqrt{t_i} - T\right)^2 \Sigma\left(M_{wti} - \overline{M_{wt}}\right)^2}}\right]^2$$

Where,

M_{wti} – the mass gain calculated from step 2 and step 3 in grams

t_i – the time corresponding to the mass gain M_{wti} in hours

$$\overline{M_{wt}} = \frac{\Sigma M_{wti}}{n}$$

$$T = \frac{\sum \sqrt{T}}{n}$$

Where n – the number of data points

Step 5. If the coefficient of correlation is less than 0.98, then the last 25 minutes of the data points are discarded and the coefficient of correlation is re-determined by adjusting the value of n.

Step 6. Repeat the above procedure until the coefficient of the correlation value is above 0.98 by discarding more data points.

Step 7. However, if a coefficient of correlation of 0.98 cannot be achieved, then the readings obtained for that particular specimen is discarded. Nevertheless, the number of data points required to obtain a coefficient of correlation of 0.98 is noted down.

Step 8. Using steps 4 to 6, linear regression analysis is conducted to determine the slope of the line of best fit (F) by the following equation

$$F = \frac{\sum\left(\sqrt{t_i} - T\right)\left(M_{wti} - \overline{M_{wt}}\right)}{\sum\left(\sqrt{t_i} - T\right)^2}$$

Where,

M_{wti} – the mass gain calculated from step 2 and step 3 in grams

t_i – the time corresponding to the mass gain M_{wti} in hours

$$\overline{M_{wt}} = \frac{\sum M_{wti}}{n};$$

$$T = \frac{\sum \sqrt{T}}{n}$$

Step 9. The water sorptivity of the specimen in $\left(\text{mm}/\sqrt{\text{h}}\right)$ is given by

$$S = \frac{F \times d}{M_{sv} - M_{s0}}$$

Where,

F – the slope of the best fit, in $\text{g}/\sqrt{\text{hour}}$

d – average thickness of the specimen to the nearest of 0.02 mm

M_{sv} – vacuum saturated mass of the specimen to the nearest of 0.01 g

M_{s0} – the mass of the specimen at time t_0 (start of the test) to the nearest of 0.01 g

Step 10. The above procedure is repeated for each specimen to determine average water sorptivity. The specimen that is determined as unsuitable according to the above procedure should not be used for determining the average water sorptivity.

Report

Results are reported as per Table 4.4.

TABLE 4.4
Sorptivity of Test Specimens

Specimen no	1	2	3	4	5
Specimen description					
Porosity (1 decimal point)					
Water sorptivity $mm/\sqrt{hour}$					
Water sorptivity index $mm/\sqrt{hour}$					
Range of data used in the calculation					
Specimen source					
Specimen location within the cube or core or member					
Specimen identification no					
Type of concrete including binder, water to binder ratio					
Curing history					
Surface preparation					
Unusual features such as cracks, voids, excessively chipped edges					
Test operator					
Age of concrete at the time of testing					

Student Remark

....
....
....
....

$$\frac{\textbf{\textit{Marks obtained}}}{\textbf{\textit{Total marks}}} = -$$

Instructor signature

Instructor Remark

....
....
....
....

4.5.13 Water Penetration Test

Water penetration of any concrete is to be considered as one of the durability performance tests (DIN 1048 Part 5, 1991).

Use and Significance

- Water permeates into the concrete and causes a significant amount of deterioration by transporting a notable amount of harmful species with them that are deleterious to concrete as well as reinforcement.
- This may cause unwanted expansion of concrete due to new products formed.

Required Apparatus

- Water reservoir with a regulator
- Water penetration cell
- Pressure regulator to regulate air pressure
- Concrete cube of 150 × 150 × 150 mm in dimension

Procedure

Step 1. Water is filled up to 75% of the reservoir. The water level can be seen through the glass tube mounted on the stand. Permeability cell cross-section and permeability cell test setup are shown in Figures 4.9 and 4.10.

Step 2. The specimen is placed in the penetration cell and the test is started as early as possible, preferably within 30 minutes after removing the test specimens from the curing chamber.

Step 3. The concrete specimen is placed in such a way that the cast surface is perpendicular to the test surface. After placing the concrete in position, bolts are tightened to ensure that the specimen is not overloaded by tightening the bolts.

Step 4. The valve is opened to permit the flow of water with the pressure set at 5 kg/cm^2 using the pressure regulator. This pressure is maintained for 3 days by proper regulation.

Step 5. After 3 days water pressure is released and the test specimen is removed from the cell. Using a heavy-duty hammer or compression testing machine the specimen is split into two halves.

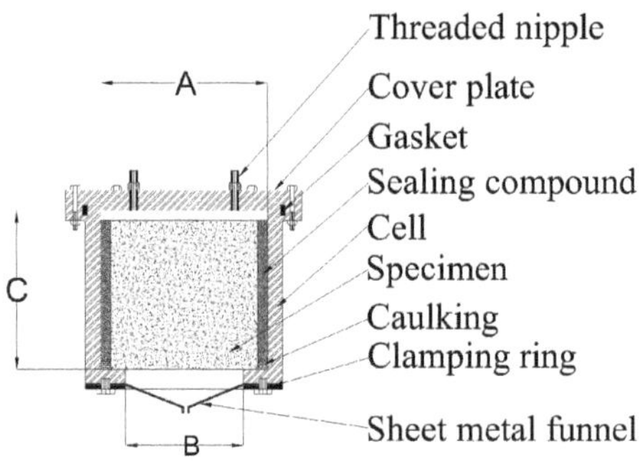

Typical details of permeability cell			
Specimen size (mm)	Dimension of cell (mm)		
	A	B	C
100	150	80	110
150	170	120	160
300	330	260	320

Section of permeability cell

FIGURE 4.9 Section of the permeability cell.

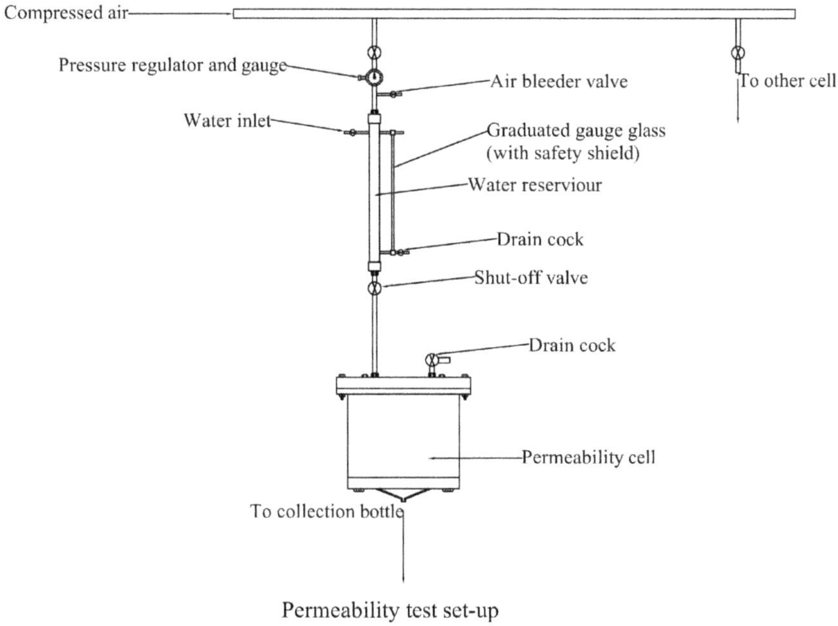

FIGURE 4.10 Permeability test setup.

Step 6. After splitting the specimen within a minute, the waterfront in the specimen is marked by a suitable marker.

Step 7. The depth of the penetration was noted for each specimen and the average depth of penetration is reported. The depth of penetration of water can be used to assess the durability parameter of the concrete.

Step 8. The more the depth of penetration of water, the less the durability of concrete. This test can be used as a potential durability test for comparing various types of concrete.

Report

Depth of penetration of water in the concrete is reported in mm.

Student Remark

....

....

....

....

$$\frac{\textbf{\textit{Marks obtained}}}{\textbf{\textit{Total marks}}} = -$$

Instructor signature

Instructor Remark

....
....
....
....

4.5.14 Standard Test for Electrical Indication of Concrete's Ability to Resist Chloride Ion Penetration (Rapid Chloride Penetration Test (RCPT))

This test method can be used to determine the electrical conductance of concrete through its resistance towards chloride ion penetration (ASTM C1202 - 19, 2019).

Test Summary

This test method consists of monitoring the amount of electric current passed through concrete for 6 hours maintained at a potential difference of 60 V (DC). Typical size of the concrete specimen will be of 50 mm thick and 100 mm diameter. A core cut from the concrete structures of similar dimension can also be used. One end of the concrete specimen is immersed in a sodium chloride solution while the other end is immersed in a sodium hydroxide solution. The resistance of the concrete to the penetration of chloride ion is related to the total charge passed in coulombs.

Use and Significance

- This test method can be used for quality control.
- It gives a qualitative relationship between the electrical current passed through concrete and its chloride ion penetration resistance.
- Care should be taken while interpreting these test results when used for surface-treated concretes.
- Other specimen sizes can also be used for testing, provided that changes need to be made on the applied voltage, etc.

Caution

- This test method can give misleading results if concrete admixed with calcium nitrate used (in general lower resistance to chloride ion penetration is obtained).
- Any concrete that is suspected of having other admixtures are recommended to undergo long-term ponding tests.
- This test method cannot be used for concrete with reinforcements, since the concrete with reinforcement placed longitudinally can act as a continuous electrical path.

Required Apparatus

- Vacuum saturation apparatus
- Separatory funnel (or other sealable, the bottom draining container with a minimum capacity of 500 ml)
- Container capable of carrying concrete specimens
- Vacuum desiccator capable of keeping the specimens entirely immersed throughout the saturation process
- Vacuum pump capable of maintaining an absolute pressure less than 50 mm Hg (6650 Pa) in a desiccator
- Vacuum gauge or manometer with an accuracy of $\pm$ 5 mm of Hg
- Coating material capable of rapid setting, non-conductive and capable of sealing side surface of concrete cores
- Paper cups, wooden spatulas and disposable brushes for mixing and applying a coating
- Specimen cutter is required if the specimens need to be cut to the appropriate size

Reagents, Materials and Test Cell

- Specimen-cell sealant – capable of sealing concrete to poly (methyl methacrylate) (PMMA) (for example, Plexiglas) against water and diluted sodium hydroxide and sodium chloride solution at temperatures up to 90 °C (e.g. synthetic silicone rubbers and silicone greases)
- Sodium chloride solution 3% by mass (reagent grade) in distilled water
- Sodium hydroxide solution 0.3 N (reagent grade) in distilled water (bring it to room temperature prior to use. The generated heat can affect the conductivity of the solution during the test)
- Filter papers – 2 numbers (90 mm diameter). These are not required if rubber gaskets are used for sealant or if the sealant can be used without overflowing from shim onto the mess
- Applied voltage cell – two symmetric PMMA chambers, each containing the electrically conductive mess and external connectors. Table 4.5 describes about the applied voltage cell parts

TABLE 4.5
Applied Voltage Cell Parts Description

Item No.	Quantity	Description	Specification
1	1	Cellblock end	Poly (methyl methacrylate)
2	4	Brass shim	0.5 mm thick
3	2	Brass screen	850 μm (No. 20) mesh
4	2	Solid copper wire	2 mm (14 gauge) nylon cladding
5	2	Ring terminal	For 2 mm (14 gauge) wire
6	2	Banana plug	6.4 mm with threaded stud

- Temperature measuring device (optional) – 0–120°C range.
- Voltage application and data readout apparatus – capable of holding DC voltage of 60 ±0.1 V across applied voltage cell over the entire range of currents with an accuracy of ± 0.1 V and current up to ± 1 mA (any of the system mentioned below can be used)
  - Voltmeter – digital (DVM), three-digit, minimum 0–99.9 V range, rated accuracy ± 0.1%
  - Voltmeter – digital (DVM), $4\frac{1}{2}$ digits, minimum 0–200 mV range, rated accuracy ± 0.1%
  - Shunt resistor – 100 mV, 10 A rating, tolerance ± 0.1%. Alternatively, a 0.01 Ω resistor, tolerance ± 0.1%, may be used. Nevertheless, care must be taken to establish very low resistance connections
  - Constant voltage power supply – 0–80 V DC, 0–2 A, capable of holding voltage constant at 60 ± 0.1 V over the entire range of currents
  - Cable – two conductors, AWG No. 11 (1.6 mm), insulated, 600 V

Test Specimen Preparation and Conditioning

- Test specimens are cast and cured according to the requirement. They are typically cured for 28 days in case of OPC concrete and 56 days and beyond in case of concrete with SCM's. Moreover, specimens are also tested for 28 days as well as 56 days for blended concrete based on the reactivity of pozzolan.
- Test specimens are cut from 100 mm diameter cylinders or 150 mm side cube specimens.
- Now water cutter is used to cut the specimens in such a way that the obtained specimen is of 100 mm diameter and 50 ± 3 mm thickness. Any burs at the end of the specimen are removed by a belt sander.
- The specimen is placed to surface dry for an hour. Hold the specimen with suitable support in such a way that the sides of the specimen can be coated with a rapid setting coat (preferably epoxy coating). The coating is allowed to cure for the required time. At the end of curing the coat won't stick to the hand. Additional curing compound is used to fill the leftover holes in the first coat and allowed to cure further.
- The cured specimens are placed with the end faces exposed in a vacuum desiccator (faces should not touch down). Vacuum desiccator is sealed and connected to a vacuum pump. The vacuum pump is started and run until a pressure of 50 mm Hg (6650 Pa), or less is reached. The vacuum is maintained for 3 hours.
- In a large sealable container, 1 litre of tap water is boiled vigorously. Once it starts to boil, remove the container, seal it tightly and place in the atmosphere until it cools down to atmospheric temperature.
- Fill the separatory funnel or a container with the de-aerated water prepared above. Open the water line stopcock while the vacuum pump is still running

and drain enough amount of water into the desiccator from the separatory funnel or the container until all the specimens are covered by the water.

- During any point of the time, no air is allowed to enter the desiccator via the stopcock.
- Once the water covers the specimen, the water stopcock line is closed and the vacuum pump was run for an additional 60 minutes. Close the vacuum line stopcock and stop the vacuum pump.
- The vacuum line stopcock is turned on to allow air to re-enter desiccator. The specimens are soaked under de-aerated water for 18 ± 2 hours.

Procedure

Step 1. The specimens are removed from the water, and the excess water is blotted off. They are transferred to a sealed can or another container, which can maintain 95% relative humidity.

Step 2. The specimen is mounted on the specimen cell by use of a low viscosity sealant or high viscosity sealant or a rubber gasket.

Step 3. 20–40 grams of cell sealant is prepared. In case of low viscosity sealant, a filter paper is placed centred over the brass screen of one cell and the low viscosity sealant is gently applied over the brass shims on the cell. The filter paper is removed and the specimen is pressed on the brass screen. The excess sealant which has flowed out of the specimen-cell boundary is smoothened or removed.

Step 4. In case of high viscosity sealant, the specimen is placed at the centre. The sealant is applied around the specimen-cell boundary.

Step 5. Moisture passage through the cell filling hole is restricted by placing a rubber plug. Similarly, the exposed face is covered by an impermeable sheet or rubber. The sealant is allowed to cure for the required time.

Step 6. Similar to step 5, the impermeable sheet on the exposed surface is removed and the same procedure as per step 4 or step 5 is followed.

Step 7. In case if a vulcanized rubber gasket of external diameter 100 mm, internal diameter 75 mm and a thickness of 6 mm is available then it is placed in each half of the specimen cell; the test specimen is inserted and the two halves of the test cell are clamped together to seal.

Step 8. Meanwhile, the filling hole near the top surface of the specimen is filled with 3% NaCl solution (this side of the cell is connected to the negative terminal of the power supply). The other filling hole nearby bottom surface is filled with 0.3 N NaOH solution (this side of the cell is connected to the positive terminal of the power supply).

Step 9. Cell banana posts are attached by lead wires. Make necessary connections, as shown in Figures 4.11 and 4.12. The power supply is turned on and the voltage is set to 60.0 ± 0.1 V. Meanwhile, the

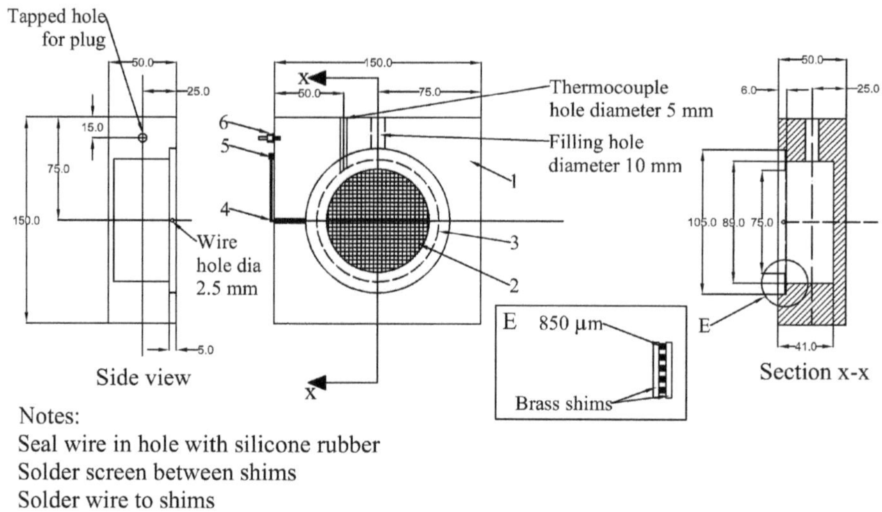

FIGURE 4.11 Applied voltage cell (construction drawing).

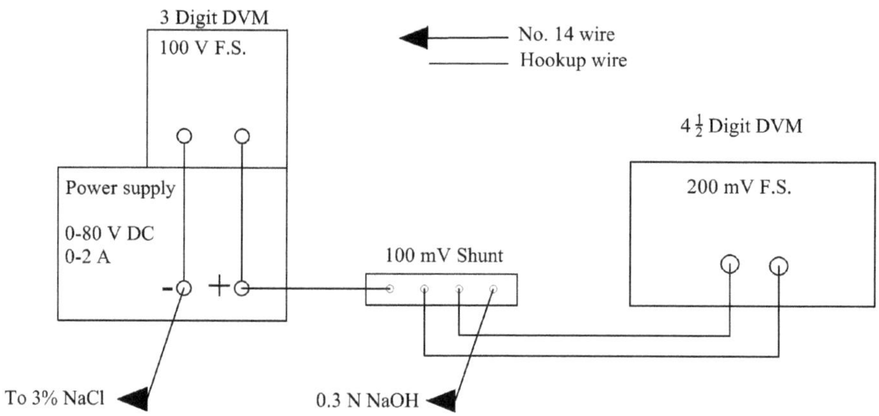

FIGURE 4.12 Electrical block diagram.

initial current reading is recorded. The power supply is turned on to ensure that the whole setup is in a temperature range of 20–25°C.

Step 10. During the test period, the surrounding temperature must be in the range of 20–25 °C. Record the reading once at least every 30 minutes. If a voltmeter is used along with a shunt resistor, then apply required corrections to convert into current.

Step 11. For the entire period of test, each filling hole is filled with the required amount of appropriate solutions.

Step 12. The test is terminated after 6 hours (except for the cases as mentioned in below note). The specimen is removed from the cell. The sealant is tripped off and the cell is rinsed thoroughly in tap water.

Note: During the test, the temperature of the solutions should not exceed 90°C. This generally happens in case of high penetrable concrete. Test that exceeds 90°C should be terminated immediately. This may cause a problem to the specimen cell and boiling of the solution in the filling hole. Moreover, the concrete specimen in the tests should be noted in the report as high chloride ion penetrability.

Observations and Calculations

Time (min)	Current (amperes)	Time (min)	Current (amperes)	Time (min)	Current (amperes)	Time (min)	Current (amperes)
30		120		210		300	
60		150		240		330	
90		180		270		360	

- Plot and draw a graph between current (amperes) and time (in seconds) and integrate the area under the curve. This area gives the charge passed through the concrete in coulombs (ampere-seconds) during the test period of 6 hours. Total charge passed through the concrete during the test period is the measure of the electrical conductance of the concrete.
- Following equation can be used if the current is recorded once in 30 minutes (trapezoidal rule).

$$Q = 900\left(I_0 + 2I_{30} + 2I_{60} + \ldots\ldots\ldots\ldots\ldots\ldots + 2I_{300} + 2I_{330} + I_{360}\right)$$

where,
Q –charge passed in coulombs
I_0 – current immediately after the voltage is applied
I_t – current at "t" minutes after the voltage is applied

- If the specimen diameter is other than 95 mm, the value for total charge passed through the concrete is adjusted by the following equation

$$Q_s = Q_x \times \left(\frac{95}{x}\right)^2$$

where,
Q_s – charge passed through a 95 mm specimen (coulombs)
Q_x – charge passed through an x mm specimen (coulombs)
x – diameter of the nonstandard specimen

TABLE 4.6
Chloride Ion Penetrability Based on Charge Passed

Charge passed (coulombs)	Chloride ion penetrability
>4000	High
2000–4000	Moderate
1000–2000	Low
100–1000	Very low
<100	Negligible

Report

Core source or specimen ………….. (in terms of location if from the site or taken from the existing structure)

Type of concrete, including binder type, the grade of concrete, water to binder ratio and other relevant data ………………………….

Curing history ………………………..

Chloride ion penetrability ……………….. (refer to Table 4.6)

Student Remark

………………………………………………………………………………………
………………………………………………………………………………………
………………………………………………………………………………………
……………………………………………………………………………………

$$\frac{\textbf{\textit{Marks obtained}}}{\textbf{\textit{Total marks}}} = -$$

Instructor signature

Instructor Remark

………………………………………………………………………………………
………………………………………………………………………………………
………………………………………………………………………………………
……………………………………………………………………………………

4.5.15 Chloride Migration Coefficient from Non-steady-state Migration Experiments (Rapid Chloride Migration Test (RCMT))

This experiment can be used to determine the chloride migration coefficient of cement-based materials (concrete/mortar) by non-steady-state migration. In this experiment migration and diffusion of molecules and ions take place, respectively, due to an externally applied electric field and chemical potential or concentration gradient (CoMSIRU, 2018c).

Test Summary

This test involves applying an external electric field across the test specimen axially to force chloride ions to migrate into the specimen. Typical specimens have a cylindrical shape of 100 mm diameter and 50 mm thickness. After the prescribed period, specimens are removed and split into two halves. In one half, spray silver nitrate solution and measure the depth until a white precipitate (silver chloride) is observed. The white precipitate gives the depth of chloride ion penetration. Chloride migration coefficient can be calculated from the chloride ion penetration depth.

Required Apparatus and Reagents

Apparatus

- Water-cooled diamond shaw
- A vacuum desiccator (capable of holding three specimens)
- A vacuum pump capable of working at a pressure less than 50 mbar (5 kPa)
- Migration setup
    - Silicone rubber sleeve (inner diameter/outer diameter 100/115 mm, about 150 mm long)
    - Clamp – diameter range 105–115 mm, 20 mm thick, stainless steel
    - Catholyte reservoir – plastic box of 370 × 270 × 280 mm (length × breadth × height)
    - Plastic support
    - Cathode – stainless steel plate about 0.5 mm thick
    - Anode – stainless steel mesh or plate with holes

Note: Other designs are acceptable, provided that during test period specimen and solution are maintained at a temperature of 20–25 °C.

- A power supply capable of maintaining 0–60 V DC that can regulate voltage with an accuracy of ± 0.1 V
- Ammeter capable of displaying current to ± 1mA
- Thermometer or thermocouple capable of reading ± 1°C
- Any suitable device for splitting the specimen
- Spray bottle
- Vernier calliper 0.1 mm accuracy
- Ruler of 1 mm accuracy
- Equipment for chloride analysis (optional any appropriate chloride test procedure can be used)

Reagents

- Distilled or de-ionized water
- Calcium hydroxide ($Ca(OH)_2$) – technical grade
- Sodium chloride (NaCl)
- Sodium hydroxide (NaOH)

- Silver nitrate ($AgNO_3$)
- Chemicals for chloride analysis as required

Test Specimen Preparation and Conditioning

- If the core specimen is obtained from the field structures, then the end faces of the core should be sawed perpendicular to the longitudinal axis of the core for a depth of 10–20 mm. Afterwards, the core is cut to 50 ± 2 mm. The face that is near to the outermost layer to the environment should be in contact with chloride solution (catholyte).
- In case of laboratory specimens, 100 × 200 mm cylinder is cut into two halves of 100 × 100 mm cylinders. Now, each 100 × 100 mm cylinder is cut into a cylinder of 50 ± 2 mm thickness. The end surface that was nearer to the first cut (the middle surface) should be exposed to the chloride solution (catholyte). Measure the dimensions of the specimen using a vernier calliper to the nearest of 0.1 mm.
- Any husk on the surfaces of the specimen is brushed off. The excess water on the surfaces is wiped off. Surface dry specimens are placed in the desiccator in such a way that both faces of the specimens are exposed.
- A vacuum pump is used to reduce the absolute pressure in the desiccator to the range of 10–50 mbar (1–5 kPa). Desiccator is maintained at this pressure for 3 hours. The container is filled with $Ca(OH)_2$ solution (excess $Ca(OH)_2$ is dissolved in de-ionized water or distilled water) to immerse all the specimens. The pressure is maintained at 10–50 mbar for a further 60 minutes. Allow the air to re-enter the container. Keep the specimens in the solution for 18 ± 2 hours.

Procedure

Step 1. Prepare catholyte and anolyte solutions. The catholyte solution is prepared by dissolving 10% by mass of NaCl in tap water (100 g of NaCl in 900 gram of water (2 N)). The anolyte solution is prepared by dissolving 12 g of NaOH in 1,000 g of distilled water or de-ionized water (0.3 N NaOH).

Step 2. The experiment is conducted at 20–25°C. Specimens and solutions are stored at 20–25°C.

Step 3. The catalytic reservoir is filled with 12 litres of prepared NaCl solution (Catholyte). The rubber sleeve is fit on the specimen and secured with two clamps, as shown in Figure 4.13. In case if the surface of the specimen is rough or a gap is found between the specimen and the rubber sleeve, silicone sealant is applied on the gaps to reduce the leakage and improve tightness.

Step 4. The specimen is placed on the plastic support as shown in Figure 4.13 in the catholyte reservoir. The sleeve above the specimen is filled with 300 ml of NaOH solution (Anolyte). The anode is immersed in the anolyte.

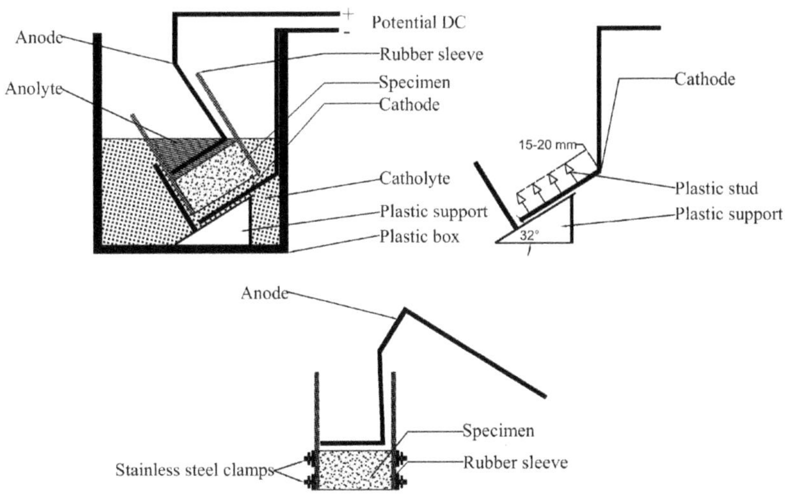

FIGURE 4.13 Migration setup arrangement.

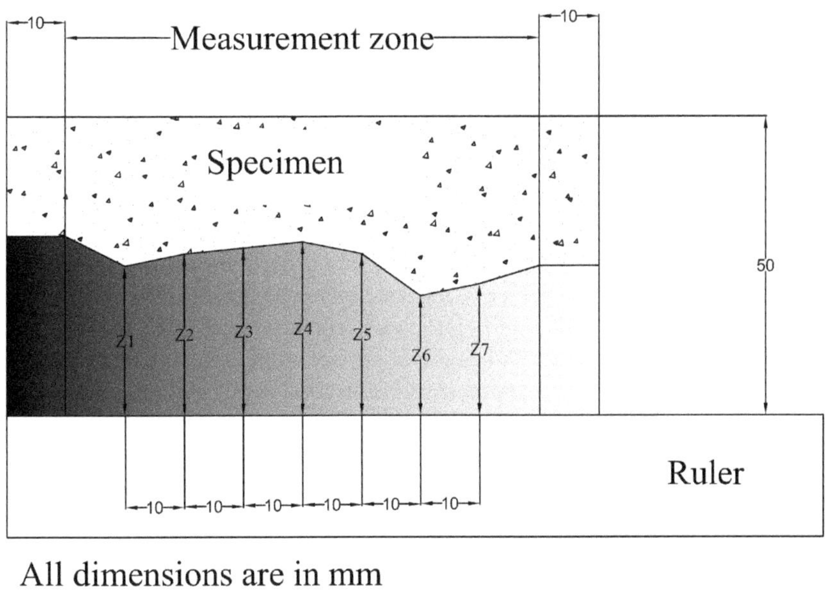

FIGURE 4.14 Illustration of chloride penetration depths measurement.

Note: For concrete with a special binder, the measured current should be corrected by multiplying a factor (approximately equal to the ratio of normal binder content to that of actual binder content) in order to use Table 4.7.

Step 5. The cathode is connected to the negative pole and the anode is connected to the positive pole of the power supply.

TABLE 4.1
Penetration Resistance and the Corresponding Time Elapsed

Initial current I30V with 30 V (mA)	Applied voltage U (after adjustment) (V)	Possible new initial current I0 (mA)	Test duration t (hour)
$I0<5$	60	$I0<10$	96
$5\leq I0<10$	60	$10\leq I0<20$	48
$10\leq I0<15$	60	$20\leq I0<30$	24
$15\leq I0<20$	50	$25\leq I0<35$	24
$20\leq I0<30$	40	$25\leq I0<40$	24
$30\leq I0<40$	35	$35\leq I0<50$	24
$40\leq I0<60$	30	$40\leq I0<60$	24
$60\leq I0<90$	25	$50\leq I0<75$	24
$90\leq I0<120$	20	$60\leq I0<80$	24
$120\leq I0<180$	15	$60\leq I0<90$	24
$180\leq I0<360$	10	$60\leq I0<120$	24
$I0\geq 360$	10	$I0\geq 120$	6

Step 6. The voltage is pre-set to 30 V, the power is turned on and the initial current is noted down through each specimen and temperature in the anolyte solution. If necessary, the voltage is adjusted, as shown in Table 4.7 to choose the appropriate test duration. The final current and temperature are recorded before the termination of the test.

Step 7. Specimens are dis-assembled and rinsed in tap water. Excess water on the surface of the specimen is wiped off. Then the specimens are split axially into two portions.

Step 8. The portion which has a split surface with less undulation is chosen for the measurement of penetration depth. Another portion is kept for chloride content analysis (optional).

Step 9. In one of the split surface (with relatively less undulation) spray 0.1 N silver nitrate solution. After 15 minutes, a white precipitate of silver chloride will be visible on the surface. The depth of penetration (*Z1*, *Z2*, *Z3*, *Z4*, *Z5*, *Z6* and *Z7*) is measured for each 10 mm interval, as shown in Figure 4.14. Each depth is measured to an accuracy of 0.1 mm.

Step 10. If the measurement front is blocked by aggregate, then measurement is moved to the nearest front in such a way that the aggregate does not interfere in the penetration depth measurement. No depth measurement is to be made at 10 mm from both the edges of the specimen to avoid edge effect (due to non-homogeneous saturation and possible leakage).

Step 11. Other split portion can be used for determination of chloride content (optional). A slice of not more than 5 mm is cut parallel to end surface exposed to chloride solution (catholyte). The thickness should always be lesser than the minimum depth of chloride penetration.

Observations and Calculations

Depth of chloride penetration ....

Non-steady state migration coefficient can be calculated by Equation (4.1).

$$D_{nssm} = \frac{RT}{zFE} \cdot \frac{X_d - \alpha\sqrt{X_d}}{t} \quad (4.1)$$

Where,

$$E = \frac{U-2}{L}$$

$$\alpha = 2\sqrt{\frac{RT}{zFE}} \cdot erf^{-1}\left(1 - \frac{2c_d}{c_0}\right)$$

D_{nssm} – non-steady state migration coefficient, m^2/s

z – absolute value of ion valence, for chloride, $z = 1$

F – Faraday constant, $F = 9.648 \times 10^4 \ \frac{J}{V \cdot mol}$

U – absolute value of the applied voltage, V

R – gas constant, $R = 8.314$ J/(Kmol)

T – the average value of initial and final temperatures in anolyte solution, °C

L – thickness of the specimen, m

X_d – the average value of the penetration depths, m

t – test duration, seconds

erf^{-1} – an inverse of the error function

c_d – chloride concentration at which the colour changes,

$c_d \approx 0.07N$ for OPC concrete

c_0 – chloride concentration in the catholyte solution,

$c_0 \approx 2N$

Since, $erf^{-1}\left(1 - \frac{2 \times 0.07}{2}\right) = 1.28$, the following simplified equation can be obtained.

$$D_{nssm} = \frac{0.0239(273+T)L}{(U-2)t}\left(X_d - 0.0238\sqrt{\frac{(273+T)LX_d}{U-2}}\right) \times 10^{-12}\ m^2/s$$

D_{nssm} – non-steady state migration coefficient $\times 10^{-12}$, m^2/s

U – absolute value of the applied voltage between two electrodes, V

T – the average value of initial and final temperatures in anolyte solution, °C

L – thickness of the specimen, mm

X_d – the average value of the penetration depths, mm
t – test duration, hour

Report

Grade of concrete and constituents of concrete (material, composition, curing age) ....

Applied voltage.........................

Initial current...........................

Final current...........................

Initial temperature.......................

Final temperature......................

Average chloride penetration depth.....................

Individual penetration depth (mm)

Migration coefficient.............

Surface chloride content.............................. (Optional)

Student Remark

....
....
....
....

$$\frac{\textit{Marks obtained}}{\textit{Total marks}} = -$$

Instructor signature

Instructor Remark

....
....
....
....

4.6 PRACTICE QUESTIONS AND ANSWERS FROM COMPETITIVE EXAMS

1. Yield of concrete is reported as
   a) 1 m^3 of concrete
   b) Per bag of cement
   c) 100 m^3 of concrete
   d) 100 bags of cement

2. What is the penetration resistance corresponding to the initial and final setting time of concrete?
   a) 2.5 MPa; 30.5 MPa
   b) 1.9 MPa; 35.5 MPa
   c) 3.5 MPa; 27.5 MPa
   d) 2.0 MPa; 29.5 MPa

3. What are the dimensions of the slump cone? Dimensions are given in following order smaller diameter, bigger diameter and height (all dimensions are in mm).
   a) 200; 100; 300
   b) 300; 200; 100
   c) 100; 200; 300
   d) 200; 300; 100

4. Two different concretes with the same slump value can be differentiated by
   a) Slump cone test
   b) Compaction factor test
   c) Vee-Bee consistency test
   d) U-box test

5. Which of the following tests is used to determine the passing ability of self-compacting concrete (SCC)?
   a) J-ring test
   b) Slump cone test
   c) Compaction factor test
   d) Vee-Bee consistency test

6. Workability of concrete is decided based on several empirical tests. Quantitatively, two different concrete can give the same magnitude of workability even they can behave differently. In such cases, how will you differentiate the workability of the concrete? What are the two crucial parameters you get from those test?
   a) Workability
   b) Plastic viscosity
   c) Yield stress
   d) Plastic viscosity and Yield stress

7. What happens to the strength of concrete if compression testing is done in wet condition rather than in dry condition?
   a) Strength of concrete increases in the wet condition when compared to dry condition.
   b) Strength of concrete decreases in the wet condition when compared to dry condition.
   c) Strength of concrete remains the same.
   d) None of the above.

8. Size effect is one of the essential factors that influence the compressive strength of concrete tested. What happens when the size of the specimen is small, considering that the shape remains the same?
   a) Compressive strength increases
   b) Compressive strength decreases
   c) Compressive strength remains the same
   d) Compressive strength increases first and then decreases

9. During durability test of concrete, carbonation is identified by splitting the specimen into two halves and spraying
   a) Phenolphthalein indicator
   b) Silver nitrate solution
   c) Potassium nitrate solution
   d) Methylene blue indicator

10. Carbonation of .... results in the reduction of pH in concrete.
   a) AFm
   b) C-S-H
   c) $Ca(OH)_2$
   d) Aft

11. The split tensile strength of M 15 grade concrete when expressed as a percentage of its compressive strength is
   a) 10–15%
   b) 15–20%
   c) 20–25%
   d) 25–30%

[IES 1995]

12. The approximate value between the strengths of cement concrete at 7 days and at 28 days is
   a) 3/4
   b) 2/3
   c) 1/2
   d) 1/3

[IES 1995]

13. Modulus of elasticity of M 25 concrete as determined by the formula of IS 456 is
   a) 124500 MPa
   b) 90125 MPa
   c) 28500 MPa
   d) 16667 MPa

[IES 1995]

14. Assertion (A): The specific surface of aggregate decreases with increases in size of the aggregate.

    Reason (R): The workability of a mix is influenced more by finer fractions than the coarse particles.

    a) Both A and R are true and R is the correct explanation of A.
    b) Both A and R are true and R is not the correct explanation of A.
    c) A is true but R is false.
    d) A is false but R is true.

[IES 1996]

15. Assertion (A): Workability of concrete is improved by air-entraining agent.

    Reason (R): Air-entraining agent increases concrete strength.

    a) Both A and R are true and R is the correct explanation of A.
    b) Both A and R are true and R is not the correct explanation of A.
    c) A is true but R is false.
    d) A is false but R is true.

[IES 1996]

16. Tensile strength of concrete is measured by
    a) Direct tension test in the universal testing machine
    b) Appling compressive load along the diameter of the cylinder
    c) Appling third point loading along on a prism
    d) Appling tensile load along the diameter of the cylinder

[IES 1996]

17. The approximate ratio of the strength of 15 cm × 30 cm concrete cylinder to that of 15 cm cube of the same concrete is
    a) 1.25
    b) 1.00
    c) 0.85
    d) 0.50

[IES 1996]

18. General shrinkage in cement concrete is caused by
    a) Carbonation
    b) Stressed due to external load
    c) Drying with starting with a stiff consistency
    d) Drying with starting with a wetter consistency

[IES 1996]

19. While concreting in cold weather where frosting is also likely, one uses
    a) High-quality Portland cement with minimum additives
    b) High alumina cement with calcium chloride additives
    c) Portland cement together with calcium chloride additives
    d) A mixture of high alumina cement and Portland cement

[IES 1996]

20. Weight batching proceeds on
    a) The assumption of the declared weight in each bag of cement
    b) Weighing the contents of each bag
    c) Accurately estimating the weight of each material to be used in each batch
    d) The assumption of correct dry weight of each size range of each material and the weight of water

[IES 1996]

21. The modulus of elasticity (E) of concrete is given by
    a) $E = 1000 f_{ck}$
    b) $E = 5700\sqrt{f_{ck}}$
    c) $E = 5500\sqrt{f_{ck}}$
    d) $E = 10000\sqrt{f_{ck}}$

[IES 1997]

22. The optimum number of revolutions over which concrete is required to be mixed in a mixer machine is
    a) 10
    b) 20
    c) 50
    d) 100

[IES 1997]

23. A splitting tensile test is performed on a cylinder of diameter D and length L. If the ultimate load is P, then splitting tensile strength of concrete is given by
    a) $\frac{P}{\pi DL}$
    b) $\frac{2P}{\pi DL}$
    c) $\frac{4P}{\pi D^3}$
    d) $\frac{4P}{\pi L^3}$

[IES 1998]

24. To make one cubic meter of 1: 2: 4 by volume concrete, the volume of coarse aggregates required is
    a) 0.95 m^3
    b) 0.85 m^3
    c) 0.75 m^3
    d) 0.65 m^3

[IES 1998]

25. The capacity of a "28 S type" concrete mixer is 0.8 m^3. For mixing one cubic meter of concrete, the quantity of cement required is 5.5 bags. In order to avoid fractional usage of cement bags, the volume of concrete (in m^3) to be mixed per batch will be
    a) 0.79
    b) 0.55
    c) 0.73
    d) 0.44

[IES 1998]

26. Batching refers to
    a) Controlling the total quantity at each batch
    b) Weighing accurately the quantity of each material for a job before mixing
    c) Controlling the quantity of each material into each batch
    d) Adjusting the water to be added in each batch according to the moisture content of the materials being mixed in the batch

[IES 1998]

27. Consider the following strengths of concrete
    1. Cube strength
    2. Cylinder strength
    3. Split tensile strength
    4. Modulus of rupture

    The correct sequence in increasing order of these strengths is
    a) 3, 4, 2, 1
    b) 3, 4, 1, 2
    c) 4, 3, 2, 1
    d) 4, 3, 1, 2

[IES 1999]

28. Consider the following statements

    Shrinkage of concrete depends upon the
    1. Relative humidity of the atmosphere
    2. Passage of time
    3. Applied stress

    Which of these statements is/are correct?
    a) 1 and 2
    b) 2 and 3
    c) 1 alone
    d) 1, 2 and 3

[IES 1999]

29. Consider the following statements

    The effect of air entrainment in concrete is to

    1. Increase resistance to freezing and thawing
    2. Improve workability
    3. Decrease strength

 Which of these statements is/are correct?

    a) 2 and 3
    b) 1 and 3
    c) 1 alone
    d) 1, 2 and 3

[IES 1999]

30. As per IS code of Practice, concrete should be cured at
    a) 5°C
    b) 10°C
    c) 27°C
    d) 40°C

[IES 2000]

31. The correct sequence of workability test(s)/method(s) in the order of their application from low to high workability is
    a) Slump test, compacting factor test and Vee-Bee consistometer
    b) Compacting factor test, Vee-Bee consistometer and slump test
    c) Vee-Bee consistometer, slump test and compacting factor test.
    d) Vee-Bee consistometer, compacting factor test and slump test

[IES 2000]

32. The ratio of direct tensile strength to modulus of rupture of concrete is
    a) 0.25
    b) 0.5
    c) 0.5
    d) 1.0

[IES 2000]

33. Assertion (A): The use of fly ash as an admixture in concrete reduces segregation and bleeding.

    Reason (R): The use of fly ash as a replacement of sand in a lean mix increases the workability and has no significant effect on drying shrinkage of concrete.

    a) Both A and R are true and R is the correct explanation of A.
    b) Both A and R are true and R is not the correct explanation of A.
    c) A is true but R is false.
    d) A is false but R is true.

[IES 2000]

34. Which one of the following types of concrete is most suitable in extreme cold climates?
    a) Air-entrained concrete
    b) Ready-mix concrete
    c) Vacuum concrete
    d) Coarse concrete

[IES 2001]

35. Match list I (workability test) with list II (measurements) and select the correct answer using the codes given below the lists:

	List I		List II
A	Slump test	1	300–500
B	Compacting factor	2	75–152
C	Vee-Bee test	3	0.80 to 0.98
D	Flow test	4	0 to 10 sec

Codes:

	A	B	C	D
a)	2	4	3	1
b)	1	3	4	2
c)	1	4	3	2
d)	2	3	4	1

[IES 2007]

36. Consider the following statements

    Curing of concrete by steam under pressure

    1. Increases the compressive strength of concrete
    2. Reduces the shear strength of concrete
    3. Increases the speed of chemical reaction

    Which of these statements is/are correct?
    a) 1, 2 and 3
    b) 1 alone
    c) 2 and 3
    d) 3 alone

[IES 2001]

37. Consider the following statements

    Ultrasonic pulse velocity test to measure the strength of concrete is

1. Used to measure the strength of wet concrete
2. Used to obtain an estimate of concrete strength of finished concrete elements
3. A non-destructive test

Which of these statements are correct?

(a) 1, 2 and 3
(b) 2 and 3
(c) 1 and 2
(d) 1 and 3

[IES 2001]

38. The length of time for which a concrete mixture will remain plastic is usually more dependent on

a) The setting time of cement than on the amount of mixing water and atmospheric temperature
b) The atmospheric temperature than on the amount of mixing water and the setting time of cement
c) The setting time of cement and the amount of mixing water than on atmospheric temperature
d) The amount of mixing water used and atmospheric temperature than on the setting time of cement

[IES 2002]

39. Match list I (admixture) with list II (action in concrete) and select the correct answer using the codes given below the lists:

	List I		List II
A	Calcium lignosulphonate	1	Anti-bleeder
B	Aluminium powders	2	Retarder
C	Tartaric Acid	3	Air-entrainer
D	Aluminium sulphate	4	Water reducer

Codes:

	A	B	C	D
a)	3	2	1	4
b)	4	3	2	1
c)	3	4	1	2
d)	4	2	3	1

[IES 2007]

40. Bleeding of concrete leads to which of the following?

1. Drying up of concrete surface.

2. Formation of pores inside.
3. Segregation of aggregate.
4. Decrease in strength.

Select the correct answer using the codes given below
- a) 1 only
- b) 1 and 2
- c) 1 and 3
- d) 2 and 4

[IES 2003]

41. Stress-strain curve of concrete is
   - a) A perfectly straight line up to failure
   - b) A straight line up to 0.002% strain value and then parabolic up to failure
   - c) Parabolic up to 0.002% strain value and then a straight line up to failure
   - d) Hyperbolic up to 0.002% strain value and then a straight line up to failure

[IES 2003]

42. Match list I (material used in individual batching of concrete) with list II (tolerance when batch weight exceeds 30% of scale capacity) and select the correct answer using the codes given below the lists:

	List I		List II
A	Cement	1	±0.3% of scale capacity
B	Water	2	±1% of scale capacity
C	Aggregates	3	±2% of scale capacity
D	Admixtures	4	±3% of scale capacity

Codes:

	A	B	C	D
a)	1	2	3	4
b)	1	2	4	3
c)	3	4	1	2
d)	4	3	1	2

[IES 2007]

43. Assertion (A): The splitting test for determining the tensile strength of concrete gives more uniform results than any other tension test.

    Reason (R): The splitting test moulds can be used for casting specimens for both compression and tension tests.
   - a) Both A and R are true and R is the correct explanation of A.
   - b) Both A and R are true and R is not the correct explanation of A.
   - c) A is true but R is false.
   - d) A is false but R is true.

[IES 2004]

44. Assertion (A): One year strength of continuously moist cured concrete is 40% higher than that of 28-day strength, while no-moisture-curing can lower the strength to about 40%.

    Reason (R): Moist curing for the first 7 to 14 days results in a compressive strength of 70–80% of that of 28th day moisture curing.

    a) Both A and R are true and R is the correct explanation of A.
    b) Both A and R are true and R is not the correct explanation of A.
    c) A is true but R is false.
    d) A is false but R is true.

[IES 2004]

45. The use of superplasticizers as admixture
    a) Increases compressive strength of concrete
    b) Permits lower water-cement ratio, thereby strength is increased
    c) Reduces the setting time of concrete
    d) Permits lower cement content, thereby strength is increased

[IES 2004]

46. The values of slump commonly adopted for the various concrete mixes are given below.
    1. Concrete for road works: 20–28 mm
    2. Ordinary RCC work: 50–100 mm
    3. Columns retaining walls: 12–25 mm
    4. Mass concrete: 75–175 mm

    Which of the pairs given above are correctly matched?
    a) 1, 3 and 4
    b) 1 and 2
    c) 3 and 4
    d) 2 and 4

[IES 2004]

47. Consider the following pairs
    1. Hand compaction of heavily reinforced sections: Low workability (0–25 mm slump)
    2. Concreting of shallow sections with vibrations: High workability (125–150 mm slump)
    3. Concreting of lightly reinforced sections like pavements: Low workability (5– 50 mm slump)
    4. Concreting of lightly reinforced section by hand or heavily reinforced sections with vibration: Medium workability (25–75 mm slump)

Which of the pairs given above are correctly matched?

a) 1 and 2
b) 2 and 3
c) 3 and 4
d) 1 and 3

[IES 2004]

48. Slump and compaction factors are two different measures of workability of concrete, for a slump of 0–20 mm, what is the equivalent range of compaction factor?
    a) 0.50–0.70
    b) 0.70–0.80
    c) 0.80–0.85
    d) 0.85–0.92

[IES 2004]

49. Consider the following statements
    Cement concrete is a/an
    1. Elastic material
    2. Visco-elastic material
    3. Visco-plastic material

    Which of these statements is/are correct?
    a) 1, 2 and 3
    b) 2 and 3
    c) 2 only
    d) 1 only

[IES 2005]

50. Why is superplasticizer added to concrete?
    1. To reduce the quantity of mixing water
    2. To increase the consistency
    3. To reduce the quantity of cement
    4. To increase resistance to freezing and thawing.

    Select the correct answer using the codes given below.
    a) 1, 2 and 4
    b) 1, 3 and 4
    c) 2 and 4
    d) 1, 4 & 3

51. On which of the following is the working principle of the concrete hammer for the non-destructive test based?
    a) Rebound deflections
    b) Radioactive waves
    c) Ultrasonic pulse
    d) Creep-recovery

[IES 2005]

52. What is the correct sequence of operations involved in concrete production?
    a) Batching-mixing-handling-transportation
    b) Mixing-batching-handling-transportation
    c) Transportation-handling-mixing-batching
    d) Handling-transportation-mixing-batching

[IES 2006]

53. What is the approximate ratio of the strength of cement concrete at 7 days to that at 28 days curing?
    a) 0.40
    b) 0.65
    c) 0.90
    d) 1.15

[IES 2006]

54. The mix design for pavement concrete is based on the
    a) Flexural strength
    b) Characteristic compressive strength
    c) Shear strength
    d) Bond strength

[IES 2006]

55. In what context is the slump test performed?
    a) Strength of concrete
    b) Workability of concrete
    c) Water-cement ratio
    d) Durability of concrete

[IES 2007]

56. Consider the following statements

    Curing of concrete is necessary because
    1. Concrete needs more water for chemical reaction
    2. It is necessary to protect the water initially mixed in concrete form being lost during evaporation
    3. Penetration of surrounding water increases the strength of concrete

Which of these statements is/are correct?

(a) 1, 2 and 3
(b) 1 and 3 only
(c) 2 only
(d) 3 only

[IES 2007]

57. Consider the following statements

Modulus of elasticity of concrete is

1. Tangent modulus
2. Secant modulus
3. Proportional to $\sqrt{f_{ck}}$
4. Proportional to $\frac{1}{\sqrt{f_{ck}}}$

Which of these statements are correct?

a) 1 and 3 only
b) 1 and 4 only
c) 2 and 3 only
d) 2 and 4 only

[IES 2007]

58. Which one of the following is employed to determine the strength of hardened existing concrete structure?

a) Bullet test
b) Kelly ball test
c) Rebound hammer test
d) Cone penetrometer

[IES 2007]

59. Which factors comprise the maturity of concrete?

a) Compressive strength and flexural strength of concrete
b) Cement content per cubic metre and compressive strength of concrete
c) Curing age and curing temperature of concrete
d) Age and aggregate content per cubic metre of concrete

[IES 2009]

60. For what reason is it taken that the nominal maximum size of aggregate may be as large as possible?

a) Larger the maximum size of aggregate, more the cement required and so higher the strength
b) Larger the maximum size of aggregate, small is the cement requirement for a particular water-cement ratio and so more economical the mix

c) Larger the maximum sizes of aggregate, lesser the voids in the mix and hence also lesser the cement required
d) Larger the maximum size of aggregate, more the surface area and better the bond between aggregates and cement, and so higher the strength

[IES 2009]
61. What is the representative geometric mean size of an aggregate sample if its fineness modulus is 3.0?
   a) 150 μm
   b) 300 μm
   c) 600 μm
   d) 12 μm

[IES 2009]
62. Which one of the following is correct regarding the most effective requirements of durability in concrete?
   a) Providing reinforcement near the exposed concrete surface
   b) Applying a protective coating to the exposed concrete surface
   c) Restricting the minimum cement content and the maximum water-cement ratio and the type of cement
   d) Compacting the concrete to a greater degree

[IES 2009]
63. Which one of the following is not required in concrete mix design?
   a) Degree of quality control of concrete
   b) Workability of concrete
   c) Characteristic compressive strength of concrete at 28 days
   d) Initial setting time of cement

[IES 2010]
64. Consider the following statements about the air-entraining admixture in concrete
   1. Improve workability
   2. Improve durability
   3. Reduce segregation during placing
   4. Decrease concrete density

   Which of these statements are correct?
   a) 1, 2, 3 and 4
   b) 1 and 2 only
   c) 2 and 3 only
   d) 3 and 4 only

[IES 2010]
65. Consider the following statements

1. Strength of concrete cube is inversely proportional to the water-cement ratio.
2. A rich concrete mix gives a higher strength than a lean concrete mix since it has more cement content.
3. Shrinkage cracks on concrete surface are due to excess water in mix.

Which of these statements is/are correct?

a) 1, 2 and 3
b) 1 and 2 only
c) 2 only
d) 2 and 3 only

[IES 2010]

66. Consider the following statements
    1. The crushing strength of concrete is fully governed by water-cement ratio.
    2. Vibration has no effect on the strength of concrete at high water-cement ratios.
    3. Workability of concrete is affected by improper grading of aggregates.

Which of these statements is/are correct?

a) 1, 2 and 3
b) 2 and 3 only
c) 2 only
d) 3 only

[IES 2010]

67. Assertion (A): Higher strength is achieved when superplasticizer is added to cement concrete mix.

    Reason (R): By adding superplasticizer, the quantity of mixing water is reduced.

    a) Both A and R are true and R is the correct explanation of A.
    b) Both A and R are true and R is not the correct explanation of A.
    c) A is true but R is false.
    d) A is false but R is true.

[IES 2010]

68. Assertion (A): In order to obtain a higher degree of workability in cement concrete, both water content and proportion of cement should be increased.

    Reason (R): Increase in water-cement ratio decreases the strength of cement concrete a mix with higher workability must have a higher proportion of cement in it.

    a) Both A and R are true and R is the correct explanation of A.
    b) Both A and R are true and R is not the correct explanation of A.
    c) A is true but R is false.
    d) A is false but R is true.

[IES 2010]

69. Assertion (A): Centrifugal pumps are not normally usable for pumping mixed concrete even if the concrete (to be pumped) can be dropped in by a hopper system.

    Reason (R): When dropping (mixed) concrete, segregation of aggregates may occur.

    a) Both A and R are true and R is the correct explanation of A.
    b) Both A and R are true and R is not the correct explanation of A.
    c) A is true but R is false.
    d) A is false but R is true.

[IES 2011]

70. Consider the following statements
    1. The compressive strength of concrete decreases with increases in the water-cement ratio of the concrete mix.
    2. Water is added to the concrete mix for hydration of cement and workability.
    3. Creep and shrinkage of concrete are independent of the concrete mix.

    Which of these statements are correct?
    a) 1 and 2 only
    b) 1 and 3 only
    c) 2 and 3 only
    d) 1, 2 and 3

[IES 2012]

71. According to the Indian Standard Specifications, concrete should be cured under a humidity of
    a) 90%
    b) 80%
    c) 70%
    d) 60%

[IES 2012]

72. Consider the following statements

    In a typical compression test with a cylindrical concrete specimen, failure is initiated by
    1. Crushing in compression
    2. Inclined shear failure
    3. Longitudinal tensile cracks

    Which of these statements is/are correct?
    a) 1 only
    b) 2 only

c) 3 only
d) 1, 2 and 3

[IES 2012]

73. If one intends to obtain the best workability of concrete, the preferred shape of aggregate is
   (a) Round
   (b) Annular
   (c) Triangular
   (d) Flinty

[IES 2012]

74. Consider the following statements as regards rheology of concrete
   1. It deals with the strength of concrete.
   2. It deals with deformation in concrete.
   3. It is the study of deformation and flow of concrete.
   4. It deals with the rate of shear and shear stress in concrete.

   Which of these are correct?
   a) 1, 2, 3 and 4
   b) 3 and 4 only
   c) 2 and 3 only
   d) 1 and 2 only

[IES 2012]

75. Consider the following for durability of well-graded concrete
   1. The environment
   2. Cover to embedded reinforcement
   3. Shape and size of the concrete member

   Which of these statements are correct?
   a) 1 and 2 only
   b) 1 and 3 only
   c) 2 and 3 only
   d) 1, 2 and 3

[IES 2012]

76. Consider the common methods related to testing of concrete
   1. Consistency
   2. Compacting factor
   3. Vee-Bee
   4. Setting time
   5. Slump

Which of these methods refer to measuring the workability of concrete?

a) 1, 2 and 3
b) 1, 2 and 5
c) 2, 3 and 4
d) 2, 3 and 5

[IES 2012]

77. Which of the following factors would greatly affect the attainment of the best possible strength of the concrete mix produced using the weigh-batcher?
    1. Moisture content in the sand and gravel
    2. Inadequate or excess use of approved admixtures
    3. Speed of rotation of the drum
    4. Non-emptying of the drum as fully as possible

    a) 1 and 4
    b) 1 and 2
    c) 2 and 3
    d) 3 and 4

[IES 2013]

78. The strength of concrete depends on
    1. Type of mortar
    2. Proportion between coarse and fine aggregates
    3. Water-cement ratio
    4. Temperature at time of mixing

    a) 1 and 2
    b) 2 and 3
    c) 2 and 4
    d) 3 only

[IES 2013]

79. Consider the following statements
    1. Strength of structural concrete is absolutely dependent upon the water-cement ratio.
    2. Increase in temperature during curing period improves the strength, especially for aluminous cement.
    3. The concern for the onset of fatigue in concrete can be overcome by increasing design loads in limit state design.
    4. Even though concrete gains strength after 28 days, such increase is not considered at design stage.

Which of these statements are correct?

a) 1, 2 and 3
b) 2, 3 and 4
c) 3 and 4 only
d) 2 and 4 only

[IES 2013]

80. The workability of concrete is assessed through
    1. Slump test
    2. Compaction factor test
    3. Setting time of cement
    4. Le-Chatelier's apparatus

    a) 1 and 2
    b) 2 and 3
    c) 3 and 4
    d) 4 and 1

[IES 2013]

81. Consider the following statements as describing the Rheological behaviour of fresh concrete
    1. Newtonian
    2. Non-Newtonian
    3. Ratio of shear stress to shear rate is constant
    4. Ratio of shear stress to shear rate depends upon the shear rate

    Which of these statements are correct?
    a) 1, 2, 3 and 4
    b) 2 and 4 only
    c) 1, 2 and 4 only
    d) 2, 3 and 4 only

[IES 2013]

82. Which of the following tests compares the dynamic modulus of elasticity of samples of concrete?
    a) Compression test
    b) Ultrasonic pulse velocity test
    c) Split test
    d) Tension test

[IES 2013]

83. UPV method in non-destructive testing for concrete is used to determine:
    1. Compressive strength
    2. Existence of voids
    3. Tensile strength
    4. Static modulus of concrete
    5. Dynamic modulus of concrete

    a) 1, 2, 3 and 4
    b) 1 and 3 only
    c) 2 and 5 only
    d) 3 and 5 only

[IES 2013]

84. Statement (I): Weight batching method does not produce concrete of required strength.

 Statement (II): Bulking of aggregates does not influence the weight batching method.

 a) Both statement (I) and statement (II) are individually true and statement (II) is the correct explanation of statement (I).
 b) Both statement (I) and statement (II) are individually true and statement (II) is not the correct explanation of statement (I).
 c) Statement (I) is true but statement (II) is false.
 d) Statement (I) is false but statement (II) is true.

[IES 2013]

85. Statement (I): An inherent weakness of finished plain concrete is the presence of micro-cracks at the aggregate interface. It is minimized by placing fibres in the concrete mix.

 Statement (II): The fibres help to transfer all incidental loads at the internal micro-cracks.

 a) Both statement (I) and statement (II) are individually true and statement (II) is the correct explanation of statement (I).
 b) Both statement (I) and statement (II) are individually true and statement (II) is not the correct explanation of statement (I).
 c) Statement (I) is true but statement (II) is false.
 d) Statement (I) is false but statement (II) is true.

[IES 2013]

86. Consider the following parameters of concrete

 1. Impermeability
 2. Compactness
 3. Durability
 4. Desired consistency
 5. Workability

 Which of the above parameters are relevant for 'water-cement ratio'?

 a) 4 and 5
 b) 1 and 2
 c) 2 and 4
 d) 3 and 5

[IES 2014]

87. Consider the following statements

 The presence of Na_2O and K_2O in concrete leads to

 1. Expansive reaction in concrete
 2. Cracking of concrete
 3. Disruption of concreter
 4. Shrinkage of concrete

Which of the above statements are correct?

a) 1 and 2 only
b) 2 and 3 only
c) 1, 2 and 3
d) 3 and 4

[IES 2014]

88. Consider the following statements concerning “elasticity of concrete”:
    1. Stress-strain behaviour concreter is a straight line up to 10% of ultimate stress.
    2. Strain determination is obtained from tangent modulus.
    3. Modulus of elasticity of concrete is also called as secant modulus.

Which of the above statements are correct?

a) 1, 2 and 3
b) 1 and 3 only
c) 1 and 2 only
d) 2 and 3 only

[IES 2014]

89. Consider the following statements

    The addition of $CaCl_2$ in concrete results in
    1. Increased shrinkage
    2. Decreased setting time
    3. Decreased shrinkage
    4. Increased setting time

Which of the above statements is/are correct?

a) 1 only
b) 2 and 3
c) 3 and 4
d) 1 and 2

[IES 2014]

90. Statement (I): Grading of concrete is based on its 28-day strength.

    Statement (II): Concrete does not gain any further strength after 28-day curing.

a) Both statement (I) and statement (II) are individually true and statement (II) is the correct explanation of statement (I).
b) Both statement (I) and statement (II) are individually true and statement (II) is not the correct explanation of statement (I).
c) Statement (I) is true but statement (II) is false.
d) Statement (I) is false but statement (II) is true.

[IES 2014]

91. Statement (I): The addition of admixture improves the workability of concrete.

    Statement (II): The addition of admixture increases the strength of concrete.

    a) Both statement (I) and statement (II) are individually true and statement (II) is the correct explanation of statement (I).
    b) Both statement (I) and statement (II) are individually true and statement (II) is not the correct explanation of statement (I).
    c) Statement (I) is true but statement (II) is false.
    d) Statement (I) is false but statement (II) is true.

[IES 2014]

92. Statement (I): Concrete of desired strength can be achieved by weight batching method.

    Statement (II): Volume-batching method does not take into account bulking of aggregates; hence concrete of desired strength cannot be achieved by volume-batching.

    a) Both statement (I) and statement (II) are individually true and statement (II) is the correct explanation of statement (I).
    b) Both statement (I) and statement (II) are individually true and statement (II) is not the correct explanation of statement (I).
    c) Statement (I) is true but statement (II) is false.
    d) Statement (I) is false but statement (II) is true.

[IES 2014]

93. Consider the following statements related to "non-destructive testing" of concrete:

    1. Indentation test is used to assess the quality of concrete.
    2. Resonant Frequency Method is based on a laboratory test.
    3. Compressive strength of concrete is estimated through Pulse Velocity Measurement.
    4. Dynamic Modulus of Elasticity is determined by a sonometer test.
    5. The thickness of concrete can be estimated by in situ Rebound Hammer Test.

Which of the above statements are correct?

a) 1, 2 and 3 only
b) 1, 2 and 5 only
c) 1, 2, 3 and 4 only
d) 1, 2, 3, 4 and 5

[IES 2015]

94. What is the amount of water required for a workable RC of mix 1: 2: 4 by weight, when w/c is 0.60 and the unit weight of concrete is 2,400 kg/m^3?
   a) 189 l
   b) 205 l
   c) 245 l
   d) 285 l

[IES 2015]

95. Assertion (A): Admixture in concrete is an essential constituent of concrete.
   Reason (R): Admixture helps in improving or modifying specific qualities in concrete.
   a) Both A and R are true and R is the correct explanation of A.
   b) Both A and R are true and R is not the correct explanation of A.
   c) A is true but R is false.
   d) A is false but R is true.

[IES 2015]

96. Assertion (A): Deadweight of a structure can be reduced by using lightweight concrete in construction.
   Reason (R): Aerated concrete, being of light weight, is used in RCC multi-storied construction.
   a) Both A and R are true and R is the correct explanation of A.
   b) Both A and R are true and R is not the correct explanation of A.
   c) A is true but R is false.
   d) A is false but R is true.

[IES 2015]

97. In a concrete mix of proportion 1: 3: 6, the actual quantity of sand, which is judged to have undergone 15% bulking, per unit volume of cement, will be
   a) 3.00
   b) 3.45
   c) 4.50
   d) 6.00

[IES 2016]

98. The rheological behaviour of concrete, when represented by shear stress vs. rate of shear, is characterized as
   a) $\tau = \tau_0 + \mu\gamma$
   b) $\tau_0 = \tau + \mu\gamma$
   c) $\frac{\tau}{\tau_0} = \mu\gamma$
   d) $\tau = \mu\gamma$

Where τ = shear stress, τ_0 = yield value, μ = at point plastic viscosity, γ = at point rate of shear

[IES 2016]

99. Which method of curing of concrete is recommendable for a rapid gain of strength of concrete?
    a) Sprinkling water
    b) Membrane curing
    c) High pressure steam curing
    d) Infrared radiation curing

[IES 2016]

100. Which of the following is appropriate as a simplified method for assessing the consistency of concrete?
    a) Compacting factor
    b) Slump test
    c) Vee – Bee test
    d) Kelly ball test

[IES 2016]

101. Which of the following are relatable to autoclaved aerated concrete?
    1. Light weight
    2. Strong
    3. Inorganic
    4. Non – toxic

    a) 1, 2 and 3 only
    b) 1, 2 and 4 only
    c) 3 and 4 only
    d) 1, 2, 3 and 4

[IES 2016]

102. The workability of concrete becomes more reliable depending on
    1. Aggregate-cement ratio
    2. Time of transit
    3. Grading of the aggregate

    a) 1 only
    b) 2 only
    c) 3 only
    d) 1, 2 and 3

[IES 2016]

103. Statement (I): Water containing less than 2,000 ppm of dissolved solids can generally be used satisfactorily for making concrete.

Statement (II): The presence of any zinc, manganese, tin, copper or lead reduces the strength of concrete considerably.

a) Both statement (I) and statement (II) are individually true and statement (II) is the correct explanation of statement (I).
b) Both statement (I) and statement (II) are individually true and statement (II) is not the correct explanation of statement (I).
c) Statement (I) is true but statement (II) is false.
d) Statement (I) is false but statement (II) is true.

[IES 2016]

104. Statement (I): Though a non-elastic material, concrete exhibits a linear relationship between stress and strain at low values of stress.

Statement (II): The modulus of elasticity of concrete is dependent on the elastic properties of aggregate and on curing.

a) Both statement (I) and statement (II) are individually true and statement (II) is the correct explanation of statement (I).
b) Both statement (I) and statement (II) are individually true and statement (II) is not the correct explanation of statement (I).
c) Statement (I) is true but statement (II) is false.
d) Statement (I) is false but statement (II) is true.

[IES 2016]

105. Consider the following statements

1. If more water is added to concrete for increasing its workability, it results in a concrete of low strength.
2. No slump is an indication of good workable concrete.
3. Higher the slump of concrete, lower will be its workability.
4. The workability of concrete is affected by water content as well as water-cement ratio.

Which of the above statements are correct?

a) 1 and 3 only
b) 2 and 3 only
c) 1 and 4 only
d) 2 and 4 only

[IES 2017]

106. Pozzolana used as an admixture in concrete has which of the following advantages?

1. It improves workability with a lesser amount of water.
2. It increases the heat of hydration and so concrete sets quickly.
3. It increases the resistance of concrete to attack by salts ad sulphates.

4. It leaches out calcium hydroxide.

Select the correct answer using the codes given below
- a) 1, 2 and 3 only
- b) 1, 2 and 4 only
- c) 1, 3 and 4 only
- d) 2, 3 and 4 only

[IES 2017]
107. Consider the following particulars in respect of a concrete mix design

	Weight	Specific gravity
Cement	400 kg/m^3	3.2
Fine aggregates	------	2.5
Coarse aggregates	1040 kg/m^3	2.6
Water	200 kg/m^3	1.0

What shall be the weight of fine aggregate?
- a) 520 kg/m^3
- b) 570 kg/m^3
- c) 690 kg/m^3
- d) 1,000 kg/m^3

[IES 2017]
108. Consider the following statements regarding Cyclopean concrete
1. Size of aggregates is more than 150 mm
2. Size of aggregates is less than 150 mm
3. High slump
4. High temperature rise due to heat of hydration

Which of the above statements are correct?
- a) 1 and 3 only
- b) 1 and 4 only
- c) 2 and 3 only
- d) 2 and 4 only

[IES 2017]
109. Consider the following statements:
1. Rich mixes are less prone to bleeding than lean ones.
2. Bleeding can be reduced by increasing the fineness of cement.

Which of the above statements is/are correct?
- a) 1 only
- b) 2 only
- c) Both 1 and 2

d) Neither 1 nor 2

[IES 2018]

110. The yield of concrete per bag of cement for a concrete mix proportion of 1: 1.5: 3 (with adopting 2/3 as the coefficient) is

a) 0.090 m^3
b) 0.128 m^3
c) 0.135 m^3
d) 0.146 m^3

[IES 2018]

111. Consider the following statements

1. The workability of concrete increases with the increase in the proportion of water content.
2. Concrete having small-sized aggregates is more workable than that containing large seized aggregate.
3. For the same quantity of water, rounded aggregates produced a more workable concrete mix as compared to angular and flaky aggregates.
4. A concrete mix with no slump shown in the slump cone test indicates its very poor workability.

Which of the above statements are correct?

a) 1, 2 and 3 only
b) 1, 2 and 4 only
c) 1, 3 and 4 only
d) 2, 3 and 4 only

[IES 2018]

112. Statement (I): Rapid method of concrete mix design will take 3 days for trials.

Statement (II): This rapid method depends on curing the concrete in warm water at or above 55°C.

a) Both statement (I) and statement (II) are individually true and statement (II) is the correct explanation of statement (I).
b) Both statement (I) and statement (II) are individually true and statement (II) is not the correct explanation of statement (I).
c) Statement (I) is true but statement (II) is false.
d) Statement (I) is false but statement (II) is true.

[IES 2018]

113. Statement (I): RMC is preferably used in the construction of large projects.

Statement (II): RMC is adaptable to achieve any desired strength of concrete, with simultaneous quality control.

a) Both statement (I) and statement (II) are individually true and statement (II) is the correct explanation of statement (I).

b) Both statement (I) and statement (II) are individually true and statement (II) is not the correct explanation of statement (I).
c) Statement (I) is true but statement (II) is false.
d) Statement (I) is false but statement (II) is true.

[IES 2018]

114. Which of the statements are wholly correct regarding broken brick aggregate usable in concretes?
   1. Broken brick aggregate is obtained by crushing waste bricks, and it has a density varying between 1,000 kg/m^3 and 1,200 kg/m^3.
   2. Such aggregate is usable in concrete for the foundation in light buildings, floorings and walkways.
   3. Such aggregate may also be used in lightweight reinforced concrete floors.

   a) 1 and 2 only
   b) 2 and 3 only
   c) 1 and 3 only
   d) 1, 2 and 3

[IES 2019]

115. In handling air-entraining admixtures the beneficial amount of entrained air depends upon certain factors like
   1. Type and quantity of air-entraining agent.
   2. Water-cement ratio of the mix
   3. Strength of aggregates
   4. Extent of compaction of concrete

   a) 1, 2 and 3 only
   b) 1, 2 and 4 only
   c) 1, 3 and 4 only
   d) 1, 2, 3 and 4

[IES 2019]

116. Which of the following methods will help in reducing segregation in concrete?
   1. Not using vibrator to spread the concrete.
   2. Reducing the continued vibration
   3. Improving the cohesion of a lean dry mix through addition of a further small quantity of water.

   a) 1, 2 and 3
   b) 1 and 2 only
   c) 1 and 3 only

d) 2 and 3 only

[IES 2019]

117. On an average, in a 125 mm slump, the concrete may lose about (in first one hour)
   a) 15 mm of slump
   b) 25 mm of slump
   c) 40 mm of slump
   d) 50 mm of slump

[IES 2019]

118. Permeability of concrete is studied towards providing for, or guarding against, which of the following features?
   1. The penetration by materials in solution may adversely affect the durability of concrete; moreover, aggressive liquids attack the concrete.
   2. In the case of reinforced concrete, ingress of moisture and air will result in the corrosion of steel, leading to an increase in the volume of steel, resulting in cracking and spalling of the concrete cover.
   3. The moisture penetration depends on permeability and if the concrete can become saturated with water it is less vulnerable to frost action.

   a) 1, 2 and 3
   b) 1 and 2 only
   c) 1 and 3 only
   d) 2 and 3 only

[IES 2019]

119. Which one of the following methods or techniques will be used for placing of concrete in dewatered caissons or cofferdams?
   a) Tremie method
   b) Placing in bags
   c) Prepacked concrete
   d) In the dry practice

[IES 2019]

120. The minimum cement content (kg/m^3) for a pre-specified strength of concrete (using standard notations) premised on "free water cement ratio" will be
   a) $1-\frac{C}{1000S_C}-\frac{W}{1000}$
   b) $\frac{\text{Water content}}{\text{Water cement ratio}}$
   c) Water content × water-cement ratio
   d) $\frac{100F}{C+F}$

[IES 2019]

121. Polymer concrete is most suitable for
    a) Sewage disposal works
    b) Mass concreting works
    c) Insulating exterior walls of an air-conditioned building
    d) Road repair works

[IES 2020]

122. Air permeability method is used to determine
    a) Soundness of cement
    b) Setting time
    c) Fineness of cement
    d) Resistance of cement

[IES 2020]

123. Compaction of concrete helps in
    a) Segregation of aggregates
    b) Removal of excess water
    c) Increase of density
    d) Addition of required air voids

[SSC JE 2004]

124. Concrete attains major part of its strength in
    a) One week
    b) Two weeks
    c) Four weeks
    d) Five weeks

[SSC JE 2004]

125. M20 grade concrete implies that strength of 15 cm cubes at 28 days of curing shall be
    a) 20 kg/cm^2
    b) 8 kg/cm^2
    c) 20 N/mm^2
    d) 8 N/mm^2

[SSC JE 2004]

126. The strength and quality of concrete depend upon
    a) Grading of the aggregates
    b) Surface area of the aggregates
    c) Surface texture of the aggregates
    d) All the above

[SSC JE 2008]

127. The concrete having a slump of 6.5 cm is said to be
   a) Dry
   b) Earth moist
   c) Semi-plastic
   d) Plastic

[SSC JE 2008]

128. Separation of water, sand and cement from a freshly mixed concrete is known as
   a) Bleeding
   b) Creeping
   c) Segregation
   d) Flooding

[SSC JE 2008]

129. Separation of coarse aggregates from mortar during transportation is known as
   a) Bleeding
   b) Creeping
   c) Segregation
   d) Shrinkage

[SSC JE 2009]

130. The correct proportion of ingredients of concrete depends upon
   a) Bulking of sand
   b) Water content
   c) Absorption
   d) All of the above

[SSC JE 2009]

131. The ratio of various ingredients (cement, sand, aggregate) in concrete of grade M20 is
   a) 1: 2: 4
   b) 1: 3: 4
   c) 1: 1.5: 3
   d) 1: 1: 2

[SSC JE 2009]

132. To prevent segregation, the maximum height for placing concrete is
   a) 100 cm
   b) 125 cm
   c) 150 cm
   d) 200 cm

[SSC JE 2009]

133. While compacting the concrete by a mechanical vibrator, the slump should not exceed
    a) 2.5 cm
    b) 5.0 cm
    c) 7.5 cm
    d) 10 cm

[SSC JE 2009]

134. Characteristic strength of concrete is measured at:
    a) 14 days
    b) 28 days
    c) 91 days
    d) 7 days

[SSC JE 2010]

135. Slump test is used for
    a) Strength
    b) Durability
    c) Workability
    d) Consistency

[SSC JE 2010]

136. Shrinkage in concrete can be reduced by using
    a) Low water-cement ratio
    b) Less cement in the concrete
    c) Proper concrete mix
    d) All the above

[SSC JE 2010]

137. The operation of removing humps and hollows of the uniform concrete surface is known as
    a) Floating
    b) Screeding
    c) Trowelling
    d) Finishing

[SSC JE 2010]

138. Which of the following methods may be used for getting a more workable concrete?
    a) Increasing cement content
    b) Decreasing water-cement ratio
    c) Using angular aggregates in place of rounded ones
    d) Reducing the size of aggregates

[SSC JE 2010]

139. The purpose of concrete compaction is to
   a) Increase the density
   b) Increase the weight
   c) Increase the voids
   d) Decrease the setting time

[SSC JE 2011]

140. The test strength of the sample is taken as the average of the strength of
   a) 2 specimens
   b) 3 specimens
   c) 4 specimens
   d) 5 specimens

[SSC JE 2011]

141. The static modulus of elasticity (E_C) of concrete for short-term loading may be derived as
   a) $E_C = 4800\sqrt{f_{ck}}$
   b) $E_C = 5000\sqrt{f_{ck}}$
   c) $E_C = 5200\sqrt{f_{ck}}$
   d) $E_C = 5500\sqrt{f_{ck}}$

[SSC JE 2011]

142. Minimum grade of concrete for moderate environmental exposure condition should be
   a) M 25
   b) M 30
   c) M 15
   d) M 20

[SSC JE 2012]

143. Workability of concrete is directly proportional to
   a) Grading of aggregate
   b) Water: Cement ratio
   c) Aggregate: Cement ratio
   d) Time of transit

[SSC JE 2012]

144. The bottom diameter, top diameter and the height of the steel mould used for slump test are, respectively,
   a) 20 cm, 30 cm and 10 cm
   b) 10 cm, 30 cm and 20 cm
   c) 20 cm, 10 cm and 30 cm
   d) 10 cm, 20 cm and 30 cm

[SSC JE 2012]

145. The addition of $CaCl_2$ in concrete results in
   (i) Increased shrinkage
   (ii) Decreased setting time
   (iii) Decreased shrinkage
   (iv) Increased setting time
      a) only (i)
      b) only (i) and (ii)
      c) only (i) and (iv)
      d) only (iv)

[SSC JE 2012]

146. The concrete mix design is achieved as per
   a) IS: 10262
   b) IS: 13920
   c) IS: 383
   d) IS: 456

[SSC JE 2012]

147. As per IS: 456 – 2000, the organic content of water used for making concrete should *not* be more than
   a) 200 mg/L
   b) 250 mg/L
   c) 100 mg/L
   d) 150 mg/L

[SSC JE 2012]

148. The grade of concrete M20 means that the characteristic compressive strength of 15 cm cubes after 28 days is not less than
   a) 10 N/mm^2
   b) 15 N/mm^2
   c) 20 N/mm^2
   d) 25 N/mm^2

[SSC JE 2013]

149. Maximum admissible water-cement ratio for mild environmental exposure should be
   a) 0.55
   b) 0.50
   c) 0.45
   d) 0.40

[SSC JE 2013]

150. The increase in the strength of concrete with time is
   a) Linear

b) Non- linear
c) Asymptotic
d) All of the above

[SSC JE 2013]
151. Air entrainment in the concrete increases
a) Workability
b) Strength
c) The effect of temperature variation
d) The unit weight

[SSC JE 2013]
152. Separation of coarse aggregates from concrete during transportation is known as
a) Bleeding
b) Creeping
c) Segregation
d) Evaporation

[SSC JE 2014]
153. Water-cement ratio is measured .... of water and cement used per cubic meter of concrete.
a) Volume by volume
b) Weight by weight
c) Weight by volume
d) Volume by weight

[SSC JE 2014]
154. Concrete cubes are prepared, cured and tested according to Indian Standard code number
a) IS: 515
b) IS: 516
c) IS: 517
d) IS: 518

[SSC JE 2014]
155. The workability of concrete is directly proportional to
(i) Time of transit
(ii) Water-cement ratio
(iii) Grading of aggregate
(iv) Strength of aggregate
(v) Aggregate-cement ratio

a) (iii), (iv), (v)
b) (i), (ii), (iv)

c) (ii), (iii), (v)
d) (ii), (iii)

[SSC JE 2014]
156. An admixture which causes early setting and hardening of concrete is called a/an
a) Air-entraining agent
b) Workability admixture
c) Accelerator
d) Retarder

[SSC JE 2014]
157. The permanent deformation of concrete with time under steady load is called
a) Visco-elasticity
b) Viscosity
c) Creep
d) Relaxation

[SSC JE 2014]
158. To prevent sulphate attack in concrete, for preparing concrete mix, water pH must be within
a) 7–10
b) 4–6
c) 5–7
d) 6–9

[SSC JE 2014]
159. Segregation in the concrete occurs when
a) Cement gets separated from the mixture due to excess water
b) Cement fails to give adequate binding quality
c) Water is driven out of concrete at a faster rate
d) Coarse aggregates try to separate out from the finer material

[SSC JE 2014]
160. Poisson's ratio of cement concrete is about
a) 0.28
b) 0.50
c) 0.40
d) 0.15

[SSC JE 2015]
161. Shrinkage in concrete increases its
a) Bond strength
b) Compressive strength

c) Flexural strength
d) Tensile strength

[SSC JE 2015]

162. The Indian standard mix design for fly ash and cement concrete recommends water content
   a) To increase by 3–5%
   b) To reduce by 15%
   c) To increase by 15%
   d) To reduce by 3–5%

[SSC JE 2015]

163. The leaching action in concrete is the example of
   a) Decomposition
   b) Creeping
   c) Crystallization
   d) Chemical reaction

[SSC JE 2015]

164. Slump test for concrete is carried out to determine
   a) Strength
   b) Durability
   c) Workability
   d) Water content

[SSC JE 2015]

165. Green concrete may be made by adding
   a) Iron hydroxide
   b) Barium manganite
   c) Iron oxide
   d) Chromium oxide

[SSC JE 2015]

166. The minimum number of test specimens required for finding the compressive strength of concrete is
   a) 3
   b) 5
   c) 6
   d) 9

[SSC JE 2017]

167. If the compaction factor is 0.95, the workability of concrete is
   a) Very low
   b) Low
   c) Medium
   d) High

[SSC JE 2017]

168. The workability of concrete can be improved by
    a) More sand
    b) More cement
    c) More fine aggregate
    d) Fineness of coarse aggregate

[SSC JE 2017]

169. Density of concrete
    a) Increases with a decrease in the size of aggregate
    b) Independent of the size of aggregate
    c) Increases with increase in the size of aggregate
    d) All options are correct

[SSC JE 2017]

170. Strength of concrete increases with
    a) Increase in water-cement ratio
    b) Decrease in water-cement ratio
    c) Decrease in size of aggregate
    d) Decrease in curing time

[SSC JE 2017]

171. Which of the following proportion of the ingredients of concrete mix is not in conformation to arbitrary method of proportioning?
    a) 1: 1: 2
    b) 1: 2: 4
    c) 1: 3: 6
    d) 1: 4: 10

[SSC JE 2017]

172. Concrete gains strength due to
    a) Chemical action of cement with coarse aggregate
    b) Hydration of cement
    c) Evaporation of water
    d) All options are correct

[SSC JE 2017]

173. Maximum shrinkage takes place in concrete after drying for
    a) 28 days
    b) Three months
    c) Six months
    d) One year

[SSC JE 2017]
174. Under constant load the Creep strain in concrete is
  a) Time dependent
  b) Temperature dependent
  c) Moisture dependent
  d) None of these

[SSC JE 2017]
175. Lightweight concrete is prepared by
  a) Using light aggregate
  b) Formation of air voids in cement by omitting sand
  c) Formation of air voids in cement paste by the substances causing foam
  d) All options are correct

[SSC JE 2017]
176. Addition of calcium chloride in concrete results in
  a) Increased strength
  b) Reduction in curing period
  c) Retardation of loss of moisture
  d) All options are correct

[SSC JE 2017]
177. Modulus of elasticity for concrete improves by
  a) Shorter curing period
  b) Age
  c) Higher W.C. ratio
  d) All options are correct

[SSC JE 2017]
178. Too wet concrete may cause
  a) Segregation
  b) Lower density
  c) Weakness of concrete
  d) All options are correct

[SSC JE 2017]
179. The curing period is minimum for concrete using
  a) Rapid hardening cement
  b) Low heat cement
  c) Ordinary Portland cement
  d) Slag cement

[SSC JE 2017]

180. If 50 kg of fine aggregates and 100 kg of coarse aggregates are mixed in a concrete whose water-cement ratio is 0.6, the weight of water required for a harsh mix is
   a) 8 kg
   b) 10 kg
   c) 12 kg
   d) 14 kg

[SSC JE 2017]

181. In a mix, if the desired slump is not obtained, the adjustment for each concrete slump difference is made by adjusting water content by
   a) 0.25%
   b) 0.5%
   c) 0.75%
   d) 1%

[SSC JE 2017]

182. In case of hand mixing of concrete, the extra cement to be added is
   a) 5%
   b) 10%
   c) 15%
   d) 20%

[SSC JE 2017]

183. If the slump of concrete mix is 60 mm, its workability is
   a) Low
   b) Medium
   c) High
   d) Very high

[SSC JE 2017]

184. Segregation is responsible for
   a) Honeycomb concrete
   b) Porous layers in concrete
   c) Sand streaks in concrete
   d) All the options are correct

[SSC JE 2017]

185. The durability of concrete is affected by
   a) Cinder
   b) Vinegar
   c) Alcohol
   d) Both cinder and vinegar

[SSC JE 2017]

186. Slump test is a measure of
  a) Tensile strength
  b) Compressive strength
  c) Impact value
  d) Consistency

[SSC JE 2017]

187. Shrinkage of concrete is directly proportional to
  a) Cement content
  b) Sand content
  c) Aggregate content
  d) Temperature of water

[SSC JE 2017]

188. The compressive strength of 100 mm cube as compared to 150 mm cube is always
  a) Less
  b) More
  c) Equal
  d) None of these

[SSC JE 2017]

189. An approximate ratio of the strength of the cement concrete of 7 days to that of 28 days is
  a) 0.65
  b) 0.85
  c) 1
  d) 1.15

[SSC JE 2017]

190. The test most suitable for concrete of very low workability is
  a) Slump test
  b) Compaction factor test
  c) Vee-Bee test
  d) All options are correct

[SSC JE 2017]

191. Water-cement ratio is generally expressed in volume of water required per
  a) 10 kg
  b) 20 kg
  c) 30 kg
  d) 50 kg

[SSC JE 2017]

192. Low temperature during concrete laying
   a) Increases strength
   b) Decreases strength
   c) Has no effect on strength
   d) Depends on other factors

[SSC JE 2017]

193. The entrained air in concrete
   a) Increases workability
   b) Decreases workability
   c) Increases strength
   d) None of these

[SSC JE 2017]

194. Poisson's ratio for concrete
   a) Increases with richer mix
   b) Decreases with richer mix
   c) Remains constant
   d) None of these

[SSC JE 2017]

195. The strength and quality of concrete depends on
   a) Aggregate shape
   b) Aggregate grading
   c) Surface area of the aggregate
   d) All options are correct

[SSC JE 2017]

196. Pick the incorrect statement from the following.
   a) A rich mix of concrete possesses higher strength than that a lean mix of desired workability with excessive quantity of water.
   b) The strength of concrete decreases as the water-cement ratio increases.
   c) If the water-cement ratio is less than 0.45, the concrete is not workable and causes a honeycombed structure.
   d) Good compaction by mechanical vibrations, increases the strength of concrete.

[SSC JE 2017]

197. Pick up the correct statement from the following.
   a) The concrete gains strength due to the hydration of cement.
   b) The concrete cured at a temperature below gains strength up to 28 days.
   c) The concrete does not set at freezing point.
   d) All options are correct.

[SSC JE 2017]

198. Pick the correct statement from the following.
   a) Higher workability indicates an unexpected increase in the moisture content.
   b) Higher workability indicates a deficiency of sand.
   c) If the concrete mix is dry, the slump is zero.
   d) All options are correct.

[SSC JE 2017]

199. The impurity of mixing water which affects the setting time and strength of concrete is
   a) Sodium sulphates
   b) Sodium chlorides
   c) Sodium carbonates and bicarbonates
   d) Calcium chlorides

[SSC JE 2017]

200. The increased cohesiveness of concrete makes it
   a) Less liable to Segregation
   b) More liable to segregation
   c) More liable to bleeding
   d) More liable for surface scaling in frosty weather

[SSC JE 2017]

201. After casting, an ordinary cement concrete on drying
   a) Expands
   b) Either expands or shrinks
   c) Shrinks
   d) None of these

[SSC JE 2017]

202. Inert material of a cement concrete mix is
   a) Water
   b) Cement
   c) Aggregate
   d) None of these

[SSC JE 2017]

203. A type of concrete in which dry, coarse aggregate are first packed to have the least voids and then the cement sand mortar is injected under pressure to fill all the voids, resulting in a very dense concrete, is known as
   a) Pre-packet concrete
   b) Vacuum concrete
   c) No fines concrete
   d) Aerate concrete

[SSC JE 2017]

204. For protection from frost, concrete should be
    a) Dense
    b) Free from cracks
    c) Adhesion between mortar and aggregate should be perfect
    d) All the options are correct

[SSC JE 2017]

205. For a satisfactory workable concrete with a constant WC ratio, increase in aggregate-cement ratio
    a) Increases the strength of concrete
    b) Decreases the strength of concrete
    c) No effect on the strength of concrete
    d) None of these

[SSC JE 2017]

206. Addition of sugar in concrete results in
    a) Increase in setting time by about 1 hour
    b) Increase in setting time by about 4 hour
    c) Decrease in setting time by about 1 hour
    d) Decrease in setting time by about 4 hour

[SSC JE 2017]

207. Strength of concrete shows an increase with
    a) Decrease in the rate of loading
    b) Increase in rate of loading
    c) Unaffected by the rate of loading
    d) Depends on the application of load

[SSC JE 2017]

208. An approximate value of shrinkage strain in concrete is
    a) 0.03
    b) 0.003
    c) 0.0003
    d) 0.00003

[SSC JE 2017]

209. For given water content, workability decreases if the concrete aggregates contain an excess of
    a) Thin particles
    b) Flat particles
    c) Elongated particles
    d) All options are correct

[SSC JE 2017]

210. On which of the following does the correct proportion of ingredients of concrete depend?
   a) Bulking of sand
   b) Water content
   c) Absorption and workability
   d) All options are correct

[SSC JE 2018]

211. Which of the below is an example of plasticizer?
   a) Hydroxylated carboxylic acid
   b) Flouro silicate
   c) Gypsum
   d) Surkhi

[SSC JE 2018]

212. If x, y and z are fineness moduli of coarse, fine and combined aggregates, respectively, the percentage (P) of fine aggregates to combined aggregate is

I. $P = \frac{z-x}{z-y} \times 100$

II. $P = \frac{x-z}{z-y} \times 100$

III. $P = \frac{x-z}{z+y} \times 100$

IV. $P = \frac{x+z}{z-y} \times 100$

   a) I only
   b) II only
   c) III only
   d) IV only

[SSC JE 2018]

213. Which of the following represents the correct expression for maturity (M) of the concrete sample?
   a) $M = \Sigma(\text{Time} \times \text{Temperature})$
   b) $M = \Sigma\left(\frac{\text{Time}}{\text{Temperature}}\right)$
   c) $M = \Sigma\left(\sqrt{\frac{\text{Time}}{\text{temperature}}}\right)$
   d) $M = \Sigma(\text{Time} + \text{Temperature})$

[SSC JE 2018]

214. If the effective working time is 7 hours and per batch time of concrete is 3 minutes, the output of concrete mixer (in litres) of 150 litre capacity is
   a) 15,900
   b) 16,900
   c) 17,900
   d) 18,900

[SSC JE 2018]

215. Grading of sand causes great variation in
   a) Workability of concrete
   b) Strength of concrete
   c) Durability of concrete
   d) All options are correct

[SSC JE 2018]

216. What is the range of slump (mm) of the concrete which used as the mass concrete?
   a) 10–15
   b) 20 –50
   c) 50–75
   d) 75–110

[SSC JE 2018]

217. The shuttering of a hall measuring 4 m× 5 m can be removed after
   a) 5 days
   b) 7 days
   c) 10 days
   d) 14 days

[SSC JE 2018]

218. No shrinkage occurs if the concrete is placed in a relative humidity of
   a) 1
   b) 0.85
   c) 0.7
   d) 0.5

[SSC JE 2018]

219. Compaction factor for heavily reinforced section with vibration is:
   a) < 0.75
   b) 0.75–0.85
   c) 0.85–0.92
   d) > 0.92

[SSC JE 2018]

220. If the internal dimensions of a warehouse are 15 m × 5.6 m and the maximum height of piles is 2.70 m, then what is the maximum number of bags to be stored in two piles?
   a) 1,500
   b) 2,000
   c) 2,200
   d) 3,000

[SSC JE 2018]

221. Sand requiring a high water-cement ratio belongs to
   a) Zone I
   b) Zone II
   c) Zone III
   d) Zone IV

[SSC JE 2018]

222. What is the gel space ratio of a sample of concrete, if the concrete is made with the 600 g of cement with the water-cement ratio of 0.65?
   a) 0.012
   b) 0.432
   c) 0.678
   d) 0.874

[SSC JE 2018]

223. For concrete tunnel linings, transportation of concrete is done by which of the following?
   a) Pans
   b) Wheel borrow
   c) Containers
   d) Pumps

[SSC JE 2018]

224. CRRI charts are used to obtain a relationship between the strength of concrete and
   a) Water-cement ratio
   b) Workability
   c) Grading of aggregate
   d) Fineness modulus

[SSC JE 2018]

225. To obtain a very high strength, concrete, use very fine-grained
   a) Granite
   b) Magnetite
   c) Barite
   d) All option are correct

[SSC JE 2018]

226. The dynamic modulus of elasticity of sample of concrete is compared in
   a) Compression test
   b) Split test
   c) Tension test
   d) Ultrasonic pulse velocity test

[SSC JE 2018]

227. .... is added to make white concrete.
   a) Fly ash
   b) Metakaolin
   c) Rice husk
   d) Figments

[SSC JE 2018]

228. .... are used to press mortar and spread it uniformly.
   a) Trowel
   b) Aluminium rod
   c) Floats
   d) Brush

[SSC JE 2018]

229. If the size of the specimen used to test the compressive strength of concrete is decreased, then compressive strength of concrete will
   a) Decrease
   b) Is not affected
   c) First decrease then increase
   d) Increase

[SSC JE 2018]

230. M20 grade of concrete is obtained from the nominal mix
   a) 1: 2: 4
   b) 1: 3: 6
   c) $1: 1\frac{1}{2} : 3$
   d) 1: 1: 2

[AE 2007]

231. The concrete slump recommended for beams and slabs is
   a) 25–50 mm
   b) 25–75 mm
   c) 30–125 mm
   d) 50–100 mm

[AE 2007]

232. The process of proper and accurate measurement of all concrete materials, uniformly of proportions and aggregates grading is called
   a) Proportioning
   b) Grading
   c) Mixing
   d) Batching

[AE 2008]

233. The concrete in which preliminary tests are performed for designing the mix is called
   a) Rich concrete
   b) Controlled concrete
   c) Lean concrete
   d) Ordinary concrete

[AE 2008]

234. Insufficient quantity of water
   a) Makes the concrete mix hard
   b) Makes the concrete mix unworkable
   c) Both A and B
   d) Causes segregation in concrete

[AE 2008]

235. For heat and sound insulation purposes, we shall use
   a) Vacuum concrete
   b) Air-entrained concrete
   c) Sawdust concrete
   d) Both A and B

[AE 2008]

236. The cement concrete in which high compressive stresses are artificially induced before it actually use is called
   a) Plain cement concrete
   b) Reinforced cement concrete
   c) Pre-stressed cement concrete
   d) Lime concrete

[AE 2008]

237. The steel mould used for slump test is in the form of a
   a) Cube
   b) Cylinder
   c) Frustum of a cone
   d) None of these

[AE 2010]

238. The concrete having a slump of 6.0 cm is said to be
  a) Medium
  b) Low
  c) Very low
  d) None of these

[AE 2010]

239. The minimum water-cement ratio to obtain workable concrete is
  a) 0.60
  b) 0.55
  c) 0.50
  d) 0.40

[AE 2010]

240. The minimum number of days required after which the side formwork of concrete column can be removed is
  (a) 14
  (b) 10
  (c) 5
  (d) 1

[AE 2010]

241. The compressive strength of first class bricks should not be less than
  a) 7 N/mm^2
  b) 10.5 N/mm^2
  c) 12 N/mm^2
  d) 15 N/mm^2

[AE 2013]

242. The minimum number of days required to strip off the side from work of RC beams after casting of the concrete is about
  a) 1
  b) 5
  c) 10
  d) 14

[AE 2013]

243. Creep in concrete is undesirable particularly in
  a) RCC columns
  b) Continuous beams
  c) Pre-stressed structures
  d) All of the above

[AE 2013]
244. If the slump of concrete mix is 70 mm, its workability is considered to be
   a) Very low
   b) Low
   c) Medium
   d) High

[AE 2013]
245. The separation of coarse aggregates from concrete during its transportation is
   a) Segregation
   b) Shrinkage
   c) Bleeding
   d) Creeping

[AE 2013]
246. In the slump test, in determining the workability of concrete, the slump is expressed in
   a) Cubic centimetres/hour
   b) Square centimetres/hour
   c) Centimetres
   d) Hours

[AE 2013]
247. A concrete is said to be workable if
   a) It shows signs of bleeding
   b) It shows signs of segregation
   c) It can be easily mixed, placed and compacted
   d) It is the form of a paste

[AE 2013]
248. Workability of concrete mix having a very low water-cement ratio should be obtained by
   a) Flexural strength test
   b) Slump test
   c) Compaction factor test
   d) Any of the above

[AE 2013]
249. More water should not be added in the concrete mix, as it will increase
   a) Strength
   b) Durability
   c) Water-cement ratio
   d) All of the above

[AE 2013]

250. The property of fresh cement concrete, in which the water in the mix tends to rise to the surface while placing and compacting is called
   a) Segregation
   b) Bleeding
   c) Creep
   d) Bulking

[AE 2014]

251. The slump (in mm) required in the trench fill in situ piling will be
   a) 25–75
   b) 50–100
   c) 75–100
   d) 100–150

[AE 2014]

252. The standard rate of application of loading to be adopted in the concrete cube testing is
   a) 14 N/mm^2 per minute
   b) 10 N/mm^2 per minute
   c) 7 N/mm^2 per minute
   d) 5 N/mm^2 per minute

[AE 2014]

253. The surface of members in tidal zones will come under which category of environmental exposure condition?
   a) Moderate
   b) Severe
   c) Very severe
   d) Extreme

[AE 2015]

254. The correct sequence of workability test methods in the order of their application from low to high workability is
   a) Slump test, compacting factor and Vee-Bee consist meter
   b) Compacting factor, Vee-Bee consist meter and slump test
   c) Vee-Bee consist meter, slump test and compacting factor
   d) Vee-Bee consist meter, compacting factors and slump test

[AE 2015]

255. Consider the following statements.

   Pozzolana used as admixtures in concrete has the following advantages
   1. It improves workability with a lesser amount of water.
   2. It increases the heat of hydration and so sets the concrete quickly.
   3. It increases resistance to attack by salts and sulphates
   4. It leaches calcium hydroxide.

Select the correct answer
a) 1 and 3
b) 1, 2 and 4
c) 2, 3 and 4
d) 1, 2, 3 and 4

[AE 2015]
256. Water-cement ratio in concrete is the ratio of
a) Volume of water to volume of cement
b) Volume of water to the weight of cement
c) Weight of water to the weight of cement
d) Weight of water to the volume of cement

[AE 2016]
257. In pavement concrete quality, a value of compacting factor suggested varies
a) 0.75–0.80
b) 0.60–0.85
c) 0.9–1.0
d) 0.5–0.6

[AE 2016]
258. Other conditions being equal, for workable mixes, the strength of concrete varies as
a) a direct function of water/cement ratio
b) a direct function of dilution of cement paste
c) an inverse function of the strength of coarse aggregate
d) An inverse function of water/cement ratio

[AE 2016]
259. If the compaction factor is 0.95, the workability of concrete is
a) Very low
b) Low
c) Medium
d) High

[AE 2016]
260. The test suitable for a concrete of very low workability is
a) Vee-Bee test
b) Slump test
c) Flow test
d) Compacting factor test

[AE 2016]

261. The split tensile strength of M20 grade concrete when expressed in percentage of the compressive strength is
   a) 10–15%
   b) 15–20%
   c) 20–25%
   d) 25– 30%

[AE 2018]

262. In the concrete mix design, the adopted maximum size of aggregate is 20 mm and water corresponding to the maximum size of aggregate is 186 kg per cubic metre of concrete. Required slump is 150 mm and pumping is used to transport the concrete. The corrected water content is ……… kg/m^3 as per IS 10262.
   a) 208.23
   b) 208.45
   c) 208.32
   d) 208.95
263. A concrete specimen is completely sealed to avoid movement of moisture and it is kept in a highly polluted place. CO_2 concentration in the place is very high. Although it is adequately sealed, shrinkage is observed and it is known as Shrinkage.
   a) Thermal
   b) Autogenous
   c) Plastic
   d) None of the above
264. Carbonation of concrete leads to two major problems in concrete: ……..…. and ……………………..…..
   a) Corrosion
   b) Carbonation shrinkage
   c) Both a and b
   d) None of the above
265. Three commonly used test methods to test flowability and workability self-compacting concrete are
   a) U-Box test
   b) J-Ring Test
   c) V-funnel Test
   d) All the above
266. Windbreaks and sunshades are used to control ………....……. shrinkage.
   a) Plastic
   b) Autogenous
   c) Carbonation
   d) None of the above
267. In concrete mix design, the water content and w/c (water-to-cement ratio) are fixed as 186 kg/m^3 and 0.39, respectively. What is the final cement content………………………………….in kg/m^3.

a) 450
b) 500
c) 400
d) 425

268. ……….... is the hydrated product mostly present in ITZ and also influence on ITZ characteristics.
a) AFt
b) AFm
c) CH
d) C-S-H

269. M50 concrete was designed as per IS mix design standard and the concrete was used in a construction site. During the construction process, cube specimens were cast at the site and compressive strength, as well as tensile strength, so they were determined after 28 days of curing at site. Assume compressive strength and tensile strength are achieved as expected as per standard. Hence, what is the difference between compressive strength and tensile strength ……………………………….….. in MPa.
a) 43
b) 44
c) 45
d) 46

270. ……………………….. special concrete has zero cement.
a) Pozzocrete
b) Papercrete
c) Alkali Activated concrete
d) Ferrocrete

1.	b	2.	c	3.	c	4.	c	5.	a	6.	d	7.	b	8.	a	9.	b	10.	c
11.	a	12.	b	13.	c	14.	b	15.	c	16.	b	17.	c	18.	d	19.	b	20.	d
21.	b	22.	b	23.	b	24.	b	25.	c	26.	c	27.	a	28.	a	29.	d	30.	c
31.	d	32.	b	33.	c	34.	a	35.	d	36.	a	37.	b	38.	d	39.	b	40.	d
41.	c	42.	a	43.	b	44.	b	45.	b	46.	b	47.	c	48.	b	49.	d	50.	a
51.	a	52.	a	53.	b	54.	a	55.	b	56.	c	57.	c	58.	c	59.	c	60.	b
61.	c	62.	c	63.	d	64.	a	65.	a	66.	b	67.	a	68.	a	69.	b	70.	a
71.	a	72.	b	73.	a	74.	a	75.	d	76.	d	77.	b	78.	b	79.	d	80.	a
81.	b	82.	b	83.	c	84.	d	85.	a	86.	d	87.	c	88.	b	89.	d	90.	c
91.	b	92.	c	93.	c	94.	a	95.	d	96.	b	97.	b	98.	a	99.	c	100.	d
101.	d	102.	d	103.	b	104.	b	105.	c	106.	c	107.	c	108.	b	109.	c	110.	b
111.	c	112.	c	113.	a	114.	d	115.	b	116.	a	117.	d	118.	b	119.	d	120.	b
121.	a	122.	c	123.	c	124.	a	125.	c	126.	d	127.	d	128.	a	129.	c	130.	d
131.	c	132.	c	133.	b	134.	b	135.	c	136.	d	137.	b	138.	a	139.	a	140.	b
141.	b	142.	a	143.	a	144.	c	145.	b	146.	a	147.	a	148.	c	149.	a	150.	b
151.	a	152.	c	153.	d	154.	b	155.	d	156.	c	157.	c	158.	d	159.	d	160.	d
161.	a	162.	d	163.	a	164.	c	165.	c	166.	a	167.	d	168.	b	169.	c	170.	b
171.	d	172.	b	173.	a	174.	a	175.	d	176.	b	177.	b	178.	d	179.	a	180.	b
181.	d	182.	b	183.	b	184.	d	185.	d	186.	d	187.	a	188.	b	189.	a	190.	c

191.	d	192.	b	193.	a	194.	a	195.	d	196.	c	197.	a	198.	c	199.	c	200.	a
201.	c	202.	c	203.	a	204.	a	205.	b	206.	b	207.	b	208.	c	209.	d	210.	d
211.	a	212.	b	213.	a	214.	d	215.	d	216.	b	217.	b	218.	a	219.	c	220.	d
221.	d	222.	c	223.	d	224.	a	225.	a	226.	d	227.	b	228.	c	229.	d	230.	c
231.	c	232.	d	233.	b	234.	c	235.	b	236.	c	237.	c	238.	b	239.	d	240.	d
241.	b	242.	a	243.	c	244.	c	245.	a	246.	c	247.	c	248.	c	249.	c	250.	d
251.	d	252.	a	253.	d	254.	d	255.	a	256.	c	257.	a	258.	d	259.	d	260.	a
261.	c	262.	c	263.	b	264.	c	265.	d	266.	a	267.	a	268.	c	269.	c	270.	c

REFERENCES

Assistant Engineer (AE). 2020. Tamil Nadu Public Service Commission (TNPSC), Tamil Nadu, India. http://www.tnpsc.gov.in/previous-questions.html (accessed September 18, 2020).

ASTM C1202 - 19. (2019). *Standard test method for electrical indication of concrete's ability to resist chloride ion penetration.* .American Society for Testing and Materials. West Conshohocken, Pennsylvania.

CoMSIRU. (2018a). *Durability index testing procedure manual.* Cape Town.

CoMSIRU. (2018b). *Durability index testing procedure manual.* Cape Town.

CoMSIRU. (2018c). *Durability index testing procedure manual.* Cape Town.

DIN 1048 Part 5. (1991). *Testing concrete – Testing of hardened concrete.* Berlin.

EFNARC. (2002a). *Specification and guidelines for self-compacting concrete.* Farnham. UK.

EFNARC. (2002b). *Specification and guidelines for self-compacting concrete.* Farnham. UK.

EFNARC. (2002c). *Specification and guidelines for self-compacting concrete.* Farnham. UK.

Engineering Service Exam (ESE). 2020. Union Public Service Commission, New Delhi, India. https://www.upsc.gov.in/examinations/previous-question-papers (accessed September 18, 2020).

Graduate Aptitude Test in Engineering (GATE). 2020. GATE Office, Chennai, India. http://gate.iitm.ac.in/gate2019/previousqp18.php (accessed September 18, 2020).

IS 1199 Part 2. (2018a). *Fresh concrete – Methods of sampling, testing and analysis: Determination of consistency of fresh concrete.* Bureau of Indian Standards. New Delhi.

IS 1199 Part 2. (2018b). *Fresh concrete – Methods of sampling, testing and analysis: Determination of consistency of fresh concrete.* Bureau of Indian Standards. New Delhi.

IS 1199 Part 2. (2018c). *Fresh concrete – Methods of sampling, testing and analysis: Determination of consistency of fresh concrete.* Bureau of Indian Standards. New Delhi.

IS 1199 Part 3. (2018). *Fresh concrete – Methods of sampling, testing and analysis: Determination of density of fresh concrete.* Bureau of Indian Standards. New Delhi.

IS 1199 Part 7. (2018). *Fresh concrete – Methods of sampling, testing and analysis: Determination of setting time of concrete by penetration resistance.* Bureau of Indian Standards. New Delhi.

IS 516. (2018). *Methods of tests for strength of concrete.* Bureau of Indian Standards. New Delhi.

IS 5816. (2018). *Splitting tensile strength of concrete – method of test.* Bureau of Indian Standards. New Delhi.

Junior Engineer (JE), Staff Selection Commission (SSC), New Delhi, India. https://ssc.nic.in/Portal/SchemeExamination (accessed September 18, 2020).

5 Bricks

5.1 INTRODUCTION

Similar to stone, brick is also one of the materials used for masonry. In fact, brick is the most widely used building material for masonry in India. Burnt bricks and unburnt bricks are made up of clay, lime and sand, cement mixtures, etc. It can be handled very conveniently because of their shape and weight. Standard bricks have dimensions of $190 \times 90 \times 90$ mm. In the case with mortar, brick size is $200 \times 100 \times 100$ mm, with average weight ranging from 30 N to 35 N.

5.2 CHEMICAL COMPOSITION AND THEIR RELEVANCE

Table 5.1 shows chemical composition, percentage range and their properties. To identify the quality of brick initial preliminary investigation can be done by visual observation. Good bricks will have uniform colour, shape and size. With upper brick surface will be plane without any cracks.

5.3 BRICKS: MANUFACTURING PROCESS

Manufacturing of brick is a four-step process. In the first step, the soil is excavated and cleaned for impurities and laid in level with the ground and exposed to the weather for a few months. Then, the exposed soil is tempered and blended with other materials in a pug mill. Then the moulding is carried out either manually or mechanically. Moulding is carried out mechanically if the quantity of production is large. There are two types of machine moulders: (1) Plastic clay machine and (2) dry clay machine.

Burning of bricks is an essential process in the manufacturing of burnt clay bricks. Based on the regulation of fire, the burning of bricks can be done by two methods. They are (1) kiln burning and (2) clamp burning. Table 5.2 shows the comparison between clamp burning and kiln burning.

5.3.1 Intermittent Kilns

Intermittent kilns are further divided into two types: (1) up draught kiln and (2) down draught kiln. In the intermittent kiln, each operation such as loading, burning, cooling and unloading are carried out separately.

5.3.2 Continuous Kilns

In the continuous kiln, each operation is carried out continuously without any intermittent disturbances. Continuous kilns are further divided into three types: (1) bull trench kiln; (2) Hoffman kiln; and (3) tunnel-type kiln.

TABLE 5.1
Chemical Composition, Percentage Range and Properties

Chemical composition	Percentage	Property
Alumina	20–30	Increases plasticity for easy moulding
Silica	50–60	It enhances the durability by preventing warping, shrinkage and cracking
Lime	10	Prevents shrinkage; Excess can cause melting during the burning process (flux)
Iron oxide	<7	It fuses silica and lime for better strength
Magnesia	<1	May induce yellow colour if more than 1%; it prevents shrinkage

TABLE 5.2
Comparison between Clamp Burning and Kiln Burning

Parameters	Clamp burning	Kiln burning
Fuel	Locally available fuels such as grass, ash or wood	Coal and biomass
Fire regulation	Not controlled	More controlled
Quality	Non-uniform	Uniform and good
Types	Intermittent kilns and Continuous kilns	–
Cost	Less	More

5.4 CLASSIFICATION OF BRICKS

Bricks are classified into three types: (1) first-class bricks; (2) second-class bricks; and (3) third-class bricks based on certain characteristics. Table 5.3 gives a classification of bricks based on their characteristics.

5.5 BRICK MASONRY

Bricks are laid in a particular bond and cemented with the help of mortar. Generally, there are four types of bonds. These four types of bonds and their short descriptions are as follows.

1. Stretcher bond: In a brick of 190×90×90 mm, the 190×90 mm face is called as a stretcher. This type of bond is useful for the construction of half-brick-thick partition walls.
2. Header bond: 90×90 mm face in brick is called a header. One-brick thick wall can be constructed by this type of bond.
3. English bond: This bond is the strongest bond of all, with alternative courses consisting of header and stretcher. It is used for the construction of walls of all thicknesses.

4. Flemish bond: Each course consists of alternate header and stretcher. The alternative course starts with a stretcher and header.

TABLE 5.3
Classification of Bricks and Their Common Characteristics

Classification characteristics	First-class bricks	Second-class bricks	Third-class bricks
Compressive strength (MPa)	10	7	–
Water absorption (%) at 24 hours	12–15	16–20	≯25
Colour due to burning	Red, cherry and copper	Slightly irregular	Under burnt or over burnt; efflorescence visible
Shape and size	Uniform	Slightly irregular	Not uniform
Edges and faces	Sharp and square; parallel	Slightly irregular	Not uniform
Ringing test	Clear ringing sound	Clear ringing sound	Unclear sound

5.6 CLOSERS AND BATS

Cut bricks that are used in the construction of brick masonry are called as closers and bats. There are four types of closers and three types of bats.

1. King closer – The brick is cut in such a way that the width of one end is half that of the full brick, while width at the other end is equal to full width.
2. Queen closer – The brick is cut in half of the width of the brick so that the width of the cut brick is half that of the full brick.
3. Bevelled closer – Brick is cut along the stretcher face in such a way that the width of one end is equal to the full width of the brick and with at the other end is half that of the width of full brick.
4. Mitred closer – In this type of closer, one end is cut splayed in such a way that the full width of the brick taken off, while the angle of splay varies between 45° and 60°. Consequently, one face of the stretcher is of full length and the other face is smaller in length.

Bats are cut along the width of the bricks. There are three types of bricks:

1. Half bat – If the length of the brick is half the length of the full brick then it is called as a half bat.
2. Three bat – If the length of the brick is three-quarter of the length of the full brick then it is called as three bat
3. Bevelled bat – If the width has been bevelled then the bat is called as a bevelled bat.

5.7 TESTING PROCEDURES

Testing of bricks can be divided into field tests and laboratory tests.

5.7.1 Field Test

Strength – Dropping a brick from a height of 1 m should not break the brick into pieces. If two good bricks have collided to each other, they should produce a metallic ringing sound.

Hardness – Resistance against nail impression on scratching by nail can be used as an indication for hardness.

Colour – Randomly selected bricks can be checked for their uniform colour.

No sign of efflorescence or staining should be there on immersing brick in water for 24 hours.

5.7.2 Laboratory Test

Laboratory tests such as compressive strength, water absorption and efflorescence tests are conducted on bricks to evaluate the quality of the brick.

1. Compression test (IS: 3495 Part 1) – Minimum three or five numbers of whole bricks are immersed in water for 24 hours at room temperature. Then the frog in the brick is filled with mortar and kept inside damp jute bag for another 24 hours. Further, the bricks are placed in clean water for 3 days. Now the specimens are placed with 3 mm thick plywood sheets and loaded at a uniform rate of 14 N/mm^2 per minute until failure. The load at the failure is noted down.
2. Water absorption of bricks (IS: 3495 Part 2) – The specimen is dried in an oven at a temperature of 105° C until it reaches a constant mass. The specimen is immersed in water and placed at a temperature of $27 \pm 2°$ C for 24 hours. The specimen is removed and wiped and weighed.
3. Efflorescence test (IS: 3495 Part 3) – Deposition of soluble salts on the surface of the porous media due to absorption or evaporation of water is called efflorescence. Bricks made up of sand should not consist of more than 0.1% of chloride and sulphate. The magnesium content of the brick should not exceed 0.05%. The brick is tested for efflorescence by placing its end in a porcelain or glass container of 150 mm diameter that contains 25 mm depth of water at room temperature. The brick is placed until 25 mm water is completely evaporated or absorbed. The above process is repeated, and the deposition of efflorescence in the exposed area is reported.

Other tests that are conducted in the laboratory are dimension test as per IS: 1077 and warping test as per IS: 3495 Part 4.

Other clay bricks – (1) Burnt clay hollow bricks (IS: 3952); (2) Burnt clay paving brick (IS: 3583); and (3) Burnt clay perforated bricks (IS: 2222)

Fly-ash-based unburnt bricks – Recent development of fly-ash-based unburnt bricks along with a small percentage of cement/lime as a binder is more popular due to their sustainable nature. Any supplementary cementitious material is mixed with cement to produce unburnt bricks.

5.7.3 Field Test to Assess the Quality of Bricks

Bricks are considered to be very important in construction applications. Care should be taken to choose appropriate bricks with good quality in the field. In order to understand the quality of bricks, there are certain field tests through which the quality of the bricks can be assessed (IS 3495 Part 1, 2016).

Test Summary

- Field tests to understand the quality of bricks can be done by following checks.
- Good-quality bricks should be well burnt with uniform shape, size and colour.
- The colour of the brick is generally deep red or copper.
- Brick should be homogeneous in texture, free from flaws and cracks.
- Edges should be square, sharp and straight edge.
- On striking two good bricks, a clear metallic sound can be observed. On striking with each other, they should not break.
- When dropped from a height of 1.2–1.5 m on to the ground, they should not break.
- Water absorption should not be more than one-sixth of its weight after immersing the bricks in water for a period of an hour.

Summary

Important characteristics to look for during a quality field check are:

- Shape, size and colour
- Dimensional accuracy
- Homogeneity
- Strength
- Water absorption

5.7.4 Compressive Strength of Bricks

The quality of the masonry brick structure largely depends upon the compressive strength of bricks. As per BIS standard, bricks should have a compressive strength of at least 3.5 N/mm^2. Moreover, based on the class of brick, the compressive strength of an individual brick should not be less than 20% of the prescribed compressive strength.

Test Summary

This test procedure is used to determine the compressive strength of bricks.

Use and Significance

- Brick masonry structure can be used as a load-bearing structure, and in such cases, it is mandatory to understand and check the quality of bricks used in the test.

Required Apparatus

- A compressive testing machine of minimum capacity 1,000 kN (100 tonnes) with an accuracy of 10 kN (1 tonne)
- 3 mm thickness 3-ply plywood sheets
- Cement
- Coarse sand
- Mixing tool
- Enamel tray

Sampling

- Five bricks are sampled from a lot as a test specimen.

Testing

Step 1. Sampled bricks are placed inside a water bath for 24 hours and maintained at a temperature of 25–29°C. After 24 hours, bricks are taken out and placed outside in a vertical position to drain out the excess moisture on the surface.

Step 2. Mortar is prepared at a cement-to-fine-aggregate ratio of 1:3. Fine aggregates used in the mortar pass 3 mm and down. This mortar is used to fill the frog of the brick. Gypsum can also be used to obtain a highly smooth surface finish.

Step 3. Filling of the frog is a mandate in order to obtain a uniform surface for loading during the determination of compressive strength.

Step 4. Bricks prepared by the above step are placed under a wet jute bag for 24 hours.

Step 5. After 24 hours, the brick specimens are immersed in a water bath for 3 days and the moisture is wiped off with no races on the surface of the brick.

Step 6. The length and width of all the specimens are measured to an accuracy of 1 mm at three different places and averaged to get a mean of length and breadth.

Step 7. A 3-ply plywood sheet is placed on the base plate. Brick is placed with a flat face facing down and mortar face facing up. Another 3-ply plywood sheet is placed on the top face of the brick and centred between the top and bottom plate of the compression testing machine.

Step 8. The load is applied axially and uniformly at a rate of 14 N/mm^2/minute. The reading is observed when the failure of the specimen occurs. In a similar way, readings are observed for all other test specimens. Table 5.4 shows the class designation and the required average compressive strength.

TABLE 5.4
Class Designation and Minimum Average Compressive Strength

	Average compressive strength not less than	
Class designation	**N/mm²**	**kgf/cm²**
35	35	350
30	30	300
25	25	250
20	20	200
17.5	17.5	175
15	15	150
12.5	12.5	125
10	10	100
7.5	7.5	75
5	5	50
3.5	3.5	35

Calculation: $\text{Compressive strength of brick} = \dfrac{\text{Maximum load at failure } (N)}{\text{Area of the loaded portion } (\text{mm}^2)}$

Note: At least three or five brick specimens are performed and the average is considered to be the mean value.

Observations and Calculations

Details	**Sample 1**	**Sample 2**	**Sample 3**
Length (mm)			
Breadth (mm)			
Height (mm)			
Compressive load at failure (kN)			
Compressive strength (N/mm²)			
Average Compressive strength (N/mm²)			

Caution: For any class of brick, the compressive strength of the brick should not be higher than or lesser than 20% of the actual strength of the class. If a brick specimen passes the above test, then the lot is considered to be satisfactory for used for the intended purpose.

Report

- Average compressive strength of brick
- Whether the individual compressive strength passes 20% limitation

Student Remark

....
....
....
....

$$\frac{\textbf{\textit{Marks obtained}}}{\textbf{\textit{Total marks}}} = -$$

Instructor signature

Instructor Remark

....
....
....
....

5.7.5 Water Absorption in Bricks

Water absorption test can be used as a test in order to understand the ability of brick to allow water penetration inside a brick masonry structure. This test can give an overview of the quality of bricks used during construction (IS 3495 Part 2, 2016).

Test Summary

Porosity and distribution of voids in the brick determine water uptake by bricks. When bricks come in contact with water or moisture, it results in the uptake of water by capillary force.

After immersing the bricks in water for 24 hours, the uptake of water by bricks should not be more than a certain percentage based on provisions according to their class designation.

Use and Significance

- In general, bricks can uptake a large amount of water due to their high porosity and distribution of voids. Uptake of water more than a certain percentage can result in the deterioration of the brick's masonry structure and the associated structures in which the bricks are used.
- To reduce deterioration of brick masonry structure, the uptake of water by bricks needs to be limited. This can be done by careful investigation of water uptake by bricks.

Required Apparatus

- Weight balance (0.1% of the accuracy of the weight of the specimen)
- Oven capable of maintaining a temperature of $110 \pm 5°C$

Sampling

Five bricks are sampled from a lot as test specimens.

Testing

Step 1. Test specimens are pre-conditioned by drying the specimen. If the specimen is relatively dry, then the drying can be accomplished within 48 hours. If the specimen has moisture, then the drying may take more time based on the requirement.

Step 2. Specimens are placed in an oven and heated at 110±5°C for 24 hours. Alternatively, specimens are kept in the oven for a time at which the specimen reaches a constant mass.

Step 3. After drying for the specified time, bricks are taken out and allowed to cool at room temperature for 4 or more hours with proper separation between each brick.

Step 4. If the cooling is done by some mechanical means such as an electric fan, then the cooling time can be reduced to less time.

Step 5. Specimens that are cooled are weighed. The weight of the dry specimen is taken as w_1 (kg).

Caution: Cooling of the specimen needs to be proper. The warm specimen should not be used. Ensure that the specimen is marked with permanent marks so that during any part of experimentation readings of corresponding specimens are noted down.

Step 6. Dry specimen is immersed in a water bath and a temperature of 27±2°C is maintained for 24 hours.

Step 7. After 24 hours, specimens are taken out of the water bath and are placed vertically as well as separately. A damp cloth is used to wipe out the specimen.

Step 8. Now, again weigh the specimen within 3 minutes. Record the weight of specimens and note it down as w_2 (kg).

Observations and Calculations

Details	Sample 1	Sample 2	Sample 3
Initial weight (w_1)			
Final weight (w_2)			
Percentage absorption of water $\frac{w_2 - w_1}{w_1} \times 100$			

$$\text{Percentage absorption of water} = \frac{w_2 - w_1}{w_1} \times 100$$

Report

- Average water absorption in percentage
- Quality of bricks is satisfactory or not (see Table 5.5.)

TABLE 5.5
Water Absorption as per BIS

Class designation of bricks	Water absorption
Until 12.5	Not more than 20%
Above 12.5	Not more than 15%

Student Remark

....
....
....
....

$$\frac{\textit{Marks obtained}}{\textit{Total marks}} = -$$

Instructor signature

Instructor Remark

....
....
....
....

5.7.6 Efflorescence in Bricks

Materials used in the manufacture of bricks should be free from or should have a limited quantity of harmful salts (lesser than the permissible limit) such as sulphates, sodium and potassium (IS 3495 Part 3, 2016).

Test Summary

This test procedure is used to determine efflorescence in brick. The porosity and distribution of voids in the brick govern efflorescence. When bricks come in contact with water or moisture, it results in the uptake of water by capillary force. The presence of such harmful salts results in the formation of crystalline products that can be identified as white patches on the surface of the bricks.

Use and Significance

- Harmful salts such as sulphates, sodium and potassium present in bricks dissolve when comes in contact with water. When the bricks that contain such salts are exposed directly to the environment or given a covering treatment, serious problems such as disruption of the surface take place due to the formation of crystalline salts on the surface of the brick. This will weaken the layers in contact with the brick as surface protection.

Required Apparatus

- A shallow flat bottom tray
- Distilled water with beaker
- Graduated steel scale in mm

Sampling

Five bricks are sampled from a lot as a test specimen.

Testing

Step 1. Take an enamel tray and fill with distilled water to a height of 25 mm. Bricks are placed vertically in the enamel tray. The temperature of the testing area should be well between 18°C and 30°C and also adequately ventilated.

Step 2. Bricks are left in the enamel tray until entire water from the tray is absorbed by the test specimens.

Step 3. When the brick appears to be dry, the same amount of distilled water is poured in the tray and water is allowed to evaporate.

Step 4. Bricks are examined once the second evaporation takes place. Then the efflorescence is observed in the bricks as per the description and the extent of deposits as shown in Table 5.6.

Note: A lot is considered to be satisfactory if four out of five bricks have efflorescence value less than moderate as per the description in Table 5.6.

Observations and Calculations

Details	Specimen 1	Specimen 2	Specimen 3
Length (mm)			
Breadth (mm)			
Height (mm)			
Surface area (mm^2)			
Efflorescence area (mm^2)			
Percentage efflorescence (%)			
Extent of efflorescence			

TABLE 5.6
Description and Extent of Deposits as per BIS

Description	Extent of deposits
Nil	No noticeable deposit of efflorescence
Slight	Thin salt area covering 10% of the area
Moderate	Heavy deposit up to 50% of the area covered with salt; No powdering and flaking
Heavy	Heavy deposit more than 50% of the area covered with salt; No powdering and flaking
Serious	Heavy deposit; powdering or flaking observed

Report

- Extent of efflorescence in bricks
- Quality of bricks is satisfactory or not

Student Remark

....
....
....
....

$$\frac{\textbf{\textit{Marks obtained}}}{\textbf{\textit{Total marks}}} = -$$

Instructor signature

Instructor Remark

....
....
....
....

5.7.7 Warpage in Bricks

Bricks can be manufactured based on requirements. After the process of moulding, it is of utmost importance to check the dimensional changes within a brick. Due to an inappropriate moulding process or due to some other reasons, the shape of the brick can twist or bend out of the plane. Tolerance should be followed to limit the out of plane change in the shape of the brick (IS 3495 Part 4, 2016).

Test Summary

This test procedure is used to determine excessive bending or twisting in manufactured brick due to the change in heat or moisture during the process of moulding.

Use and Significance

- For the construction of brick masonry structures and other structures, bricks should have perfect shape and size. If the shape and the size of the bricks are irregular, then the alignment of structures that involves bricks will be distorted. In order to understand the tolerance limit, it is of significant importance to check the dimensional variability within a brick.

Required Apparatus

- A steel ruler of 0.5 mm precision
- A straight steel edge
- A steel wedge of 60 mm in length, 15 mm in width and starting at 0 mm height in one end and 15 mm height at the other end, graduation with a precision of 0.5 mm is marked both on the slope and on the base
- A flat surface of 300 mm × 300 mm in area and a thickness of 0.02 mm

Sampling

Five bricks are sampled from a lot as a test specimen. Dust adhering to the surface of the brick is removed.

Testing

Step 1. The warpage of bricks can be both convex and concave. Convexity and concavity of bricks are decided based on the surface without the frog.

Step 2. Concave and convex warpage can be measured both on edges as well as on surfaces of the bricks.

Step 3. Concave warpage can be measured by placing the straight edge or steel wedge along the length or the diagonal across the surface. A location is selected by determining the point of greatest departure from straightness using a straightedge or wedge. The maximum departure of the straightness from the surface of the specimen is recorded to the nearest of 1 mm precision by using a steel ruler of 0.5 mm precision.

Step 4. In a similar manner, the concave warpage on the edges can also be determined by placing the straightedge or wedge across the ends of the concave edge.

Step 5. Similar to the determination of concave edges and concave surfaces, the convexity of the edges and surfaces can be determined by placing the straightedge or wedge parallel to the edge or surface.

Step 6. The maximum deviation of edges and surfaces from straightness is recorded by using a steel ruler.

Step 7. An average of two corners is considered for an edge, whereas an average of four corners is considered for a surface. Moreover, both are considered in the case of concave warpage and convex warpage.

Report

- Higher of the distance in concave and convex is considered to be warpage.

Student Remark

....

....

....

....

$$\frac{\textbf{\textit{Marks obtained}}}{\textbf{\textit{Total marks}}} = -$$

Instructor signature

Instructor Remark

....

....

....

....

5.8 PRACTICE QUESTIONS AND ANSWERS FROM COMPETITIVE EXAMS

1. During compressive strength testing of bricks, end platen effect .... the compressive strength of bricks.
   a) Increases
   b) Decreases
   c) Remains the same
   d) None of the above

2. In water absorption test for tiles, tiles are kept under boiling water condition
   a) To increase the rate of water absorption of tiles
   b) To decrease the rate of water absorption of tiles
   c) To open up the micro-sized pores to facilitate increased water absorption
   d) None of the above

3. 3-point loading test on tiles results in higher values than the 4-point loading test?
   a) True
   b) False
   c) Remains the same
   d) None of the above

4. Why are AAC blocks more preferred than burnt clay bricks in high rise buildings?
   a) Reduces the dead weight of the structure
   b) Reduces the temperature variability of the room
   c) Both a and b
   d) None of the above

5. What is the purpose of adding lime in the manufacture of bricks?
   a) Increases plasticity
   b) Enhances durability
   c) Prevents shrinkage
   d) Increases strength

6. Why are bricks placed vertically during efflorescence test?
   a) To enhance the area of contact with water
   b) To increase capillary action
   c) To prevent water from entering from sides
   d) None of the above

7. During drying of bricks in water absorption test, bricks are placed in an oven for a specific time period. On what basis is the period decided?
   a) Until a constant mass is reached

b) Until a constant bulk density is reached
c) On the day basis
d) None of the above

8. Concavity and convexity of bricks are determined by which of the following tests?
   a) Efflorescence test
   b) Warpage test
   c) Ringing test
   d) Ball-mill test

9. Which of the following gives plasticity to the bricks?
   a) Silica
   b) Lime
   c) Alumina
   d) Magnesia

10. What is the role of addition of iron oxide in the brick?
    a) Prevents shrinkage
    b) Increases strength
    c) Enhances durability
    d) Increases plasticity

11. Bull's trench kiln is used in the manufacture of
    a) Lime
    b) Cement
    c) Bricks
    d) None of these

[GATE 2016 FN; 1 mark]

12. As per Indian standards for bricks, the minimum acceptable compressive strength of any class of burnt clay bricks in dry state is
    a) 10.0 MPa
    b) 7.5 MPa
    c) 5.0 MPa
    d) 3.5 MPa

[GATE 2016 AN; 1 mark]

13. Consider the following statements

    P. Walls of one-brick thickness are measured in square metres.

    Q. Walls of one-brick thickness are measured in cubic metres.

    R. No deduction in the brickwork quantity is made for openings in walls up to 0.1 m^2 area.

    S. For the measurement of excavation from the borrow pit in a fairly uniform ground, dead men are left at suitable intervals.

For the above statements, the correct option is
   a) P – False, Q – True, R – False, S – True
   b) P – False, Q – True, R – False, S – False
   c) P – True, Q – False, R – True, S – False
   d) P – True, Q – False, R – True, S – True

[GATE 2017 AN; 2 marks]
14. A good brick should not absorb water by weight more than
   a) 10%
   b) 20%
   c) 25%
   d) 30%

[IES 1995]
15. For good bonding in bricks
   a) All bricks need not be uniform in size
   b) Bats must be used in alternate course only
   c) The vertical joints in alternate course should fall in plumb
   d) Cement mortar used must have surkhi as an additive

[IES 1995]
16. King closers are related to
   a) Doors and windows
   b) King post truss
   c) Queen post truss
   d) Brick masonry

[IES 1996]
17. A good brick, when immersed in a water bath for 24 hours, should not absorb more than
   a) 20% of its dry weight
   b) 30% of its saturated weight
   c) 10% of its dry weight
   d) 20% of its saturated weight

[IES 1996]
18. For one cubic metre of brick masonry, the number of modular bricks needed is
   a) 400 or less
   b) 400–450
   c) 500–550
   d) 600–650

[IES 1997]

19. In brick masonry
    a) Mortar strength should match brick strength
    b) Mortar strength should exceed brick strength
    c) Mortar strength should exceed mortar strength
    d) The strength of masonry and brick are independent

[IES 1998]

20. Consider the following statements:
    1. About 25% of alumina in brick earth imparts the plasticity necessary for moulding bricks into required shape.
    2. Iron pyrite present in brick earth preserves the form of the bricks at high temperatures.
    3. Presence of weeds in brick earth makes the bricks unsound.

    Which of these statements are correct?
    a) 1 and 2
    b) 1 and 3
    c) 2 and 3
    d) 1, 2 and 3

[IES 1999]

21. The most important purpose of frog in a brick is to
    a) Emboss manufacture's name
    b) Reduce the weight of brick
    c) Form keyed joint between brick and mortar
    d) Improve insulation by providing "hollows"

[IES 1999]

22. Bricks are burnt at a temperature range of
    a) 500–700°C
    b) 700–900°C
    c) 900–1200°C
    d) 1200–1500°C

[IES 1999]

23. A king closer is a
    a) Full brick
    b) $\frac{3}{4}$ brick
    c) Longitudinally $\frac{1}{2}$ brick
    d) Crosswise $\frac{1}{2}$ brick

[IES 2000]

24. Brick masonry walls and columns of a building are to be protected from the earthquake. The earthquake proofing is done by providing
    a) Cross walls
    b) Less openings
    c) Under-reamed piles
    d) A steel band at comers above windows below ceiling

[IES 2000]

25. When the comer of a brick is removed along the line joining midpoints of adjoining sides, the portion left is called
    a) Closer
    b) Squint brick
    c) Queen closer
    d) King closer

[IES 2000]

26. Which one of the following procedures is applied to determine the soundness of bricks?
    a) Immersing the brick under water for 16 hours and determining the quantity of water absorbed by the brick.
    b) Immersing the brick under water for 24 hours and determining its expansion using Le Chatlier's apparatus.
    c) Taking two bricks, hitting one against the other and observing whether they break or not and the type of sound produced while hitting.
    d) Scratching the brick by finger nail and noting whether any impression is made or not.

[IES 2001]

27. The number of bricks required per cubic metre of brick masonry is
    a) 400
    b) 450
    c) 500
    d) 550

[IES 2001]

28. The bricks which are extensively used for basic refractories in furnaces are
    a) Chrome bricks
    b) Silimanite bricks
    c) Magnesite bricks
    d) Forsterite bricks

[IES 2001]

29. The given figure shows the Tee junction in brick masonry, which is known as

   a) English bond
   b) English cross-bond
   c) Flemish bond
   d) Double Flemish bond

[IES 2001]

30. The minimum compressive strength of first-class bricks should be
   a) 5 MPa
   b) 7.5 MPa
   c) 9 MPa
   d) 10 MPa

[IES 2001]

31. Match list I (constituents of bricks) with list II (corresponding influence) and select the correct answer using the codes given below the lists:

	List I		List II
A	Alumina	1	Colour of brick
B	Silica	2	Plasticity recovery for moulding
C	Magnesia	3	Reacts with silica during burning and causes particles to unite together and development of strength
D	Lime stone	4	Preserves the form of brick at high temperature and prevents shrinkage

Codes:

	A	B	C	D
a)	2	1	4	3
b)	3	4	1	2
c)	2	4	1	3
d)	3	1	4	2

[IES 2002]

32. In some brick masonry walls, patches of whitish crystals were found on the exposed surfaces, also chipping and spalling of brick took place from the same walls. Which among the following are the causes of these defects?
    1. Settlement of foundation
    2. Overloading of the walls
    3. Sulphate attack
    4. Efflorescence

Select the correct answer from the codes given below

a) 1 and 2
b) 2 and 3
c) 2 and 4
d) 3 and 4

[IES 2002]

33. Consider the following stages in the manufacturing of bricks
    1. Weathering
    2. Moulding
    3. Tempering

The correct sequence of these stages in the manufacturing of the bricks is

a) 1, 2, 3
b) 2, 3, 1
c) 1, 3, 2
d) 3, 2, 1

[IES 2002]

34. Consider the following statements:

    Bricks are soaked in water before use in masonry work:
    1. To remove dust
    2. To remove air voids
    3. So that they do not absorb water from cement mortar

Which of these statements is/are correct?

a) 1, 2 and 3
b) 1 only
c) 2 and 3
d) 3 only

[IES 2003]

35. Consider the following characteristics with respect to brick:
    1. Minimum compressive strength = 175 (Standard units)
    2. Minimum absorption in 24 hours, (in % of dry weight) = 12
    3. Very little efflorescence
    4. Tolerance in dimensions = ±8%

As per India standards classification, a brick with the characteristics given above is termed as

a) H-I
b) F-II
c) L-II
d) H-II

[IES 2004]

36. Refractory bricks resist
    a) High temperature
    b) Chemical action
    c) Dampness
    d) All of the above

[IES 2005]

37. For ornamental work, which type(s) of bricks is/are preferred?
    1. Silica bricks
    2. Silica lime bricks
    3. Bricks produced in autoclaves

Select the correct answer using the codes given below:

a) 1 and 3 only
b) 2 and 3 only
c) 1 only
d) 2 only

[IES 2007]

38. Which one of the following is the nominal size of a standard modular brick?
    a) 25 cm × 13 cm × 8 cm
    b) 25 cm × 10 cm × 8 cm
    c) 20 cm × 10 cm × 10 cm
    d) 20 cm × 15 cm × 10 cm

[IES 2007]

39. Consider the following statements:
    1. Soil containing more than 30% of calcium hydroxide is used for the manufacture of sand lime brick.
    2. Carbon brick is made from crushed coke bonded with tar.

Which of these statements is/are correct?
- a) 1 only
- b) 2 only
- c) Both 1 and 2
- d) Neither 1 nor 2

[IES 2008]

40. For high-class brick masonry, which are the proper bricks?
- a) Refractory bricks
- b) Jhumb bricks
- c) Bull nose bricks
- d) Modular bricks

[IES 2008]

41. Modular bricks are of nominal size $20 \times 10 \times 10$ cm and 20% of the volume is lost in the mortar between joints. Then what is the number of modular bricks required per cubic metre of brickwork?
- a) 520
- b) 500
- c) 485
- d) 470

[IES 2009]

42. In order to achieve a safe compressive strength of 20 kg/cm^2 in brick masonry, what should be the suitable range of crushing strength of bricks?
- a) 35 kg/cm^2 to 70 kg/cm^2
- b) 70 kg/cm^2 to 105 kg/cm^2
- c) 105 kg/cm^2 to 125 kg/cm^2
- d) More than 125 kg/cm^2

[IES 2009]

43. Consider the following statements:
    1. Strength of brick masonry is influenced by the type of mortar.
    2. Brick masonry with lime mortar achieves full strength earlier than cement mortar masonry.
    3. Mortar strength decides the strength of masonry.

Which of these statements is/are correct?
- a) 1, 2 and 3

b) 1 only
c) 1 and 3 only
d) 3 only

[IES 2010]

44. The average compressive strength of a burnt clay brick is less than 12.5 MPa. The allowable rating of efflorescence is
a) Moderate
b) Serious
c) Heavy
d) Zero

[IES 2010]

45. Consider the following statements:
1. Bricks lose their strength by 25% when soaked in water.
2. Minimum crushing strength of brick in buildings should be 3.5 kg/cm^2
3. The size of modular type brick is 20 cm × 10 cm × 10 cm including mortar thickness.

Which of these statements are correct?
a) 1, 2 and 3
b) 1 and 2 only
c) 1 and 3 only
d) 2 and 3 only

[IES 2010]

46. Consider the following statements
Perforated bricks are preferred in construction since
1. They are lighter
2. They are stronger than class I bricks
3. They have heat-insulating properties
4. They are cheaper and need less mortar

Which of these statements are correct?
a) 1, 2, 3 and 4
b) 2 and 3 only
c) 1 and 3 only
d) 3 and 4 only

[IES 2012]

47. The standard size of a brick is
a) 20 cm × 10 cm × 10 cm
b) 19 cm × 9 cm × 9 cm
c) 18 cm × 9 cm × 9 cm
d) 18 cm × 10 cm × 10 cm

[IES 2012]

48. As per IS classification, the minimum compressive strength of a first-class brick should be
    a) 75 kg/cm^2
    b) 100 kg/cm^2
    c) 125 kg/cm^2
    d) 150 kg/cm^2

[IES 2013]

49. When provided with alternating courses of all headers and all stretchers, the front elevation of such brick masonry is designated as
    a) English bond
    b) Single Flemish bond
    c) Double Flemish bond
    d) Rat-trap bond

[IES 2013]

50. IS Code specifies that the compressive strength of brick shall be determined by keeping the brick
    1. On edge
    2. On bed
    3. After soaking in water for 2 hours
    4. After soaking in water for 24 hours

    a) 1 and 2
    b) 2 and 3
    c) 3 and 4
    d) 2 and 4

[IES 2013]

51. Statement (I): Bricks are soaked in water before using in brick masonry for removing dirt and dust.

    Statement (II): Bricks are soaked in water before using in brick masonry so that bricks do not absorb moisture from the bonding cement mortar.
    a) Both statement (I) and statement (II) are individually true and statement (II) is the correct explanation of statement (I).
    b) Both statement (I) and statement (II) are individually true and statement (II) is not the correct explanation of statement (I).
    c) Statement (I) is true but statement (II) is false.
    d) Statement (I) is false but statement (II) is true.

[IES 2014]

52. Statement (I): Brick masonry in mud mortar is weak in strength.
    Statement (II): Cement mortar enhances the strength of the bricks relative to mud mortar.
    a) Both statement (I) and statement (II) are individually true and statement (II) is the correct explanation of statement (I).
    b) Both statement (I) and statement (II) are individually true and statement (II) is not the correct explanation of statement (I).
    c) Statement (I) is true but statement (II) is false.
    d) Statement (I) is false but statement (II) is true.

[IES 2014]

53. Consider the following statements:
    1. Brickwork will have high water tightness.
    2. Brickwork is preferred for monument structures.
    3. Brick resist fire better than stones.
    4. Bricks of good quality shall have thin mortar bonds.

    Which of the above statements are correct?
    a) 1 and 2
    b) 3 and 4
    c) 2 and 3
    d) 1 and 4

[IES 2014]

54. The relation between the strength of brick masonry f_w, the strength of bricks f_b, and the strength of mortar f_m is given by (where K_w is a coefficient based on the layout of the bricks and the joints)

    a) $f_w = \sqrt{K_w \frac{f_b}{f_m}}$

    b) $f_w = K_w \sqrt{\frac{f_b}{f_m}}$

    c) $f_w = \sqrt{K_w f_b f_m}$

    d) $f_w = K_w \sqrt{f_b f_m}$

[IES 2015]

55. As per IS 3102–1965, for F1 class bricks, the percentage water absorption after 24 hours of immersion in cold water shall not exceed
    a) 20%
    b) 12%

c) 25%
d) 5%

[IES 2015]

56. Consider the following statements related to autoclave bricks:
    1. Less water absorption compared to other bricks.
    2. Noise reduction.
    3. It is cheap compared to other types of bricks.
    4. Requirement of bulk volume of mortar in joints being relatively less compared to other types of bricks masonry.
    5. Not recommended for outer walls.

Which of the above statements are relevant to the use of “autoclave” bricks?
a) 1, 2 and 4
b) 1, 3 and 5
c) 2, 3 and 4
d) 2, 4 and 5

[IES 2015]

57. Assertion (A): Fly ash bricks are used in construction as alternatives to burnt clay bricks.
    Reason (R): Fly ash bricks are lighter in weight and are stronger than burnt clay bricks.
    a) Both A and R are true and R is the correct explanation of A.
    b) Both A and R are true and R is not the correct explanation of A.
    c) A is true but R is false.
    d) A is false but R is true.

[IES 2015]

58. Assertion (A): Strength of brick wall is dependent on the type of bricks and the mortar used.
    Reason (R): Slenderness ratio of masonry decides the strength of the wall, and also mortar type to be used.
    a) Both A and R are true and R is the correct explanation of A.
    b) Both A and R are true and R is not the correct explanation of A.
    c) A is true but R is false.
    d) A is false but R is true.

[IES 2015]

59. Efflorescence of bricks is due to
    a) Excessive burning of bricks
    b) High silt content in brick clay
    c) High porosity of bricks
    d) Soluble salts present in parent clay

[IES 2016]

60. Disintegration of brick masonry walls is primarily due to
    1. Efflorescence
    2. Magnesium sulphate in bricks
    3. Calcined clay admixtures
    4. Kankar nodules

Which of the above statements are correct?
- a) 1, 2 and 3 only
- b) 1, 2 and 4 only
- c) 3 and 4 only
- d) 1, 2, 3 and 4

[IES 2016]

61. Statement (I): In general, bricks cannot be used in industrial foundations.
    Statement (II): Heavy duty bricks can withstand higher temperatures.
    - a) Both statement (I) and statement (II) are individually true and statement (II) is the correct explanation of statement (I).
    - b) Both statement (I) and statement (II) are individually true and statement (II) is not the correct explanation of statement (I).
    - c) Statement (I) is true but statement (II) is false.
    - d) Statement (I) is false but statement (II) is true.

[IES 2016]

62. Statement (I): In multi-storeyed constructions, burnt clay perforated bricks are used to reduce the cost of construction.
    Statement (II): Perforated bricks are economical and they also provide thermal insulation.
    - a) Both statement (I) and statement (II) are individually true and statement (II) is the correct explanation of statement (I).
    - b) Both statement (I) and statement (II) are individually true and statement (II) is not the correct explanation of statement (I).
    - c) Statement (I) is true but statement (II) is false.
    - d) Statement (I) is false but statement (II) is true.

[IES 2016]

63. Consider the following statements
    1. IS 3583 refers to Burnt clay paving bricks.
    2. IS 5779 refers to Burnt clay soling bricks.
    3. IS 3952 refers to Burnt clay hollow bricks.
    4. IS 2222 refers to Burnt clay lay bricks.

Which of the above statements are correct?
- a) 1, 2 and 3 only
- b) 1, 2 and 4 only

c) 3 and 4 only
d) 1, 2, 3 and 4

[IES 2017]

64. Consider the following statements regarding refractory bricks in furnaces:
    1. The furnace is fired at temperatures more than 1,700°C.
    2. Silica content in the soil should be less than 40%.
    3. Water absorption of bricks should not exceed 10%.
    4. Chrome bricks are known as basic bricks.

    Which of the above statements are correct?
    a) 1 and 2 only
    b) 2 and 4 only
    c) 1 and 3 only
    d) 3 and 4 only

[IES 2017]

65. Statement (I): Mud bricks can be completely replaced by Fly ash lime – gypsum bricks in a building.

    Statement (II): Useful fertile soil is used in manufacturing mud bricks, causing high CO_2 release in the atmosphere.
    a) Both statement (I) and statement (II) are individually true and statement (II) is the correct explanation of statement (I).
    b) Both statement (I) and statement (II) are individually true and statement (II) is not the correct explanation of statement (I).
    c) Statement (I) is true but statement (II) is false.
    d) Statement (I) is false but statement (II) is true.

[IES 2017]

66. Statement (I): Contemporarily, even in high-rise buildings, ordinary brick is being replaced by glass blocks for load-bearing walls.

    Statement (II): Bricks have high thermal conductivity and are not heat insulators.
    a) Both statement (I) and statement (II) are individually true and statement (II) is the correct explanation of statement (I).
    b) Both statement (I) and statement (II) are individually true and statement (II) is not the correct explanation of statement (I).
    c) Statement (I) is true but statement (II) is false.
    d) Statement (I) is false but statement (II) is true.

[IES 2017]

67. When the deposit of efflorescence is more than 10% but less than 50% of the exposed area of the brick, the presence of efflorescence is
    a) Moderate
    b) Slight

c) Heavy
d) Serious

[IES 2020]

68. Clay and silt content in a good brick earth must be at least
a) 20%
b) 50%
c) 35%
d) 70%

[SSC JE 2007]

69. The standard size of a modular brick is
a) 18 cm × 18 cm × 18 cm
b) 19 cm × 9 cm × 9 cm
c) 20 cm × 10 cm × 10 cm
d) 21 cm × 11 cm × 11 cm

[SSC JE 2008]

70. Crushing strength of first-class bricks should not be less than
a) 35 kg/cm^2
b) 70 kg/cm^2
c) 100 kg/cm^2
d) 150 kg/cm^2

[SSC JE 2009]

71. The size of modular brick is
a) 10 × 10 × 9 cm
b) 19 × 9 × 9 cm
c) 22.5 × 10 × 8.5 cm
d) 22.5 × 8.0 × 9 cm

[SSC JE 2010]

72. King closers are related to
a) Doors and windows
b) King post truss
c) Queen post truss
d) Brick masonry

[SSC JE 2011]

73. The water absorption for good brick should not be more than
a) 10% of its dry weight
b) 15% of its dry weight
c) 10% of its saturated weight
d) 15% of its saturated weight

[SSC JE 2012]

74. Clay bricks are made of earth having
    a) Nearly equal proportion of silica and alumina
    b) Nearly equal proportions of alumina silica and lime
    c) 35–70% silica and 10–20% alumina
    d) 10–20% silica and 35–70% alumina

[SSC JE 2012]

75. The plasticity to mould bricks in suitable shape is contributed by
    a) Alumina
    b) Lime
    c) Magnesia
    d) Silica

[SSC JE 2013]

76. The crushing strength of a first class brick is
    a) 3.5 N/mm^2
    b) 10.5 N/mm^2
    c) 5.5 N/mm^2
    d) 7.5 N/mm^2

[SSC JE 2013]

77. Strength-based classification of brick is made on the basis of
    a) IS: 3101
    b) IS:3102
    c) IS: 3495
    d) IS: 3496

[SSC JE 2014]

78. The compressive strength of common building bricks should not be less than
    a) 3.5 N/mm^2
    b) 5.5 N/mm^2
    c) 7.5 N/mm^2
    d) 10.5 N/mm^2

[SSC JE 2015]

79. The number of standard bricks in one cubic metre of brick masonry is
    a) 300
    b) 500
    c) 700
    d) 1,000

[SSC JE 2015]

80. The standard size of a masonry brick is

a) 18 cm × 8 cm × 8 cm
b) 18 cm × 9 cm × 9 cm
c) 19 cm × 9 cm × 9 cm
d) 19 cm × 8 cm × 8 cm

[SSC JE 2015]

81. The indentation provided in the face of the brick is called
    a) Frog
    b) Pallet
    c) Strike
    d) None of these

[SSC JE 2017]

82. A layer of dry bricks put below the foundation concrete, in the case of soft soils, is called
    a) Soling
    b) Shoring
    c) DPC
    d) None of these

[SSC JE 2017]

83. Hollow bricks are generally used with the purpose of
    a) Reducing the cost of construction
    b) Providing insulation against heat
    c) Increasing the bearing area
    d) Ornamental look

[SSC JE 2017]

84. The shape of the brick gets deformed due to rain water falling on hot brick. This defect is known as
    a) Chuffs
    b) Bloating
    c) Nodules
    d) Lamination

[SSC JE 2017]

85. The standard size of brick as per Indian standards is
    a) 20 cm × 10 cm × 10 cm
    b) 23 cm × 12 cm × 8 cm
    c) 19 cm × 9 cm × 9 cm
    d) 18 cm × 9 cm × 9 cm

[SSC JE 2017]

86. A brick masonry could fail due to
    a) Rupture along a vertical joint in poorly bonded walls

b) Shearing along a horizontal plane
c) Crushing due to overloading
d) Any of these

[SSC JE 2017]
87. Which of the following statements is correct?
a) Excess of alumina in the clay makes the brick brittle and weak.
b) Excess of alumina in the clay makes the brick crack and warp on drying.
c) Excess of alumina in the clay leaves high power deposit on the brick.
d) Excess of alumina in the clay improves impermeability and durability of the brick.

[SSC JE 2017]
88. The portion of the brick without a triangular corner is equal to half the width and half of the length is called
a) Closer
b) Queen closer
c) King closer
d) Squint brick

[SSC JE 2017]
89. When a brick is cut into two halves longitudinally, one part is called
a) King closer
b) Cornice brick
c) Queen closer
d) Voussoir

[SSC JE 2017]
90. The red colour obtained by the bricks is due to the presence of
a) Lime
b) Silica
c) Manganese
d) Iron oxide

[SSC JE 2017]
91. Which of the following is good for making the bricks?
a) Silted soil
b) Weathered clay
c) Soil
d) None of these

[SSC JE 2017]

92. The process of mixing clay, water and other ingredients to make bricks is known as
    a) Tempering
    b) Kneading
    c) Pugging
    d) Moulding

[SSC JE 2017]

93. Excess of silica in the clay
    a) Makes the brick brittle and weak
    b) Makes the brick crack and warp on drying
    c) Changes the colour of the brick from red to yellow
    d) Improves the impermeability and durability of the brick

[SSC JE 2017]

94. The term frog means
    a) An apparatus to lift the stone
    b) A depression on a face of brick
    c) Vertical joint in a brick work
    d) Soaking brick in water

[SSC JE 2017]

95. The minimum compressive strength of 2nd class bricks should be
    a) 70 kg/cm^2
    b) 90 kg/cm^2
    c) 100 kg/cm^2
    d) 120 kg/cm^2

[SSC JE 2017]

96. A pug mill is used for
    a) Softening brick earth
    b) Moulding brick earth
    c) Tempering brick earth
    d) Providing brick earth

[SSC JE 2017]

97. A bull nose brick is not used for
    a) Rounding off sharp corners
    b) Pillars
    c) Decoration purpose
    d) Arches

[SSC JE 2017]

98. The defect that is caused by falling of rain water on the hot surfaces of the bricks is known as
    a) Bloating
    b) Chuffs
    c) Cracks
    d) Lamination

[SSC JE 2018]

99. Which one of the following bricks is suitable for high-class brick masonry?
    a) Bullnose brick
    b) Jhumb brick
    c) Modular bricks
    d) Under burnt bricks

[SSC JE 2018]

100. In the composition of good bricks, the total content of silt and clay by weight should not be less than
    a) 20%
    b) 30%
    c) 50%
    d) 75%

[SSC JE 2018]

101. Refractory bricks are generally used to resist
    a) Chemical action
    b) Dampness
    c) High temperature
    d) Weathering action

[SSC JE 2018]

102. What is the percentage content of silica in a good-quality brick earth?
    a) 20–30%
    b) 30–40%
    c) 40–50%
    d) 50–60%

[SSC JE 2018]

103. Which of the following defect appears due o presence of alkalies in the bricks?
    a) Bloating
    b) Black core
    c) Cracks
    d) Efflorescence

[SSC JE 2018]
104. For which of the following is an ideal warehouse provided?
   a) Water proof masonry walls
   b) Water proof roof
   c) Few windows which remain generally closed
   d) All of these

[SSC JE 2018]
105. The water absorption (expressed in percentage) for burnt clay perforated bricks should not be more than
   a) 5
   b) 15
   c) 25
   d) 35

[SSC JE 2018]
106. What is the thickness of one-and-half-brick wall made up of standard modular brick?
   a) 20
   b) 30
   c) 40
   d) 50

[SSC JE 2018]
107. Which of the following is the most important characteristic of the alumina in the brick earth?
   a) Maintain plasticity
   b) Increase strength of bricks
   c) To manufacture impermeable
   d) Reduce wrapping when heated

[SSC JE 2018]
108. What is the thickness (cm) of a two-brick wall made up of standard modular brick?
   a) 9
   b) 10
   c) 20
   d) 40

[SSC JE 2018]
109. The expected out turn (cubic metre) of reinforced brickwork per mason per day is
   a) 1
   b) 3

c) 5
d) 10

[SSC JE 2018]
110. Efflorescence in bricks causes due to
a) Excessive burning of bricks
b) High content of silt in brick clay
c) High porosity of the bricks
d) Presence of soluble salt in parent clay

[SSC JE 2018]
111. Which one of the following minerals is responsible for red colour in bricks?
a) Iron oxide
b) Lime
c) Magnesia
d) Silica

[SSC JE 2018]
112. Tempering is the process used in the manufacture of
a) Bricks
b) Bitumen
c) Cement
d) Paints

[SSC JE 2018]
113. Which of the following is burnt in the Hoffman's kiln during the process of manufacturing?
a) Bitumen
b) Brick
c) Clinker
d) Varnishes

[SSC JE 2018]
114. The crushing strength off a first class brick is
a) 30 kg/cm^2
b) 55 kg/cm^2
c) 75 kg/cm^2
d) 105 kg/cm^2

[AE 2007]
115. The standard size of bricks as per Indian Standard is
a) $20\ cm \times 10\ cm \times 10\ cm$
b) $23\ cm \times 12\ cm \times 8\ cm$
c) $19\ cm \times 9\ cm \times 9\ cm$
d) $18\ cm \times 9\ cm \times 9\ cm$

[AE 2007]
116. Expansion joints in masonry walls are provided in wall lengths greater than
a) 10 m
b) 20 m
c) 30 m
d) 40 m

[AE 2007]
117. Which of the following scaffolding is suitable for stone masonry?
a) Single scaffolding
b) Double scaffolding
c) Cantilever scaffolding
d) Suspended scaffolding

[AE 2007]
118. The hardest stone available till now is
a) Diamond
b) Carborundum
c) Quartz
d) Marble

[AE 2007]
119. The head room in a staircase should not be less than
a) 3.5 m
b) 3 m
c) 2.1 m
d) 2 m

[AE 2007]
120. The arrangement of supporting an existing structure by providing supports underneath is known as
a) Shoring
b) Jacking
c) Pilling
d) Underpinning

[AE 2007]
121. A type of bond into a brick masonry consisting of alternate course of headers and stretchers are called
a) English bond
b) Flemish bond
c) Stretching bond
d) Heading bond

[AE 2007]
122. Herring-bone bond is used for
   a) Walls having thickness more than four bricks
   b) Architectural finish to the face work
   c) Environmental panels in brick flooring
   d) All of these

[AE 2007]
123. A brick laid with its length perpendicular to the face of the wall is called as
   a) Header
   b) Stretcher
   c) Bond
   d) Course

[AE 2008]
124. The construction of a temporary structure required to support an unsafe structure is called
   a) Underpinning
   b) Scaffolding
   c) Shoring
   d) Jacking

[AE 2008]
125. The stone masonry of finely dressed stones laid in cement or lime is
   a) Random rubble masonry
   b) Coursed rubble masonry
   c) Dry masonry
   d) Ashlar masonry

[AE 2008]
126. The vertical members fixed between steps and handrail are known as
   a) Balusters
   b) Strings
   c) Newel posts
   d) Soffits

[AE 2008]
127. The number of standard bricks required for 1 m^3 of brick masonry is
   a) 400
   b) 500
   c) 425
   d) 550

[AE 2008]

128. The standard size of masonry brick is
   a) 180 mm × 80 mm × 80 mm
   b) 190 mm × 90 mm × 90 mm
   c) 200 mm × 100 mm × 100 mm
   d) 210 mm × 110 mm × 110 mm

[AE 2008]

129. A good brick, when immersed in a water bath for 24 hours, should not absorb more than
   a) 20% of its dry weight
   b) 30% of its saturated weight
   c) 10% of its dry weight
   d) 20% of its saturated weight

[AE 2010]

130. Refractory bricks are specially manufactured to
   a) Withstand high temperature
   b) Withstand high crushing pressure
   c) Have high insulation against sound
   d) None of these

[AE 2010]

131. For rectangular foundation of width b, the eccentricity of load should not be greater than
   a) $\frac{b}{3}$
   b) $\frac{b}{4}$
   c) $\frac{b}{5}$
   d) $\frac{b}{6}$

[AE 2010]

132. Formation of whitish deposit on the bricks due to the presence of excess salts is called
   a) Efflorescence
   b) Disintegration
   c) Warping
   d) Floating

[AE 2010]
133. The size of a step commonly adopted for residential building is
   a) 25 cm × 16 cm
   b) 17 cm × 15 cm
   c) 30 cm × 13 cm
   d) 35 cm × 10 cm

[AE 2010]
134. Stucco plastering is used in
   a) Excellent finish
   b) X-ray rooms
   c) Sound proofing
   d) All of the above

[AE 2013]
135. Refractory bricks resist
   a) Chemical action
   b) Shocks and vibration
   c) Dampness
   d) High temperature

[AE 2013]
136. A stretcher bond is usually used for
   a) Half-brick wall
   b) One-and-half-brick wall
   c) Two-brick wall
   d) One-brick wall

[AE 2013]
137. A series of steps without any platform or landing is called
   a) Soffit
   b) Flight
   c) Pitch
   d) Nosing

[AE 2013]
138. A levelled horizontal mortar joint in a masonry wall is called
   a) Wall joint
   b) Bed joint
   c) Cross joint
   d) Bonded joint

[AE 2013]

139. A wall constructed to withstand the pressure of an earth filling is
   a) Parapet wall
   b) Sloping wall
   c) Buttress
   d) Retaining wall

[AE 2013]

140. When a brick is cut into two halves longitudinally, one part is called
   a) King closer
   b) Queen closer
   c) Cornice brick
   d) Bat

[AE 2013]

141. The window used with the objects of providing light and air to the enclosed space below the pitched roof is called
   a) Dormer window
   b) Corner window
   c) Bay window
   d) Clerestory window

[AE 2013]

142. A bat is the portion of a
   a) Wall between facing and backing
   b) Wall not exposed to weather
   c) Brick cut across the width
   d) Brick cut in such a manner that its one long face remains uncut

[AE 2013]

143. A type of bond in a brick masonry in which each course consists of alternate leaders and stretchers is called
   a) English bond
   b) Flemish bond
   c) Stretching bond
   d) Heading bond

[AE 2013]

144. Plastering material that is supplied in the form of rolls is
   a) Glass
   b) Cork
   c) Linoleum
   d) Mosaic

[AE 2014]

145. A sloping member which supports the steps in a stair is
    a) Stringer
    b) Carriage
    c) Flight
    d) Landing

[AE 2014]

146. Out of all the rules and specifications, choose the most important requirement with respect to brick masonry.
    a) All brick should be of uniform shape and size.
    b) The mortal joints should be as thin as possible.
    c) The vertical joints should not be contiguous and should be staged in a consecutive layer.
    d) The entire wall height should be constructed without a stoppage in between.

[AE 2014]

147. Masons who went to ensure the best quality and best workmanship in brick masonry will insist on using only
    a) Single-layer scaffolding with area supports on the wall itself
    b) Double-layer scaffolding which does not require the wall for its support
    c) Steel scaffolding instead of timber scaffold
    d) Moving scaffold

[AE 2014]

148. A type of scaffolding that can be provided on the side of a busy street without obstructing the traffic on the road is
    a) Single scaffolding
    b) Mason's scaffolding
    c) Ladder scaffolding
    d) Needle scaffolding

[AE 2014]

149. According to Indian standard institution, the size of modular brick excluding mortar thickness is
    a) 19 cm × 9 cm × 7 cm
    b) 19 cm × 9 cm × 9 cm
    c) 19 cm × 7 cm × 9 cm
    d) 20 cm × 10 cm × 10 cm

[AE 2015]

150. In a hostel construction at your college campus, different classes of bricks were used. Average compressive strength of first-class bricks was observed as 68% of the minimum value prescribed in the Indian standard. Moreover,

the ratio of the strength of first-class bricks to the strength of second class bricks was observed as 1.21. What is the strength of second class bricks .... in MPa?

a) 5.62
b) 5.95
c) 5.26
d) 5.92

151. Excess of lime in brick earth leads to .... and excess of magnesia leads to ...., respectively.

a) Melting of bricks during burning and Yellow colour bricks
b) Yellow colour bricks and melting of bricks during burning
c) Shrinkage of bricks and melting of bricks during burning
d) None of the above

1.	a	2.	c	3.	a	4.	c	5.	c	6.	b	7.	a	8.	b	9.	c	10.	b
11.	c	12.	d	13.	d	14.	b	15.	c	16.	d	17.	a	18.	c	19.	a	20.	b
21.	c	22.	c	23.	b	24.	d	25.	d	26.	c	27.	c	28.	c	29.	d	30.	d
31.	c	32.	d	33.	c	34.	d	35.	d	36.	d	37.	b	38.	c	39.	b	40.	d
41.	d	42.	d	43.	c	44.	a	45.	a	46.	c	47.	b	48.	b	49.	a	50.	d
51.	d	52.	b	53.	b	54.	d	55.	a	56.	a	57.	a	58.	b	59.	d	60.	b
61.	b	62.	a	63.	a	64.	c	65.	a	66.	d	67.	a	68.	b	69.	b	70.	c
71.	b	72.	d	73.	b	74.	c	75.	a	76.	b	77.	b	78.	a	79.	b	80.	c
81.	a	82.	a	83.	b	84.	a	85.	c	86.	d	87.	b	88.	b	89.	c	90.	d
91.	b	92.	b	93.	a	94.	b	95.	a	96.	c	97.	d	98.	b	99.	c	100.	c
101.	c	102.	d	103.	d	104.	d	105.	b	106.	b	107.	a	108.	d	109.	a	110.	d
111.	a	112.	a	113.	b	114.	d	115.	c	116.	d	117.	b	118.	a	119.	c	120.	d
121.	a	122.	d	123.	a	124.	c	125.	d	126.	a	127.	b	128.	b	129.	a	130.	a
131.	d	132.	a	133.	a	134.	a	135.	d	136.	b	137.	b	138.	b	139.	d	140.	b
141.	a	142.	c	143.	a	144.	c	145.	a	146.	c	147.	b	148.	d	149.	b	150.	a
151.	a																		

REFERENCES

Assistant Engineer (AE). 2020. Tamil Nadu Public Service Commission (TNPSC), Tamil Nadu, India. http://www.tnpsc.gov.in/previous-questions.html (accessed September 18, 2020).

Engineering Service Exam (ESE). 2020. Union Public Service Commission, New Delhi, India. https://www.upsc.gov.in/examinations/previous-question-papers (accessed September 18, 2020).

Graduate Aptitude Test in Engineering (GATE). 2020. GATE Office, Chennai, India. http://gate.iitm.ac.in/gate2019/previousqp18.php (accessed September 18, 2020).

IS 3495 Part 1. (2016). *Methods of tests of burnt clay building bricks: Determination of compressive strength*. Bureau of Indian Standards. New Delhi.

IS 3495 Part 2. (2016). *Methods of tests of burnt clay building bricks: Determination of water absorption*. Bureau of Indian Standards. New Delhi.
IS 3495 Part 3. (2016). *Methods of tests of burnt clay building bricks: Determination of efflorescence*. Bureau of Indian Standards. New Delhi.
IS 3495 Part 4. (2016). *Methods of tests of burnt clay building bricks: Determination of warpage*. Bureau of Indian Standards. New Delhi.
Junior Engineer (JE), Staff Selection Commission (SSC), New Delhi, India. https://ssc.nic.in/Portal/SchemeExamination (accessed September 18, 2020).

6 Lime

6.1 INTRODUCTION

Lime was used as a chief construction material for binding before the invention of cement. Lime is also used in current practices such as in repair work of monuments, in the stabilisation process, as an ingredient in unburnt bricks and paints, etc.

6.2 PROCESS OF MANUFACTURING

Similar to the primary source that is used for the manufacture of cement, calcium carbonate acts as a chief source for the manufacturing of lime. Calcium carbonate is heated at a higher temperature of 800°C. Decarbonation takes place that results in the formation of quicklime. Then the slacking of quicklime is prepared by the sprinkling of water over quicklime. Vertical kilns are used for burning the limestone. The system of the kiln can be a tunnel or flare shaped and either continuous or intermittent.

Based on the arrangement of fuel and limestone, the kiln can be classified as mixed-feed and separate-feed. Slacking of lime is completed within 10 minutes, once quicklime is sprinkled with water. Slacked lime is sieved by 3.5 mm IS sieve, and it can be used for many applications such as whitewashing, plastering, making mortars and lime putty. The process of slacking was done previously in the site; nowadays, it is done in a factory. Slacking of lime requires water equivalent to 32% of the weight of Calcium oxide.

6.2.1 Impurities in Lime

Some of the major impurities present in lime are

1. Magnesium carbonate – Slacking and setting processes are delayed due to its presence. On the other hand, it imparts up to 5% higher strength.
2. Clay – The presence of clay enhances the lime hydraulic property. An average of 10–30% of clay is added to impart hydraulicity to the lime, although it makes the lime insoluble in water.
3. Silica – The addition of silica results in poor cementing and hydraulicity.
4. Some of the different impurities are components of iron, carbonates, sulphates and alkalis.

6.2.2 Classification of lime

Based on the purpose of lime used in construction IS: 712 classifies lime as A, B, C, D, E and F.

1. Class A – Extremely hydraulic, used for making mortar, concrete for construction and foundation works, i.e. structural purpose
2. Class B – Semi-hydraulic, used for flooring, concrete in the ordinary construction, plaster undercoat and masonry mortar
3. Class C – Fat lime, used for finishing coat in plastering whitewashing and with pozzolana can be used for mortar
4. Class D – Dolomite, used for finishing coat in plastering and whitewashing
5. Class E – Kankar mortar, used for masonry mortar, plastering and whitewashing
6. Class F – Siliceous dolomite, used for undercoat and finishing coat of plaster

6.3 TESTING OF LIME

The quality of lime can be checked by conducting both lab tests and field tests. Some of the field tests are as follows:

1. Workability test – A handful of mortar mixed in the required proportion, say lime:sand in 1:3 proportion, is taken and thrown on the surface where it is used. Area and the quantity covered by the mortar are recorded. This quantity estimates the workability of mortar.
2. Impurity test – A known quantity of lime is mixed with water and the residue is dried in sun heat for 8 hours and weighed.
3. Ball test – Stiff lime is made into a ball and left for 6 hours and placed in a basin with water. The type of lime can be identified by expansion and disintegration.

6.3.1 Lab tests

1. Fineness test (IS: 6932 Part-4) – Sample is placed in sieves that are arranged from coarse to fine from top to bottom. 100 g of hydrated lime is placed on top of the sieve and washed gently for 30 minutes by passing a moderate jet of water. Then the residue in each sieve is dried at 100°C until the constant mass is achieved and weighed.
2. Setting time test – Vicat's apparatus is used to determine the initial and final setting time of hydrated lime.
3. Soundness test (IS: 6932 Part-9) – Le Chatelier's apparatus is used to identify the expansion or unsoundness or disintegration. This test is to identify the quality of the lime.

4. Volume yield of quicklime (IS: 6932 Part-6) – Consistency of the lime putty is adjusted in such a way that the slump obtained is 13 mm. Consistency can also be measured by a viscometer. Then, density vessel is used to weigh a known volume of this putty.
5. Strength test (IS: 6932 Part-7) – Both compression and transverse loading are performed on a lime mortar specimen. The compression test is done in the compression testing machine by placing a mortar specimen and loading at a rate of 150 N/min till the specimen breaks. Similarly, mortar specimen is placed in two rollers with a thin roller precisely in the middle of the specimen and the load is transferred uniformly at a rate of 150 N/min till the specimen breaks.

6.3.2 Precautions in storage and handling lime

Due care should be given for the storage of lime as they are extremely reactive under moisture. Sometimes air slacking can take place due to the moist atmosphere. This can be avoided by proper, compact insulated storage. Slacking of lime is highly exothermic and releases an enormous amount of dust and heat. Precautions such as wearing goggles, glows, respirators and skin cream can protect the people working with lime mortar.

6.4 TESTING PROCEDURES

6.4.1 Lime reactivity

Slacking characteristics of pulverised lime is used to identify the reactivity of quicklime.

Test Summary

This test method is used to identify the reactivity of un-slaked lime (quicklime), either by acid neutralisation (which produces a specific reactivity value (RDIN value)) or by measuring the rise in temperature in 3 minutes according to EN: 459-2019 (BS EN 459 Part 2, 2019).

Use and Significance

- Overall degree of reactivity of un-slaked lime can be identified and can be used to determine the total slaking period.
- Un-slaked lime is suitable for use as a binder.

Required Apparatus

- Oven capable of maintaining 1000°C
- Sieve of 0.5 mm size
- Calcinator that can maintain the temperature at 950°C
- Mechanical mixer or stirrer that can mix at a speed of 300 ± 10 rpm

- Thermostatically isolated vacuum flask or container of 100 mL capacity with an internal diameter of 77 mm and an internal height of 235 mm with a plastic lid capable of placing thermometer and sample
- Thermometer from 0°C to 100°C with a temperature division of 0.5°C
- Stopwatch
- Weighing balance
- Feeding vessel
- Graduated cylinder

Procedure

Slacking rate test or reactivity

Step 1. About 600 mL of distilled water is placed in a container and maintained at a temperature of 20°C.

Step 2. The thermometer and the stirrer are inserted. The stirrer is rotated at a speed of 300 ± 10 rpm. The temperature of the container is maintained at 20°C.

Step 3. By means of a feeding vessel, the prepared sample of weight 150 ± 0.1 g is added in the container with water, and simultaneously stopwatch is switched on.

Step 4. For every 30 seconds, the slaking time is recorded. Readings are recorded until the last three consecutive readings recorded are of not more than 0.5°C change.

Step 5. The time at which the first consecutive reading at which the temperature rise is not more than 0.5°C recorded is considered to be the end of the slacking period.

Step 6. If the slaking period exceeds 10 minutes, then one or two readings are recorded until the maximum temperature is reached.

Step 7. The slacking curve is drawn, with an increase in temperature (°C) on the y-axis and time (minutes) on the x-axis. The time at which the slacking temperature is reached can be determined from the graph.

Step 8. The reactivity index value (RDIN) is calculated by dividing 2400 (temperature rise of 40°C (from 20°C to 60°C) × 60 seconds/minute) by time in seconds required for the temperature rise to occur.

Step 9. An average of three repeats is considered to obtain reasonable repeat values.

Residue on slacking test

Step 1. Once the slacking test is completed, the lime solution is sieved through a 300 μm IS sieve.

Step 2. The residue collected on the IS sieve is heated in an oven at a temperature of 100°C for 3 hours and dried at room temperature.

Step 3. The cooled residue is brushed and the coarser particles remained in the sieve are weighed. Grit is calculated as a percentage of the original mass of lime.

Caution**:** Care should be taken while handling quicklime. Quicklime in the vessel should be mixed thoroughly through the test.

Observations and calculations

Step 1. The slaking period is the time at which the first of three consecutive readings is taken (where the temperature rise is not more than 0.5°C).

Step 2. The difference between the periods at which the temperature rises from 20°C to 60°C can be calculated by noting down the final time at which the temperature reaches 60°C.

Step 3. Lime slacking rate can be calculated as per observations from Table 6.1.

Step 4. Lime slacking period for the residue can be calculated as per observations from Table 6.2.

TABLE 6.1
Lime Slacking Rate

Time (minute)	0.0	0.5	1.0	1.5	2.0	2.5	3.0	3.5	4.0	4.5	5.0
Temperature (°C)											
Time (minute)	5.5	6.0	6.5	7.0	7.5	8.0	8.5	9.0	9.5	10.0	10.5
Temperature (°C)											

TABLE 6.2
Observation for Lime Slacking Period and Residue

	Control specimen		
Sample no.	**1**	**2**	**3**
Temperature (°C)			
Time taken for 40°C rise in temperature (minutes)			
Temperature rise in 3 minutes (°C)			
Total temperature rise (°C)			
Mass of sample m_1 (g)			
Mass of dried residue m_2 (g)			
Residue on slacking $\frac{m_2}{m_1} \times 100$			

TABLE 6.3
Recommended Criteria for Lime Reactivity According to ASTM

Lime reactivity	Criteria	Recommendation
High	1. Temperature increase of 40°C in 3 minutes 2. Total slacking period less than 10 minutes	Acceptable
Medium	1. Temperature increase of 40°C in 3–6 minutes 2. Total slacking period between 10 minutes and 20 minutes	Chemical analysis of the sample is required
Low	1. Temperature increase of 40°C in 6 minutes or more 2. Total slacking period more than 20 minutes	Chemical analysis of the sample is required

Result

***Caution*:** The fineness of the lime to be used is to be identified, since one of the major physical properties that influence the reactivity of the lime is its fineness.

Specification: Table 6.3 shows the recommended criteria for lime reactivity, according to ASTM.

Student Remark

....
....
....
....

$$\frac{\textit{Marks obtained}}{\textit{Total marks}} = -$$

Instructor signature

Instructor remark

....
....
....
....

6.4.2 Soundness test

Soundness test is performed in order to determine the quality of lime against its disintegration.

Test Summary

This test method is used to sample and investigate the slacking property of quicklime by accelerating the slacking process by means of applied heat. Using Le-Chatlier's apparatus expansion of the lime is checked (IS 6932 Part 9, 2019).

Use and Significance

- Presence of excess of over burnt lime in hydrated lime can be identified.
- Soundness test is indicative of the appearance of potential surface defects in plastering applications.

Required Apparatus

- Weighing balance
- Le Chatelier's apparatus
- Glass plates
- Water bath – temperature controlled
- Ruler
- Porcelain dish for mixing
- Graduated cylinder
- Trowel
- IS sieve of 850 μm

Procedure

Step 1. Test sample is proportioned in a ratio of 1:3:12 (Cement: hydrated lime: standard sand).

Step 2. The test sample is mixed with 12% of the mass of water (taken by means of the dry sample).

Step 3. Three well-greased moulds are placed over a non-porous glass plate with the sample filled in the moulds maintaining the edges of the moulds to be visible.

Step 4. A non-porous glass plate is placed over the moulds with a weight placed above the glass plate, as shown in Figures 6.1 and 6.2.

Step 5. Initial indicator point (D_1) is noted down after an hour and the moulds are transferred to a damp air cupboard for 48 hours.

Step 6. Moulds are removed from the cupboard and placed in a water bath (boiling vigorously) for 3 hours. Ensure that the moulds are not immersed entirely in the boiling water.

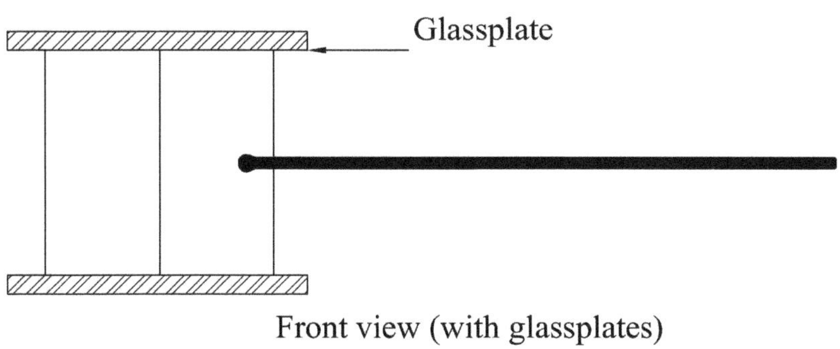

FIGURE 6.1 Le Chatelier's apparatus (front view).

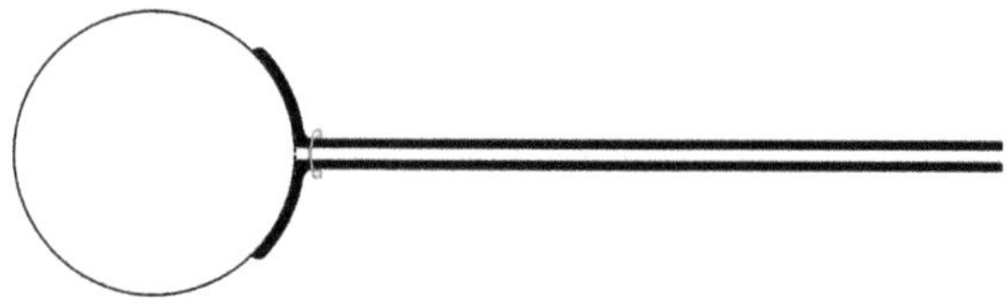

Top view (without glassplates)

FIGURE 6.2 Le Chatelier's apparatus (top view).

Step 7. Moulds are allowed to cool and the distance between the indicator is measured (D_2).

Step 8. The difference between (D_2) and (D_1) gives the total expansion of the sample due to cement and lime. To get the expansion of lime, the initial expansion of 1 mm is subtracted from the difference between (D_2) and (D_1).

Caution: Samples should be handled with care since hydrated lime and quicklime have a higher affinity to react with moisture and CO_2.

Observations and Calculations

	Trail 1	Trial 2	Trail 3
Initial distance between indicators (D_1) (mm)			
Final distance between indicators (D_2) (mm)			
Expansion of lime ($D_2 - D_1 - 1$) (mm)			

Result

Specification

According to IS, the net expansion of lime should not be more than 10 mm (IS 6932 Part 9, 2019).

Student Remark

....

....

....

....

$$\frac{\textit{Marks obtained}}{\textit{Total marks}} = -$$

Instructor signature

Instructor remark

....

....

……………………………………………………………………………………………
……………………………………………………………………………………………

6.5 PRACTICE QUESTIONS AND ANSWERS FROM COMPETITIVE EXAMS

1. Fineness of lime is determined by
   a) Blaine's air permeability apparatus
   b) Le Chatelier's apparatus
   c) Vicat's apparatus
   d) None of the above

2. Soundness of lime is determined by
   a) Le Chatelier's apparatus
   b) Vicat's apparatus
   c) Compression testing machine
   d) None of the above

3. The primary source for manufacturing of lime is
   a) Tri-calcium silicate
   b) Calcium carbonate
   c) Di-calcium silicate
   d) All the above

4. How quicklime is formed
   a) Decarbonation of calcium hydroxide
   b) Decarbonation of magnesium hydroxide
   c) Decarbonation of calcium carbonate
   d) None of the above

5. On sprinkling water on quicklime which of the following is formed
   a) Slacked lime
   b) Hydraulic lime
   c) Calcium carbonate
   d) Fat lime

6. The percentage of water required to slack quicklime is
   a) 22% by weight
   b) 32% by weight
   c) 42% by weight
   d) None of the above

7. Presence of magnesium carbonate in lime results in
   a) Delayed setting process
   b) Enhanced hydraulic property

c) Poor cementing property
d) None of the above

8. Which of the following results in enhanced hydraulic property in lime
a) Magnesium carbonate
b) Silica
c) Clay
d) None of the above

9. Stiff lime is made into a ball and left for 6 hours and placed in a basin with water. The type of lime can be identified by expansion and disintegration. This test is called
a) Impurity test
b) Workability test
c) Both a and b
d) Ball test

10. According to ASTM, if the temperature of the lime increases to 40°C in 3 minutes and the total slacking period is less than 3 minutes, then the reactivity of the lime is
a) Low
b) High
c) Medium
d) None of the above

11. Lime mortar is generally made with
a) Quick lime
b) Fat lime
c) Hydraulic lime
d) White lime

[IES 1995]

12. High alumina cement is produced by fusing together a mixture of
a) Limestone and bauxite
b) Limestone, bauxite and gypsum
c) Limestone, gypsum and clay
d) Limestone, gypsum, bauxite, clay and chalk

[IES 1997]

13. A gauged mortar is obtained by adding which of the following ingredient(s) to cement?
a) Sand alone
b) Sand and surkhi
c) Sand and lime
d) Surkhi alone

[IES 1998]

14. Surkhi is added to lime mortar to
    a) Prevent shrinkage
    b) Decrease setting time
    c) Increase bulk
    d) Impart hydraulicity

[IES 1999]

15. Assertion (A): Lime mortar can retain its bond with masonry unit and be free from cracks.
    Reason (R): Lime mortar undergoes only negligible volume change after setting and initial shrinkage.
    a) Both A and R are true, and R is the correct explanation of A
    b) Both A and R are true, and R is not the correct explanation of A
    c) A is true but R is false
    d) A is false but R is true

[IES 2000]

16. A mortar for which both cement and lime are mixed is called
    a) Gauged mortar
    b) Cement mortar
    c) Lime mortar
    d) Lightweight mortar

[IES 2000]

17. One of the main demerits in using the lime mortar is that it
    a) Is not durable
    b) Does not set quickly
    c) Swells
    d) Is plastic

[IES 2001]

18. Why is lime added to cement slurry for the topcoat of plastering?
    a) To improve the strength of plaster
    b) To stiffen the plaster
    c) To smoothen the plaster for ease of spread
    d) To make the plaster non-shrinkable

[IES 2007]

19. Consider the following statements about lime.
    1. Calcination of limestone results in quicklime.
    2. Lime produced from a pure variety of chalk is hydraulic lime.
    3. Hydrated lime is obtained by treating quicklime with water.

    Which of the above statements are correct?
    a) 1, 2 and 3

b) 1 and 2 only
c) 2 and 3 only
d) 1 and 3 only

[IES 2017]

20. Statement (I): Lime surkhi mortar is used in the construction of Anicuts (dams) since the nineteenth century.

Statement (II): Portland cement is a recent material compared to surkhi mortar, which is best suited for hydraulic structures.

a) Both statement (I) and statement (II) are individually true and statement (II) is the correct explanation of statement (I).
b) Both statement (I) and statement (II) are individually true and statement (II) is not the correct explanation of statement (I).
c) Statement (I) is true but statement (II) is false.
d) Statement (I) is false but statement (II) is true.

[IES 2018]

21. Which one of the following limes will be used for finishing coat in plastering and whitewashing?
a) Semi-hydraulic lime
b) Kankar lime
c) Magnesium/dolomitic lime
d) Eminently hydraulic lime

[IES 2020]

22. Pick up the correct statement
a) Lime is available in free state.
b) Lime is made from gypsum.
c) Lime is made from dolomite/calcium carbonate.
d) None of the above

[SSC JE 2007]

23. The commonly used lime in whitewashing is
a) Quicklime
b) Fat lime
c) Hydraulic lime
d) All of the above

[SSC JE 2010]

24. Which of the following type of lime is used for underwater constructions?
a) Fat lime
b) Quicklime
c) Slaked lime
d) Hydraulic lime

[SSC JE 2014]

25. The purpose of the soundness test of cement is
    a) To determine the presence of free lime
    b) To determine the setting time
    c) To determine the soundproof quality of cement
    d) To determine the fineness

[SSC JE 2017]

26. The lime which has the property of setting in water is known as
    a) Fat lime
    b) Hydraulic lime
    c) Hydrated lime
    d) Quicklime

[SSC JE 2017]

27. The quicklime as it comes from the kiln is called
    a) Milk lime
    b) Hydraulic lime
    c) Lump lime
    d) Hydrated lime

[SSC JE 2017]

28. For construction of structures underwater, the type of lime used is
    a) Hydraulic lime
    b) Fat lime
    c) Quicklime
    d) Pure lime

[SSC JE 2017]

29. Soundness test of cement determines
    a) Quantity of free lime
    b) Ultimate strength
    c) Durability
    d) Initial setting

[SSC JE 2017]

30. The lime which contains a high percentage of calcium oxide is generally called
    a) Fat lime
    b) Rich lime
    c) White lime
    d) All of these

[SSC JE 2017]

31. Lime putty

a) Is made from hydraulic lime
b) Is made by adding lime to water
c) Can be used only up to three days
d) All options are correct

[SSC JE 2017]

32. The maximum percentage of chemical ingredient of cement is
    a) Magnesium oxide
    b) Iron oxide
    c) Aluminium
    d) Lime

[SSC JE 2017]

33. In a lime cement plaster, ratio 1: 1: 6 corresponds to
    a) Lime : Cement : Sand
    b) Cement : Lime : Sand
    c) Lime : Sand : Gravel
    d) Cement : Sand : Gravel

[SSC JE 2018]

34. The calcination of pure lime results in:
    a) Quick lime
    b) Hydraulic lime
    c) Hydrated lime
    d) Fat lime

[SSC JE 2018]

35. The process of adding water to lime to convert it into a hydrated lime is termed as
    a) Watering
    b) Baking
    c) Hydration
    d) Slaking

[SSC JE 2018]

36. Which of the following is fused together to prepare the high alumina cement?
    a) Bauxite and limestone
    b) Bauxite limestone and gypsum
    c) Limestone gypsum and clay
    d) Limestone, bauxite gypsum, clay and chalk

[SSC JE 2018]

37. What is the main reason to use lime in the cement slurry during the plastering in the top coat?

a) To make the surface bright
b) To harden the cement
c) To make the plaster non-shrinkable
d) To improve the workability of plaster

[SSC JE 2018]

38. In plastering the first coat is called .... and its thickness should be .... mm.
    a) Undercoat, 6–9
    b) Floating coat, 6–9
    c) Floating coat, 0–15
    d) Under coat, 10–15

[SSC JE 2018]

39. Which of the following statements is true?
    A. Mud plastering does not require curing.
    B. Mud plastering requires curing.
    C. It depends on the situation.
        a) Only A
        b) Only B
        c) Only C
        d) None of these

[SSC JE 2018]

40. .... is used to ensure that the thickness of plastering is uniform.
    a) Bull point
    b) Pivot point
    c) Bull mark
    d) Bench mark

[SSC JE 2018]

41. Plaster of Paris is obtained from the calcination of
    a) Kankar
    b) Limestone
    c) Dolomite
    d) Gypsum

[AE 2007]

42. Consider the following statements:
    Assertion (A): Pure lime takes a long time to develop adequate strength.
    Reason (R): Pure lime has slow hardening characteristics.
    a) Both A and R are true and R is the correct explanation of A
    b) Both A and R are true and R is not the correct explanation of A
    c) A is true but R is false
    d) A is false but R is true

[AE 2007]
43. Lime suitable for making motor of good strength is
   a) Hydraulic lime
   b) Fat lime
   c) Lean lime
   d) None of these

[AE 2007]
44. Le Chatelier's apparatus is used to
   a) Carry out consistency test
   b) Carry out tensile test
   c) Carry out soundness test
   d) Determine compressive strength

[AE 2008]
45. The property by virtue of which lime sets under water is known as
   a) Setting
   b) Slacking
   c) Hydraulicity
   d) Hydration

[AE 2008]
46. The process of adding water to lime to convert it into hydrated lime is termed as
   a) Watering
   b) Baking
   c) Hydration
   d) Slaking

[AE 2010]
47. The process of burning the limestone to redness in contact with air is termed as
   a) Carbonation
   b) Oxidation
   c) Hydration
   d) Calcination

[AE 2013]
48. Lime mortar is generally made with
   a) Quick lime
   b) Fat lime
   c) Hydraulic lime
   d) White lime

[AE 2013]

49. Hydraulic lime has small quantities of
    a) Silica
    b) Alumina
    c) Iron oxide
    d) All of the above

[AE 2013]

50. Hydrated lime is also called .... lime.
    a) Slaked
    b) Calcined
    c) Fat
    d) None of the above

1.	a	2.	a	3.	b	4.	c	5.	a	6.	b	7.	a	8.	c	9.	d	10.	b
11.	c	12.	a	13.	c	14.	d	15.	a	16.	a	17.	b	18.	c	19.	d	20.	b
21.	c	22.	c	23.	b	24.	a	25.	a	26.	b	27.	c	28.	a	29.	a	30.	d
31.	b	32.	d	33.	b	34.	a	35.	d	36.	a	37.	c	38.	d	39.	a	40.	c
41.	d	42.	a	43.	a	44.	c	45.	c	46.	d	47.	d	48.	c	49.	d	50.	a

REFERENCES

BS EN 459 Part 2. (2019). *Building lime: Test methods*. UK.

IS 6932 Part 9. (2019). *Methods of tests for building limes: Determination of soundness*. Bureau of Indian Standards. New Delhi.

Graduate Aptitude Test in Engineering (GATE). (2020). *GATE Office*, Chennai, India. http://gate.iitm.ac.in/gate2019/previousqp18.php (accessed September 18, 2020).

Engineering Service Exam (ESE). (2020). *Union Public Service Commission*, New Delhi, India. https://www.upsc.gov.in/examinations/previous-question-papers (accessed September 18, 2020).

Assistant Engineer (AE). (2020). Tamil Nadu Public Service Commission (TNPSC), Tamil Nadu, India. http://www.tnpsc.gov.in/previous-questions.html (accessed September 18, 2020).

Junior Engineer (JE), *Staff Selection Commission (SSC)*, New Delhi, India. https://ssc.nic.in/Portal/SchemeExamination (accessed September 18, 2020).

7 Bitumen

7.1 INTRODUCTION

Bitumen is obtained in the fractional distillation of crude petroleum as an end product. Tar is obtained from the distillation of organic substances such as wood and coal.

7.2 CHEMICAL COMPOSITION OF BITUMEN

Bitumen has a complicated and variable composition. Chemical composition changes from bitumen to bitumen and is dependent on the origin of the crude oil. For that reason, a thorough chemical analysis of bitumen is difficult. Nevertheless, bitumen is considered a high molecular weight hydrocarbon and is generally represented as $CnH_{2n+b}X_d$, where X represents elements such as sulphur, nitrogen, oxygen or trace metals.

7.3 STRUCTURE OF BITUMEN

Bitumen consists of three main components in its structure: asphaltenes, resins and oils. Asphaltenes are solid particles in the bitumen and they are dissolved in the oil. Resins are the interface between asphaltenes and oil. Asphaltenes impart strength and stiffness to the bitumen. Oil imparts viscosity and fluidity to the bitumen based on temperature. Resins impart ductility and the adhesion property to the bitumen. Asphaltenes are dispersed particles, whereas oils are dispersant. Resins are interfacial.

7.4 CLASSIFICATION OF BITUMEN BASED ON STRUCTURE

Based on structure, it is further categorised as (1) sol, (2) gel and (3) sol-gel. If asphaltenes are widely dispersed (randomly), then it is called as sol bitumen. If asphaltenes are arranged in a three-dimensional network through molecular attraction, then it is called gel bitumen. If the behaviour is intermediate, then it is called as sol-gel bitumen; it means that the asphaltenes are arranged far and wide as well in the three-dimensional network.

7.5 FACTORS INFLUENCING PROPERTIES OF BITUMEN

Temperature and manufacturing methods are the two primary factors. If temperature increases, then asphaltenes dissolve in the oil and resins, and hence bitumen becomes less viscous. If the temperature decreases, then the asphaltene becomes less soluble and bound in the ordered structure. Therefore, the viscosity of the bitumen is increased. If the temperature is decreased, lower than the glass transition temperature, then the structure of bitumen is frozen, and the material becomes rigid and brittle.

7.6 MODIFIED BITUMEN

When modifiers are added to enrich the properties of bitumen, it is known as modified bitumen. Modifiers aid in improving the bitumen quality and the following benefits can be achieved: better flexibility, improved durability, superior crack resistance, enhanced resistance to UV rays, better resistance to daily and seasonal temperature variations. There are three commonly used modifiers:

1. Styrene-butadiene rubber (SBR) (TR)
2. Styrene butadiene styrene (SBS)
3. Ethylene vinyl acetate (EVA)

SBS and SBR are available in powder form and pellet form. 20% Ethylene and 80% vinyl acetate are called EVA. EVA is available in pellet form.

7.6.1 Crumbed Rubber Modified Bitumen (CRMB)

Disposed tyres are crushed into powder form and this powder is added as a modifier in the bituminous mix to enhance properties. This bitumen is known as CRMB. For the production of CRMB, the following method is adopted. Aggregates are heated to 190–210°C. The crumbed rubber particles are directly added to the aggregates and mixed for 10 seconds prior to the addition of bitumen. The heated bitumen (145–165°C) is added to the mix. Bitumen leaning temperature is 190–210°C. Afterwards, digestion time is adopted as 60 minutes. The rubber particle absorbs light fraction oily component and becomes a gel-like structure. Therefore, the quality of bitumen is enhanced.

7.6.2 Blown Bitumen

Bitumen is heated until it becomes liquid and the air is passed under pressure to drive out the volatile compounds present in the bitumen. Therefore, bitumen becomes hard at room temperature. It is known as blown bitumen.

7.6.3 Cutback Bitumen

Generally, bitumen is heated to lower the viscosity. In cutback bitumen, an appropriate solvent is added to lower the viscosity of bitumen instead of heating. From the environmental point of view, cutback bitumen is favoured. In the cutback bitumen, 80% is residual bitumen whereas only 40–60% bitumen is followed in bitumen emulsion. When cutback bitumen is applied, then solvent material is evaporated and bitumen is available to bind the aggregates. The evaporation time is called curing time of cutback bitumen. There are three types of cutback bitumen:

1. Rapid curing cutback bitumen
2. Medium curing cutback bitumen
3. Slow curing cutback bitumen

Different types of solvents are used, and most commonly used solvents are naphtha, kerosene, gasoline and white spirit.

7.6.4 Bitumen Emulsion

At present, 20% of bitumen used worldwide is bitumen emulsion. A dispersion of bitumen particles in water that is stabilised with the help of surfactants (emulsifiers) is known as bitumen emulsion. The bitumen content in the emulsion is about 60%. There are three types of bitumen emulsion. They are rapid curing bitumen emulsion, medium curing bitumen emulsion, slow curing bitumen emulsion. In the production of bituminous emulsion, water is treated with emulsifying agents. Bitumen is crushed into tiny particles and added to water. Emulsifying agents migrate to the bitumen water interface and separate bituminous particles. Emulsifying agents are the chemicals used to keep billions and billions of bitumen drops separated from one another. Bitumen emulsions are divided into two:

1. Anionic with negatively charged globules
2. Cationic with positively charged globules

7.7 PAVEMENT BINDERS

Pavements are pathways in which the graded aggregates are combined with binders at higher temperatures to attain a suitable bituminous mix. Pavements can be classified into the flexible pavement and rigid pavements. Flexible pavements are constructed with graded aggregates that are bound together by a suitable binder (bitumen). Bitumen is one of the most complex construction materials. Commonly used binders in the flexible pavement are

1. Natural bitumen
2. Tar
3. Manufactured bitumen

Natural bitumen is further classified into two types depending upon its origin as lake bitumen or rock bitumen.

Tar is another binder used in the construction of flexible pavement. However, it is rarely preferred in current practices. Tar is obtained by the destructive distillation of coal. Due to its severe harmful gas emission effect, its usage has been proscribed by several government agencies all over the world.

Fractional distillation of crude oil gives manufactured bitumen. These are the most extensively used binders for the construction of flexible pavement. Almost 1,500 various crudes have been manufactured throughout the world. Depending upon the yield and quality, only certain crudes are used for the production of manufactured bitumen.

7.7.1 Distress in Bituminous Pavements

a. Fatigue cracking
b. Rutting
c. Bleeding
d. Permeability and Stripping
e. Weathering
f. Ageing
g. Polishing
h. Water bleeding and pumping
i. Thermal cracking
j. Shoving
k. Ravelling

   a. **Fatigue cracking:** If very stiff bitumen is used in the flexible pavement and load is heavy, then fatigue cracking may occur.
   b. **Rutting:** Permanent deformation of the pavement under the wheel path leads to cracking. This permanent deformation is called rutting.
   c. **Bleeding:** Moving up of bitumen to the top surface of the flexible pavement is called bleeding; a shiny, black surface carried by liquid asphalt migrates to the top of the road and leads to damage of the surface texture. Reasons for bleeding are (1) high bitumen content and (2) poor mix design.
   d. **Permeability and stripping:** Water infiltrates into the pavement if the permeability of bituminous concrete is high. Air diffuses into the bituminous pavement and oxidises light fractions, and as a result, hardening of bitumen also occurs. When water is permeated and available at the interface of bitumen and aggregate, it leads to stripping. Under wheel pressure, water migrates at the interface and separates the aggregates. It means that the binding between aggregate and bitumen is lost. This is called stripping.
   e. **Weathering:** Bitumen becomes more brittle when it is exposed to UV rays.
   f. **Ageing:** Bitumen becomes harder and less ductile with respect to time due to the changes in the structure and composition. This is known as ageing. Due to ageing, two significant processes occur in bitumen. They are surface hardening and hardening of the whole mass. Oxidation of some light oils proximate to the surface leads to hardening of surface bitumen. It commences with surface and progressively spreads into the bitumen. Hardening of the entire mass occurs because of the condensation of molecules.

g. **Polishing:** If soaked aggregates are used in heavy traffic conditions, then the aggregates are polished under wheel loads. If bleeding happens, the bitumen has been detached from the top surface due to continuous vehicle movement. As a result, aggregates get exposed to the traffic and get polished.
h. **Water bleeding and pumping:** Leakage of water from any adjacent water source (pipeline) on to the pavement is called water bleeding and pumping.
i. **Thermal cracking:** Due to the temperature variation between the top surface and bottom surface of bituminous pavements, cracking occurs.
j. **Shoving:** Shoving is the deformation of the pavement surface. It is commonly observed in the pavement adjacent to the traffic signals, where most braking incidents occur. It leads to the formation of a wavy surface on the top of the road. The reasons for shoving are poor mix design, braking and stopping and slippage between layers.
k. **Ravelling:** Loose aggregates that ravel from the surface or edges of the flexible pavements cause moisture damage to the road. Water may further permeate through the damaged surface and loosen aggregates, which may cause severe problems.

7.8 PHYSICAL PROPERTIES, CHEMICAL PROPERTIES AND GRADING OF BITUMEN

The quality of the pavement is widely influenced by bitumen. The chemical composition of bitumen consists mostly of hydrocarbons (hydrogen and carbon) along with a small trace of heterocyclic compounds such as sulphur, nitrogen and oxygen. Bitumen can be further sub-divided into two types as asphaltenes and maltenes based on its complex chemical composition. Maltenes can be further sub-divided into saturated hydrocarbons, aromatic hydrocarbons and resins. This separation can be carried out by the following methods.

1. Solvent extraction
2. Chromatography
3. Adsorption by finely divided solid and removal of un-adsorbed solids by the filtration
4. Molecular distillation used in conjunction with one of the other techniques

Different types of empirical test methods have evolved with time. Grading of bitumen is prepared by different methods; some of the conventional methods used are:

1. Penetration
2. Softening point
3. Viscosity
4. Ductility

5. Specific gravity
6. Flashpoint and fire point

Based on different properties, bitumen can be classified as (1) paving grade bitumen, (2) hard paving grade bitumen and (3) oxidised bitumen.

Different properties classify grading of bitumen. For example, as determined by consistency test, if the consistency of the bitumen is intermediate, then the bitumen is said to be of paving grade. Other tests include softening point or viscosity test to determine consistency at other temperature, resistance to hardening; durability test; Fraass breaking test, which determines the brittleness of bitumen at low service temperatures and penetration index that gives the temperature-dependent consistency. All the above tests are determined in order to know the usage and requirements of bitumen depending upon the wide temperature variations.

Hard pavement bitumen gives relatively higher performance than paving grade bitumen. Hard pavement bitumen is used in roads, especially in pavements with high traffic flow and where the temperature is high and intermediate.

Oxidised bitumen is used in roofing, waterproofing, adhesives and insulations.

Based on the requirements and suitability, bitumen can be graded by three different methods. They are:

1. Performance graded bitumen
2. Viscosity graded
3. Penetration graded

Standards are available based on the continuous research programmes led by the US government in 1993. A programme called Strategic Highway Research Program gave specifications based on performance graded bitumen called as super pave (superior paving performance) specifications. The Asphalt Institute introduced this specification (SP-1), which was later converted into the AASHTO M320 and ASTM D 6373 2007 specifications.

7.9 TESTING PROCEDURES

7.9.1 Softening Point of Bitumen

Softening point of bitumen is useful in understanding the degree of softening at a specific temperature. Materials such as bitumen are visco-elastic in nature and these materials do not possess a definite melting point. Because of their visco-elastic nature, these materials will have a range of melting points.

Test Summary

This test method is used to understand the softening point of the bitumen using the ring and ball experiment with the specimen immersed in distilled water. This test is a measure of temperature at which the steel ball passes through the bitumen sample in a mould and falls through a height of 25 mm when heated underwater at the specified condition of the test (IS 1205, 2019).

Use and Significance

- The softening point of visco-elastic material gradually reduces with respect to the increase in temperature. Usage of such kind of material as a building material requires caution during its service life.
- This test is also used as a quality check test to assess the uniformity of the material.
- Determining the softening point of bitumen is used to have an understanding of the temperature required for heating the bitumen for various road applications.

Required Apparatus

- Ring and ball apparatus – It consists of a metal frame having three plates
- The top plate consists of a hole to place a thermometer, the middle one consists of two big holes at the middle for placing the ring and the bottom plate is flat
- The distance between the bottom surface of the ring placed and the top surface of the bottom plate is 25 mm; cylindrical ring is tapered with a diameter of 17.5 mm at top and height of 6.4 mm
- The centring guide with three pins forms an imaginary circle of a diameter slightly more than the diameter of steel ball (7.5 mm) to enable easy movement of steel ball
- Steel ball of diameter 7.5 mm and weight 3.5 grams
- Two numbers of steel balls and centring guides
- Centring guide and ring can be placed coaxially within each other and placed in the middle plate
- A heat resistance beaker of 600 mL
- A thermometer of range 120°C and an accuracy of 0.5°C
- A hot plate connected with a heat regulator and a stirrer to maintain the uniform temperature in the bar

Sampling

Step 1. The bitumen sample is heated to a temperature of 75–100°C above the softening point. Simultaneously, an equal portion of glycerol and dextrin are mixed in a glass plate.

Step 2. The mixture is applied on the surface to avoid sticking of the bitumen to the surface of the plate.

Step 3. Once the bitumen acquires desired consistency, it is poured in the rings. The sample is allowed to cool for 30 minutes in the air. A hot short-edged knife is used to remove the excess bitumen.

Testing

Step 1. Distilled water at 5°C is poured inside a beaker. The beaker is filled in such a way that the water surface is about 50 mm above the level of the specimen.

Step 2. The ring and the ball guide are assembled and placed on the metallic frame (in the middle plate). Instantly, the whole assembly is placed inside the beaker with distilled water and left for 15 minutes.

Step 3. After 15 minutes, steel balls previously cooled to 5°C are placed on the top of the ball guide by lifting the assembly outside the beaker and the whole assembly is placed back inside the beaker immediately.

Step 4. Now, the whole assembly is placed over the hot plate with a stirrer. The hot plate is switched on and the thermometer is inserted.

Step 5. Heating should be done in such a way that the rate of rising in temperature is 5°C per minute. The rate of increase in temperature is controlled by the heat regulator.

Step 6. Bitumen starts to soften as the temperature increases. At a specific temperature the steel ball, due to its self-weight, starts falling down.

Step 7. The temperature at which the steel ball falls down and just touches the bottom plate is considered to be the softening point of the bitumen. Temperature can be recorded from the thermometer to the accuracy of 0.5°C.

Step 8. Two readings are recorded from the test and the average of the two readings is the softening point of the bitumen.

Step 9. The ring and ball apparatus and the setup is shown in Figures 7.1, 7.2, 7.3 and 7.4.

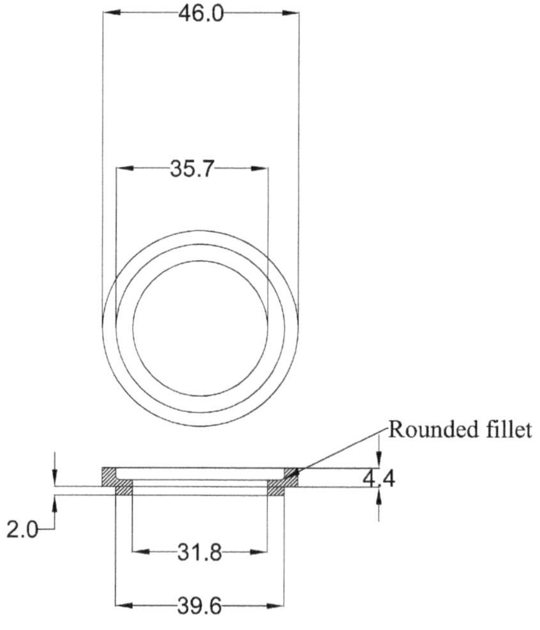

FIGURE 7.1 Shouldered ring.

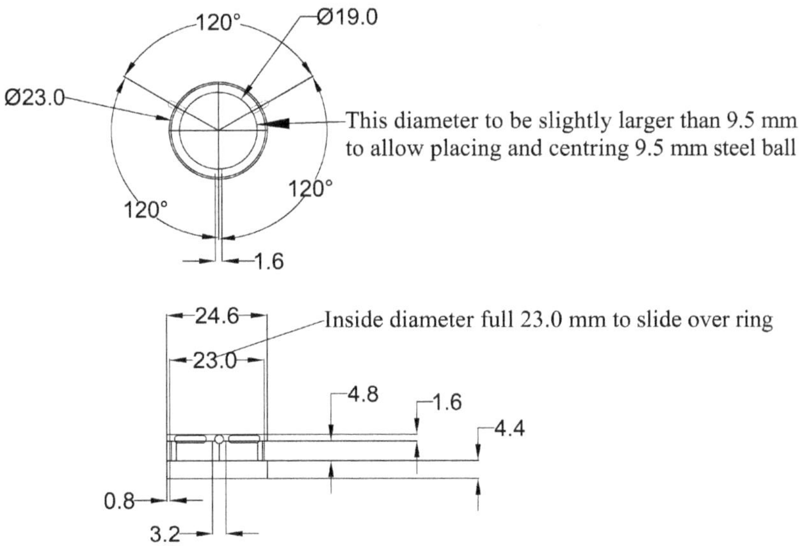

FIGURE 7.2 Ball centring guide.

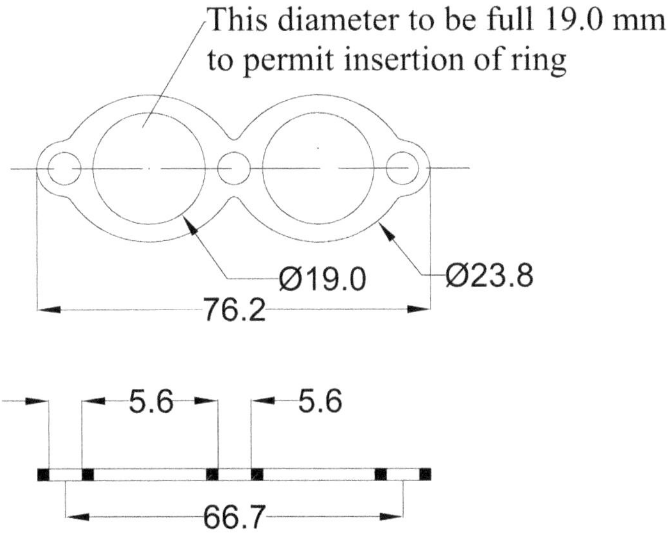

FIGURE 7.3 Ring holder.

Note: The difference between the two readings should not exceed 1°C for a temperature range of 40–60°C and 1.5°C for a range of 61–80°C. If the difference between the two readings is more than the prescribed value, then the test should be repeated. If the softening point of bitumen is expected to be more than 80°C, then the test needs to do with glycerol as a heating medium instead of water and the starting temperature should be 35°C.

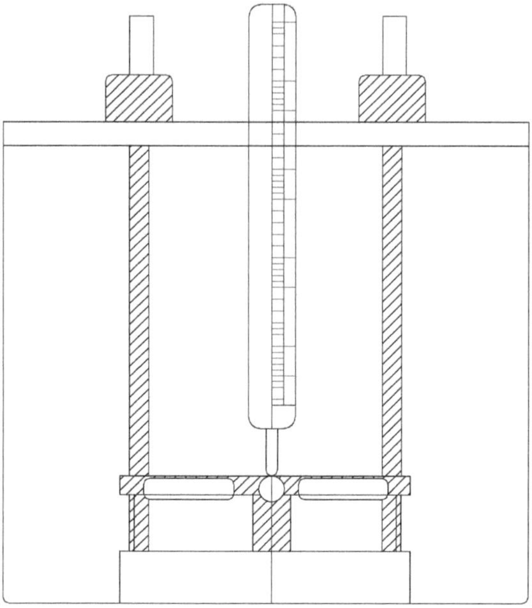

FIGURE 7.4 Ball-ring apparatus.

TABLE 7.1
Softening Point for Paving Bitumen

Grade	Softening point (°C) (min)
VG10	40
VG20	45
VG30	47
VG40	50

Report

Softening point of the bitumen is

Note: Table 7.1 shows the softening point of different grades of bitumen according to BIS (IS 73, 2018).

Student Remark

....
....
....
....

$$\frac{\textit{Marks obtained}}{\textit{Total marks}} = -$$

Instructor signature

Instructor Remark

....
....
....
....

7.9.2 Flash and Fire Point of Bitumen

Flash point of bitumen is the temperature at which the fuel oil gives off vapour that flashes when exposed to an open flame and does not catch fire. The minimum and maximum limits of flash point of bitumen are 175°C and 180°C, respectively. The fire point is the measure of the sample to support combustion.

Test Summary

This test method is used to understand the flash point and fire point of the bitumen using the Pensky-Martens closed cup apparatus or Cleveland open cup apparatus (IS 1209, 2019).

Use and Significance

- Under controlled laboratory conditions, the flash point measures the tendency of the sample to form a flammable mixture with air.
- Due to the hazardous nature of the volatile materials emitted during the combustion of bitumen, it is of utmost importance to understand the fire point of bitumen.
- Knowing the flash point and fire point will be helpful to restrict the working temperature in the field. In other words, heating of bitumen should be limited within the flash point and the fire point.

Required Apparatus

- Pensky-Martens closed cup apparatus consist of a test cup
- Test cover and shutter
- Stirrer
- Heating source
- Ignition source
- Air bath and top plate
- Cleveland open cup apparatus consists of a test cup
- Heating plate
- Test flame applicator
- Heater
- Thermometer support

- Thermometer
- Heating plate support
- Sampling and testing

Pensky-Martens Closed Cup Apparatus

Step 1. The bitumen sample is heated above the softening point, generally between 75°C and 100°C and thoroughly stirred to remove air bubbles.

Step 2. The cup, as shown in Figure 7.5, is filled with softened bitumen until the filling mark on the cup.

Step 3. Now, the thermometer is inserted, and heat is applied in a controlled manner from the heat source using a thermostat, in such a way that the increase in temperature is between 5°C and 6°C per minute.

Step 4. The stirring rate should be approximately 60 revolutions per minute.

Step 5. The test flame is lighted once the temperature observed in the thermometer is 17°C below the flash point (based on trial and error or previous experience).

Step 6. The stirring is stopped and the test flame is applied for every 1°C increase in temperature after the above point.

Step 7. At a particular temperature, the sample reaches sufficient heat and forms a flash. This temperature is called as flash point of the bitumen.

Step 8. The fire point of the bitumen is the point at which the sample catches fire and the fire stays for 5 or more seconds. This is done by continuing the flash for every 2°C increase in temperature after the flash point.

Step 9. The above experiment is repeated for at least three times and the average is taken as the flash point and fire point in the apparatus as shown in Figure 7.6.

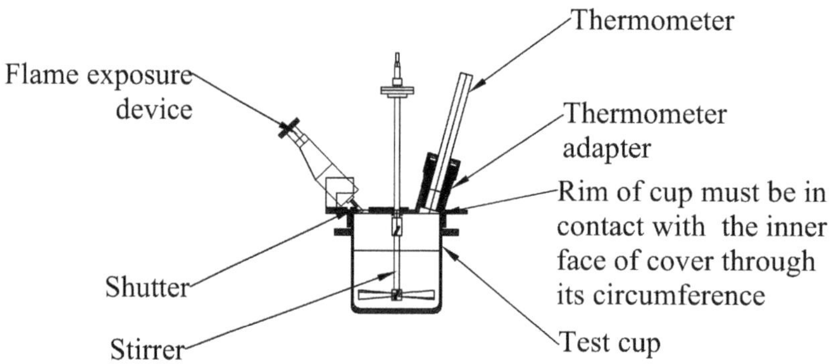

FIGURE 7.5 Pensky-Martens closed cup.

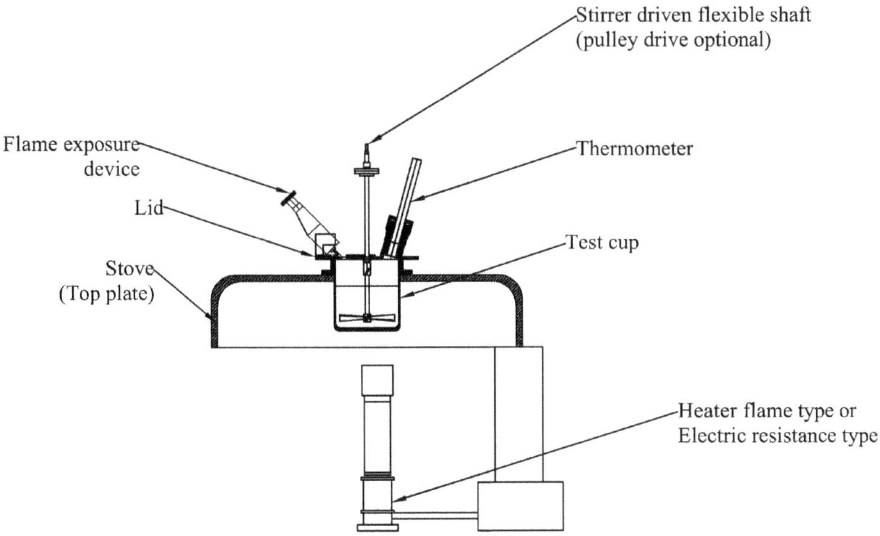

FIGURE 7.6 Pensky-Martens closed cup apparatus.

Cleveland Open Cup Apparatus

Step 1. Bitumen is heated below 100°C and above softening point and filled in a cup in such a way that the meniscus is at the filling mark of the cup, as shown in Figure 7.7. The excess bitumen is removed by suitable methods and air bubbles are removed.

Step 2. The temperature of the sample is raised at a rate of 14–17°C. Afterwards, the light flame is used and the diameter of the flame is adjusted to 3.8–5.4 mm.

Step 3. Approximately, when the temperature of the sample is 56°C below the flash point, the rate of rise in temperature is reduced to 5–6°C per minute for the last 28°C before the flash point.

Step 4. Flame is applied once, for every 2°C increase in the temperature, starting from at least 28°C below the flash point.

Step 5. The test flame is passed smoothly and continuously across the cup, at right angles or in straight lines or in the circumference of the cup having a radius of at least 150 mm. The test flame is passed at a height of 2.5 mm above the upper edge of the cup. First, the flame is moved in one direction and then in the opposite direction; the time required for passing the test flame is 1 second.

Step 6. Care must be taken to avoid disturbing the vapours from the test cup during the last 17°C by careless movement or bathing near the cup.

Step 7. The temperature at which the flash appears on the surface of the specimen is recorded from the thermometer. A bluish halo colour

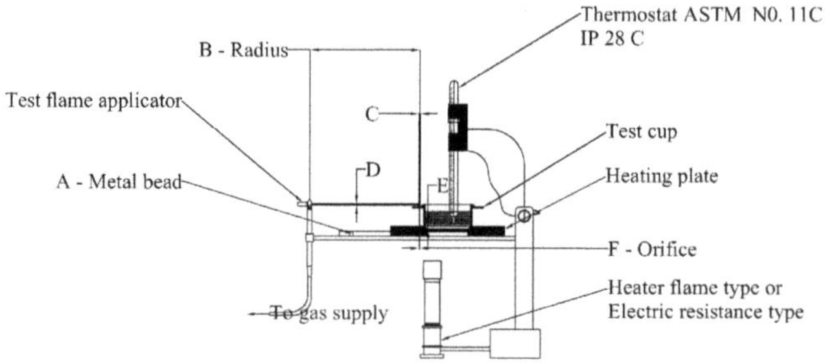

	Minimum	Maximum
A - diameter	3.2000	4
B - radius	152	nominal
C - diameter	1.6000	nominal
D	-	2
E	5.9000	6.900000
F - diameter	0.800000	nominal

FIGURE 7.7 Cleveland open cup apparatus (all dimensions are in mm).

that appears during the test should not be considered as the flash point.

Step 8. The fire point is determined by continuing the heating at a rate of 5–6°C. Heating is continued until the specimen ignites and stays flaming for 5 seconds.

Step 9. The test is repeated two more times and the average of three tests is taken as flash point and fire point.

Report

- Flash point of the bitumen is
- Fire point of the bitumen is

Student Remark

....
....
....
....

$$\frac{\textbf{\textit{Marks obtained}}}{\textbf{\textit{Total marks}}} = -$$

Instructor signature

Instructor Remark

....
....
....
....

7.9.3 Penetration Value of Bitumen

Bitumen is available in different grades. The choice of the grade of bitumen depends upon the requirement of surfacing and climatic conditions. Grading of bitumen is done based on their consistency. Consistency of a particular bitumen is determined by measuring the penetration resistance offered by bitumen.

Test Summary

The structure of bitumen influences the grade of bitumen. In order to distinguish different grades, bitumen is allowed to penetrate by means of a specific needle (based on the requirements) at specific experimental conditions (IS 1203, 2019).

Use and Significance

- This test method is used to understand the consistency. The higher the penetration value, softer the consistency.

Required Apparatus

- Penetrometer (available in manual type as well as mechanical type)
- Penetrometer consisting of a metal base
- A weight with a carrier of 100 g
- The weight with the carrier is released and returned back to its original position through a release knob
- A dial graduated at 1/10 of a millimetre is fixed along with the assembly
- A stopwatch with an accuracy of 1/10 of a second
- A steel or an aluminium container of diameter 55 mm and height 35 mm
- A straight edge
- Highly polished hard steel needle
- Water bath to maintain 25°C

Sampling

Step 1. Bitumen sample is heated to a temperature of 90°C above the softening point. The sample is continuously stirred to make a homogeneous mix and to remove air bubbles.

Step 2. After sufficient heating, the sample is poured into a container and allowed to cool for 1 hour at room temperature.

Step 3. The sample is placed in a water bath at 25°C for a period of one to one and a half hours, while ensuring that the entire sample is maintained at 25°C.

Testing

Step 1. The prepared sample is placed on the base plate of the penetrometer.

Step 2. The penetration needle is cleaned with benzene, dried and fixed in place. Ensure that the needle assembly has a free movement. If needed, the assembly is lubricated.

Step 3. The prepared specimen is placed below the needle and the needle is brought just in contact with the top surface of the bitumen sample. This can be verified by using fine adjustment and making sure that the image of the needle and the real tip of the needle coincide.

Step 4. The penetration test is performed by releasing the needle. The needle is allowed to penetrate for 5 seconds. Before starting the penetration, the initial reading is noted down. Let the reading be denoted as A.

Step 5. After 5 seconds, the final reading displayed on the dial is noted down. Let the reading be denoted as B_1.

Note: At least three different points are chosen and tests are performed in a single sample. The distance between two consecutive penetrations and the distance from the wall of the container are equal to or greater than 10 mm and the average is considered to be the mean value.

- Before taking consecutive readings, the needle is taken out and cleaned with benzene.
- Penetration value is the difference between the final reading and the initial reading. The average of three different values is considered to be the penetration value.

Report

- Penetration value of the bitumen is

Note: The grade of bitumen can be decided based on the penetration value obtained. This value can be used as a quality check. Table 7.2 shows the penetration value of different grades of bitumen, according to BIS (IS 73, 2018).

Student Remark

....

....

....

....

$$\frac{\textit{Marks obtained}}{\textit{Total marks}} = -$$

Instructor signature

TABLE 7.2
Penetration Requirement for Paving Bitumen

Paving grades	Penetration value @ 25°C minimum value
VG10	80
VG20	60
VG30	45
VG40	35

Instructor Remark

....
....
....
....

7.9.4 Ductility Test of Bitumen

This test procedure is used to determine the ductility of bitumen at a specified temperature at a specified elongation rate. This test is considered to be an important one among the quality check test for bitumen.

Test Summary

The type of bitumen to be tested is placed and held between the two ends of a briquette specimen and pulled apart at a specific rate and temperature until the breaking point to determine elongation of the bitumen (IS 1208, 2019).

Use and Significance

- This test method is used to understand the tensile properties of asphalt material by virtue of which the ductility of the asphalt material can be determined.

Required Apparatus

- Ductility apparatus consists of a rectangular tank that is provided with steel or copper liner sheet, with a built-in heater to maintain temperature and a stirrer to circulate water to maintain a uniform temperature
- Moreover, the apparatus is equipped with a clutch to control the speed at which the elongation rate need to be given
- To hold the specimen, a briquette mould made of brass is used
- The briquette mould has a base plate with two clips and removable sides
- One end of the briquette mould is fixed and the other end is movable and both pins can be removed in order to connect it with the ductility apparatus
- Temperature indicator or thermometer to measure temperature in the rectangular tank

Sampling

Step 1. A sample is prepared by heating the bitumen above the softening point of the bitumen. This temperature generally varies from 75–100°C.

Caution: A mixture of equal parts of glycerine and dextrin is prepared. The prepared mixture is applied on the base plate, interior and sides of the mould to avoid sticking of the bitumen on the surface of the briquette mould.

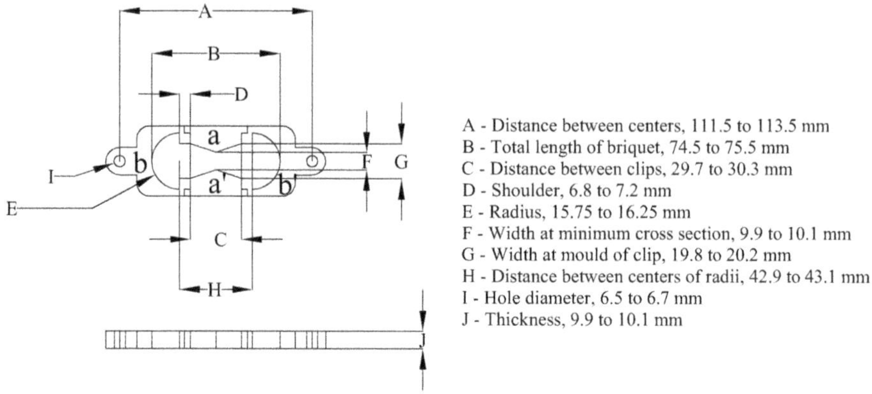

FIGURE 7.8 Mould for ductility test specimen.

Step 2. Heated bitumen is poured into the moulds, as shown in Figure 7.8. Afterwards, moulds with the sample are placed in a water bath and maintained at 27°C for a period of half an hour.

Testing

Step 1. The sides of the mould are removed and curing is conducted in the apparatus. Note down the initial reading in the apparatus or set the scale to zero. Let the initial reading be denoted as A.

Step 2. In order to set an elongation rate, clutch in the system is adjusted and geared in such a way that the elongation is performed at a rate of 50 mm/minute.

Step 3. One end of the mould is kept fixed while the other end is pulled apart.

Step 4. Once the bitumen thread of the specimen breaks, then note the final reading (B) from the scale.

Step 5. The difference between the final reading (B) and the initial reading (A) is the ductility value.

Note: The test is performed on at least three samples and the average value is considered to be the mean value.

Report

- Ductility value of the bitumen is

Note: If the specimen elongates beyond 75 cm, the apparatus is stopped and the elongation is mentioned as more than 75 cm. The ductility of the bitumen depends upon its grade. Table 7.3 shows the ductility of different grades of bitumen, according to BIS (IS 73, 2018).

TABLE 7.3
Ductility Requirement for Paving Bitumen

Grade	Minimum ductility (in cm) @ 25°C minimum value
VG10	75
VG20	50
VG30	40
VG40	25

Student Remark

....
....
....
....

$$\frac{\textit{Marks obtained}}{\textit{Total marks}} = -$$

Instructor signature

Instructor Remark

....
....
....
....

7.10 PRACTICE QUESTIONS AND ANSWERS FROM COMPETITIVE EXAMS

1. Bitumen is a .... material
   a) Elastic
   b) Viscous
   c) Visco-elastic
   d) Visco-plastic

2. What is the melting point of bitumen?
   a) 100°C
   b) 600°C
   c) 750°C
   d) Cannot be defined accurately

3. What is the purpose behind determining the softening point of bitumen?
   a) To know the penetration resistance

b) To understand the temperature required to melt for a particular application
c) To understand the temperature at which the material undergoes combustion
d) None of the above

4. What is the test apparatus used to determine the flash and fire point of the bitumen?
   a) Ring and ball apparatus
   b) Penetration resistance apparatus
   c) Cleveland open cup apparatus
   d) None of the above

5. What are the purpose of conducting flash point and fire point test for bitumen?
   a) Heating beyond this point may result in catastrophic events in the field
   b) Without heating beyond this point, bitumen cannot be used
   c) When heated at this point material undergoes a chemical reaction that results in enhanced strength
   d) None of the above

6. ………… colour that appears during the test should not be considered as flash point.
   a) Bluish halo
   b) Pinkish green
   c) Golden green
   d) Greyish yellow

7. Consistency of bitumen is determined by
   a) Flash point
   b) Penetration resistance
   c) Softening point
   d) None of the above

8. Ductility test is apparatus also used to check
   a) Elastic recovery of the bitumen
   b) Consistency of the bitumen
   c) Penetration resistance of the bitumen
   d) Melting point of the bitumen

9. Which of the following bitumen is used in airport pavements
   a) Paving grade bitumen
   b) Hard paving grade bitumen
   c) Oxidisable bitumen
   d) None of the above

10. If the softening point of bitumen is 50°C, what is the grade of bitumen?
    a) VG-20
    b) VG-30
    c) VG-40
    d) VG-50

11. Bureau of Indian Standards classifies bitumen into grades 65/25, 85/40, etc. The first and second numbers respectively refer to
    a) Softening point and penetration
    b) Penetration and softening point
    c) Flash point and penetration
    d) Flash point and softening point

[IES 1996]

12. Polyvinyl chloride (PVC) is a
    a) Thermosetting material
    b) Thermoplastic material
    c) Elasto-plastic material
    d) Rigid plastic material

[IES 1996]

13. In building construction, the place for providing a damp-proof course is at the
    a) Basement level
    b) Window sill level
    c) Lintel level
    d) Roof level

[IES 2001]

14. Statement (I): When plastering on building exteriors, more of coarser particles of sand are used in regions where seasonal rainfall is often intense and the total annual rainfall also is relatively more.

    Statement (II): Such type of "dhabbah" plastering effects the minimisation of rainfall impacts, resulting in less formation of moss and less surface discoloration but may not reduce seepage to the interior.
    a) Both statement (I) and statement (II) are individually true and statement (II) is the correct explanation of statement (I).
    b) Both statement (I) and statement (II) are individually true and statement (II) is not the correct explanation of statement (I).
    c) Statement (I) is true but statement (II) is false.
    d) Statement (I) is false but statement (II) is true.

[IES 2012]

15. Statement (I): In areas where extreme cold conditions are a regular feature, and more so particularly in winter, it is necessary to use lighter oil for automobiles than in summer.

Statement (II): Lighter in statement (I) refers to the oil density, which may be adjusted by admixtures.

a) Both statement (I) and statement (II) are individually true and statement (II) is the correct explanation of statement (I).
b) Both statement (I) and statement (II) are individually true and statement (II) is not the correct explanation of statement (I).
c) Statement (I) is true but statement (II) is false.
d) Statement (I) is false but statement (II) is true.

[IES 2018]

16. Plastic asphalt is
    a) Used as a waterproofing layer over roof
    b) A mixture of cement and asphalt
    c) A natural asphalt
    d) A refinery product

[SSC JE 2008]

17. Which of the following test is used for the bitumen?
    a) Slump test
    b) Abrasion test
    c) Penetration test
    d) Fineness test

[SSC JE 2018]

18. Which one of the following material is used as a bonding admixture?
    a) Natural rubber
    b) Synthetic rubber
    c) Organic polymers
    d) All options are correct

[SSC JE 2018]

19. The quantity of damp-proof course (DPC) is worked in
    a) m^3
    b) m
    c) m^2
    d) Lump sum

[AE 2008]

20. Plastic asphalt is
    a) Used as waterproofing layer over roof
    b) A mixture of cement and asphalt
    c) A natural asphalt
    d) A refinery product

[AE 2008]

21. For damp-proof course at plinth level the commonly adopted material is
    a) Membrane sheeting
    b) Bitumen sheeting
    c) Paint coating
    d) Mortar sheeting

[AE 2010]
[AE 2013]

22. In case of building without basement, the best position for D.P.C lies at
    a) Plinth level
    b) Ground level
    c) 15 cm above the plinth level
    d) 15 cm above the ground level

[AE 2013]

23. A semi tight material which forms an excellent impervious layer for damp-proofing is called
    a) Bitumen
    b) Bituminous felt
    c) Alumina
    d) Mastic asphalt

[AE 2014]

24. .... is the distress which is occurred at intersections due to the poor mix design of bituminous concrete.
    a) Rutting
    b) Breaking
    c) Eroding
    d) Shoving

1.	c	2.	d	3.	b	4.	c	5.	a	6.	a	7.	b	8.	a	9.	b	10.	c
11.	a	12.	b	13.	a	14.	a	15.	c	16.	b	17.	c	18.	d	19.	c	20.	b
21.	b	22.	a	23.	d	24.	d												

REFERENCES

Assistant Engineer (AE). 2020. Tamil Nadu Public Service Commission (TNPSC), Tamil Nadu, India. http://www.tnpsc.gov.in/previous-questions.html (accessed September 18, 2020).

Engineering Service Exam (ESE). 2020. Union Public Service Commission, New Delhi, India. https://www.upsc.gov.in/examinations/previous-question-papers (accessed September 18, 2020).

Graduate Aptitude Test in Engineering (GATE). 2020. GATE Office, Chennai, India. http://gate.iitm.ac.in/gate2019/previousqp18.php (accessed September 18, 2020).
IS 1203. (2019). *Methods for testing tar and bituminous materials: Determination of penetration*. Bureau of Indian Standards. New Delhi.
IS 1205. (2019). *Methods for testing tar and bituminous materials: Determination of softening point*. Bureau of Indian Standards. New Delhi.
IS 1208. (2019). *Methods for testing tar and bituminous materials: Determination of ductility*. Bureau of Indian Standards. New Delhi.
IS 1209. (2019). *Methods for testing tar and bituminous materials: Determination of flash point and fire point*. Bureau of Indian Standards. New Delhi.
IS 73. (2018). *Paving bitumen – Specification*. Bureau of Indian Standards. New Delhi.
Junior Engineer (JE), Staff Selection Commission (SSC), New Delhi, India. https://ssc.nic.in/Portal/SchemeExamination (accessed September 18, 2020).

8 Timber

8.1 INTRODUCTION

Wood is one of the widely used materials for construction. Wood generally has a high strength-to-mass ratio. Based on the engineering applications timber can be obtained from different trees.

8.2 CLASSIFICATION OF TREES

Trees can be classified into exogenous and endogenous based on their mode of growth. If the mode of growth is in the outward direction, then the trees are said to be exogenous. Exogenous trees find numerous applications in case of construction. They are further subdivided into conifers and deciduous. Some of the examples include oak, teak, deodar, chir, etc. On the other hand, if the mode of growth is in the inward direction, then the trees are said to endogenous and consist of fibrous mass in their longitudinal direction. Some of the examples include palm, cane, bamboo, etc.

8.3 MACROSTRUCTURE OF TIMBER

A cross-section of timber consists of following essential parts.

1. Pith – Innermost central part of the tree cross-section is called as pith.
2. Annual ring of heartwood – This wood covers the innermost portion surrounding the pith of the annual ring. They are highly rigid and strong, enabling their usage as a construction material.
3. Annual ring of sapwood – The annual rings that are located in between the heartwood and vascular cambium is called as sapwood.
4. Vascular cambium – Cambium layers are responsible for the production of new cells in the trees. They produce both xylem and phloem. These layers are later converted into the sapwood.
5. Medullar rays – These rays start from the pith and extend towards outward direction until the cambium layers. These rays extend in the transverse direction such a way that they impart strength to the timber and hold the cross-section of timber intact.
6. Barks – This protects the vascular cambium and phloem layer and safeguards the tree for its future growth.

8.4 TIMBER TESTING

Timber is generally identified by their species and defects present by visual examination. However, some of the essential tests are conducted in the laboratory. (1)

Moisture content determination; (2) Tensile strength parallel to grains; (3) Tensile strength perpendicular to grains; and (4) Brittleness measurement by Charpy test.

8.4.1 Defects in Timber

Defects in timber can be divided majorly into four types. Defects can occur principally due to various reasons. Defects can occur mainly, during the growing process conversion and seasoning process. Defects in any material generally reduce its quality and strength and further its appearance gets spoiled, resulting in decay.

1. **Defects due to conversion** – Planning and sawing process may produce chip mark over the surface of the finished timber; this defect is identified by a mark on the surface of the timber. Similar to the chip mark, a diagonal mark may result due to an improper sawing technique adopted. Usually, diagonal marks are identified by diagonal grains over the direction of the straight-grained structure. During section conversion, falling of tools may form section depression that results in defect called as torn grains. In the finished timber surface, the original rounded section can be identified and are called as wane.
2. **Defects due to the attack of fungi** – Frequently, fungi attack can bring about a lot of defects in timber. They can be classified into the blue stain, sap stain, brown rot, white rot, dry rot and wet rot.
3. **Defects due to natural forces** – Shakes are defects that are formed due to natural forces in timber. They are longitudinal separations that are formed between the annul rings in the wood. Based on the shape of the longitudinal separations, they are classified into cup shakes, star shakes, heart shakes and ring shakes and also they are named based on their patterns on the timber. Generally, the formation of shakes affects the shear strength of the wood to a greater extent.
4. **Defects due to seasoning** – Changes in the moisture content movement causes defects in the timber. Seasoning defects result in disruption of decoration and loosening of fixing. Some of the common seasoning defects are bows, cups, twisting and warps.

8.5 PROCESSING OF TIMBER

In order to use timber as a construction material, it needs to be appropriately processed. Timber processing involves the following steps. They are

1. **Felling of trees** – Cutting down of trees in order to obtain timber is called as the felling of trees. In general, 50–100 years is considered to be the best age for the felling of trees.
2. **Seasoning of timber** – Once the timber is cut, the first stage of processing is drying also called as seasoning. Seasoning is done to prevent the timber

from fermentation. In timber, water is present in the form of sap or in the form of moisture. Complete removal of water from the inside cell walls is called a fibre saturation point. The seasoning can be carried out either naturally or artificially. Artificial seasoning involves five methods. They are (1) Boiling; (2) Electrical seasoning; (3) Kiln seasoning; (4) Water seasoning and (5) Chemical seasoning.

3. **Timber conversion –** Shaping and sizing of timber to the desired dimension is called as timber conversion. Timber conversion can be done by following methods.
   a. **Ordinary sawing –** This is the most general, easiest and economical way of sawing. In this type of sawing, the cut is done tangentially to the annular rings.
   b. **Quarter sawing –** When the cuts are made at right angles to each other, then the sawing is called as quarter sawing.
   c. **Tangential sawing –** In this type of sawing cuts are made tangential to the annular rings as same as that of ordinary swing. Nonetheless, two tangential cuts to the annular rings are perpendicular to each other.
   d. **Radial sawing –** This a type of sawing in which cuts are made along the radial direction on the medullary rays. This type of cuts imparts more wastage. Nonetheless, they provide decorative effects.
4. **Timber preservation (IS: 401-2001) –** Preservation of timber is carried out to make the timber more resistant against insect attack and more durable in order to prolong the life of the timber. The following are some of the treatments that carried out to preserve the timber.
   - **AS CU treatment –** Timber is coated with 1 part of arsenic pent-oxide, 3 parts of copper sulphate and 4 parts of potassium and sodium dichromate mixed in water.
   - **Chemical salts –** The application of chemical salts such as copper sulphate, chlorides of sodium, mercury and zinc.
   - **Coal tar –** Sometimes, timber is coated with hot tar to preserve it.
   - **Creosote oil application –** Tar distillation gives creosote oil. This oil is applied at 0.7–1 N/mm^2 at 50° C and timber is left to stand for 1–2 hours for this proper oil penetration
   - **Oil and solignum paints –** Coating of timber with oil and solignum paints increases its resistance against water penetration.

8.6 TESTING PROCEDURES

8.6.1 Compression Failure of the Timber Block

Timber is one of the vital construction materials. Although the usage of timber as a building material is limited in India, it is one of the most used construction material in other countries due to its availability and durable nature.

Test Summary

Timber is a naturally occurring organic material with high heterogeneity and anisotropy. In this test, a standard timber specimen is loaded until failure, to understand some important characteristics of the timber to use them as a construction material. Due to high heterogeneity, characteristics of the timber will be dissimilar in different directions. In order to understand its characteristics, especially strength-related characteristics, tests are done in both parallel as well as perpendicular direction to the orientation of the grain.

In this test, the compressive strength of the timber is determined in parallel as well as perpendicular to the grains (IS 2408, 2015).

Use and Significance

- The above test method is used to check the quality of the timber to be used as a construction material. As well, this test can be used to understand the suitability of a product to be used for a particular application.

Required Apparatus

- Compression testing machine (CTM)
- Specimens of required dimensions

Sampling

Determination of compressive strength of timber requires sampling of the timber and timber specimen of definite size.

Step 1. To determine the compressive strength of the timber parallel to the grains, timber specimen of size $200 \times 50 \times 50$ mm or $80 \times 20 \times 20$ mm is to be used.

Step 2. To determine the compressive strength of the timber perpendicular to the grains, timber specimen of size $150 \times 50 \times 50$ mm or $100 \times 20 \times 20$ mm is to be used.

Caution: The surface of the loading should be smooth and parallel. So that in the course of loading, force applied is uniformly distributed across the cross-section.

Testing

Compressive strength of timber parallel to the grains

Step 1. A test specimen is placed in CTM with the grains in the timber parallel to the direction of loading. A gauge length of 150 mm is marked for a specimen with 200 mm length.

Step 2. The specimen is placed in such a way that the specimen is directly below the movable head of the specimen.

Step 3. For both the size of the specimen, loading is given at a rate of 0.6 mm per minute. An initial load of 2.5 kN or 250 kg is applied to set 200 mm specimen whereas, an initial load of 2.0 kN or 200 kg is applied to set 80 mm specimen.

Step 4. A suitable compressometer is fixed over the gauge length in order to observe and record the change in deformation with respect to the applied load.

Step 5. Readings from the compressometer are recorded until the load-carrying capacity of the specimen starts to decrease. It is advisable to remove the compressometer before the stage where the specimen fails in different modes in different planes.

Step 6. Deformation should be measured to the nearest of 0.002 mm. To obtain the required characteristics of the material, enough amount of data is obtained before the proportionality limit (at least 8–10 readings).

Step 7. For a 80×20×20 mm specimen, load at failure is noted down. In case the data is required within proportionality limit, a dial gauge of precision 0.01 mm is used and the required amount of data is recorded within the proportionality limit (at least 10–15 readings).

Step 8. In order to understand the failure pattern, the machine is run for a longer time until the load-carrying capacity of the specimen reaches zero or very minimum value.

Compressive strength of timber perpendicular to the grains

Step 1. A test specimen is placed in CTM with the grains in the timber perpendicular to the direction of loading.

Step 2. The specimen is placed in such a way that the specimen is directly below the movable head of the specimen.

Step 3. For both the size of specimen loading is given at a rate of 0.6 mm per minute. An initial load of 0.5 kN or 50 kg is applied to set 150 mm specimen while an initial load of 0.1 kN or 10 kg is applied to set 100 mm specimen.

Step 4. A suitable dial gauge is fixed over the gauge length to observe and record the change in deformation with respect to the applied load.

Step 5. Readings from the dial gauge are recorded until the load-carrying capacity of the specimen starts to decrease.

Step 6. Deformation should be measured to the nearest of 0.02 mm. To obtain the required characteristics of the material, an enough amount of data is obtained before the proportionality limit (at least 8 to 10 readings). Readings are recorded for every 2.5 mm deformation until failure occurs.

Step 7. To understand the failure pattern, the machine is run for a longer time until the load-carrying capacity of the specimen reaches zero or very minimum value.

Calculation

A plot is obtained between the recorded load and deformation. Various characteristics of timber are obtained from the following calculation, as shown in Table 8.1.

Where,

- $\boldsymbol{P}$ – the load at the limit of proportionality (kN)
- $\boldsymbol{P_m}$ – the maximum load-carrying capacity (kN)
- $\boldsymbol{P_0}$ – the load corresponding to 2.5 mm deformation (kN)
- $\boldsymbol{A}$ – the area of cross-section (mm^2)
- $\boldsymbol{L}$ – the gauge length of the specimen in case of a compressive load applied parallel to the grains of the specimen or is the total height of the specimen (mm)
- Δ – the deformation within the limit of proportionality (mm)

TABLE 8.1
Characteristics of Timber

Characteristics	Unit	Formula	Parallel to grain	Perpendicular to grain
Compressive strength at proportionality limit	N/mm^2	$\frac{P}{A}$		
Compressive strength at maximum load	N/mm^2	$\frac{P_m}{A}$		
Compressive strength at 2.5 mm	N/mm^2	$\frac{P_0}{A}$	–	
Modulus of elasticity	N/mm^2	$\frac{PL}{A\Delta}$		

Report

- In order to compare specimens of different sizes during the compression test
  - In case of specimen of size 200×50×50 mm or 80×20×20 mm a ratio of 0.98 is considered for cross-section
  - In case of specimen of size 150×50×50 mm or 100×20×20 mm a ratio of 1.07 is considered for cross-section

Student Remark

....

....

....

....

$$\frac{\textit{Marks obtained}}{\textit{Total marks}} = -$$

Instructor signature

Instructor Remark

....

....

....

....

8.6.2 Tensile Failure of Timber Block (Parallel to Grains)

Timber is one of the vital construction materials. Although the usage of timber as a building material is limited in the modern construction practices in India, it is one of the most used construction materials in other countries due to its availability and durable nature.

Test Summary

Timber is a naturally occurring organic material with high heterogeneity and anisotropy. In this test, a standard timber specimen is loaded until failure, to understand the tensile strength of timber to use them as a construction material. Due to high heterogeneity, characteristics of the timber will be different in different directions. To understand its characteristics, especially strength-related characteristics, tests are done in both parallel as well as perpendicular direction to the orientation of the grain.

In this test, the tensile strength of the timber is determined parallel to the grains.

Use and Significance

- The above test method is used to check the quality of the timber to be used as a construction material. As well, this test can be used to understand the suitability of a product to be used for a particular application.

Required Apparatus

- Universal testing machine (UTM)
- Specimens of required dimensions with suitable holding grips to avoid slippage of the specimen during testing

Sampling

Determination of tensile strength of timber requires sampling of the timber and timber specimen of definite size.

- To determine the tensile strength of the timber parallel to the grains, timber specimen of size with a central cross-section area of 7×7 mm or 5×5 mm is taken. The gauge length of 50 mm and 30 mm is considered for 7×7 mm and 5×5 mm, respectively.

Testing

Tensile strength of timber parallel to the grains

Step 1. A test specimen is placed in UTM with the grains in the timber parallel to the direction of loading. A gauge length of 50 mm or 30 mm is marked for a specimen with 7×7 mm and 5×5 mm cross-section, respectively.

Step 2. The specimen is placed in such a way that during the process of loading, slippage does not occur. The specimen is fitted with a suitable extensometer.

Step 3. For both the sizes of specimens, the rate of loading is given at a rate of 1 mm per minute.

Step 4. Readings from the extensometer are recorded until the load-carrying capacity of the specimen starts to decrease. It is advisable to remove the compressometer before the stage where the specimen fails in different modes in different planes.

Step 5. Deformation should be measured to the nearest of 0.002 mm. In order to obtain the required characteristics of the material, enough amount of data is obtained before the proportionality limit (at least 8–10 readings).

Step 6. For 80×20×20 mm specimen, load at failure is noted down. In case the data is required within proportionality limit, a dial gauge of precision 0.01 mm is used and the required amount of data is recorded within the proportionality limit (at least 10–15 readings).

Step 7. To understand the failure of a specimen, the machine is run for a longer time until the load-carrying capacity of the specimen reaches zero or very minimum value.

Calculation

A plot is obtained between the recorded load and deformation. Various characteristics of timber are obtained from the following calculation, as shown in Table 8.2.

TABLE 8.2
Characteristics of Timber

Characteristics	Unit	Formula	Parallel to grain
Tensile strength at proportionality limit	N/mm²	$\frac{P}{A}$	
Tensile strength at maximum load	N/mm²	$\frac{P_m}{A}$	
Modulus of elasticity	N/mm²	$\frac{PL}{A\Delta}$	

Where,

- $\boldsymbol{P}$ – the load at the limit of proportionality (kN)
- $\boldsymbol{P_m}$ – the maximum load-carrying capacity (kN)
- $\boldsymbol{A}$ – the area of cross-section (mm^2)
- $\boldsymbol{L}$ – the gauge length of the specimen in case of a compressive load applied parallel to the grains of the specimen or is the total height of the specimen (mm)
- Δ – the deformation within the limit of proportionality (mm)

Report

- Tensile strength of the specimen parallel to the grains

Student Remark

....
....
....
....

$$\frac{\textit{Marks obtained}}{\textit{Total marks}} = -$$

Instructor signature

Instructor Remark

....
....
....
....

8.6.3 Tensile Failure of Timber Block

Timber is one of the important construction materials. Although the usage of timber as a building material is limited in India it is one of the most used construction materials in other countries due to its availability and durable nature.

Test summary

Timber is a naturally occurring organic material with high heterogeneity and anisotropy. In this test, a standard timber specimen is loaded until failure, in order to understand some essential characteristics of the timber to use them as a construction material. Due to high heterogeneity, characteristics of the timber will be different in different directions. In order to understand its characteristics, especially strength-related characteristics, tests are done in both parallel as well as perpendicular direction to the orientation of the grain.

In this test, the tensile strength of the timber is measured perpendicular to the grains.

Use and Significance

- The above test method is used to check the quality of the timber to be used as a construction material. As well, this test can be used to understand the suitability of a product to be used for a particular application.

Required Apparatus

- Universal testing machine (UTM)
- Specimens of required dimensions with suitable holding grips to avoid slippage of a specimen during testing

Sampling

Determination of tensile strength of timber requires sampling of the timber and timber specimen of definite size.

Step 1. To determine the tensile strength of the timber perpendicular to the grains, timber specimen of size 50×50×56 mm or 20×20×60 mm is used.

Step 2. For the test specimen of size 50×50×56 mm, notches of radius 12 mm is created in 50×56 mm face. Notches are located in such a way that the centre of the notches is located at the opposite edges of the face with 6 mm inside the edges.

Step 3. For the test specimen of size 20×20×60 mm, wedge-shaped notches are created in 20×60 mm face. Notches are located in such a way that the centre of the notches is located at a distance of 28 mm inside the opposite edges. Notches extend until the corresponding edges equally on both sides with an internal angle of 30° that makes a 4 mm cut from both the corners.

Step 4. With the help of a driller, a hole with a radius of 2 mm is drilled by taking the centre of notches as a centre.

Step 5. Similarly, two holes of 2 mm radius are drilled on the slanting lines of the notches (lying precisely opposite to each other in a notch) having a centre of the hole located at a distance of 13 mm from the centre of the notch. Meanwhile, similar holes are drilled in the other notch at a similar distance.

Note: The face chosen for making the notches should have the fibres oriented in the direction notches.

Testing

Tensile strength of timber perpendicular to the grains

Step 1. A test specimen is placed in UTM provided such that suitable grips are used for holding the specimen without slipping.

Step 2. These grips are provided with cushioning springs that prevent any damage to the machine while the test specimen breaks.

Step 3. The specimen is placed in such a way that during the process of loading, slippage does not occur. The specimen is fitted with a suitable extensometer.

Step 4. For both the sizes of specimens, the rate of loading is given at a rate of 2.5 mm per minute.

Step 5. The load at which the specimen fails is recorded.

Calculation

- The maximum load at which the specimen fails divided by the area gives the tensile strength of the specimen during failure.

Report

- Tensile strength of the specimen perpendicular to the grains.

Student Remark

....
....
....
....

$$\frac{\textit{Marks obtained}}{\textit{Total marks}} = -$$

Instructor signature

Instructor Remark

....
....
....
....
....

8.7 PRACTICE QUESTIONS AND ANSWERS FROM COMPETITIVE EXAMS

1. What kinds of trees are useful in construction?
   a) Conifers
   b) Deciduous
   c) Exogenous
   d) All the above

2. Which of the following is an endogenous tree?
   a) Coconut
   b) Oak
   c) Teak
   d) Chir

3. Which of the following enables them as a good construction material in exogenous trees?
   a) Pith
   b) Annual ring of heartwood
   c) Annual ring of sapwood
   d) Vascular cambium

4. Which of the following is responsible for the production of new cells?
   a) Vascular cambium

b) Xylem
c) Phloem
d) All the above

5. Which of the following is the central portion of a tree?
   a) Vascular cambium
   b) Pith
   c) Medullar rays
   d) Barks

6. Which of the following part protects and safeguards tree against the external environment?
   a) Pith
   b) Medullar rays
   c) Sapwood
   d) Barks

7. Which among the defects is formed naturally?
   a) Chip marks
   b) Diagonal marks
   c) Blue stain
   d) Star shakes

8. Fibre saturation point is related to which among the following word?
   a) Strengthening of fibre by elongation
   b) Removal of water
   c) Stretching of fibres
   d) None of the above

9. Which of the following is a defect not caused by seasoning?
   a) Bows
   b) Cups
   c) Twisting
   d) Cup shakes

10. Which part of the tree produces xylem and phloem?
    a) Pith
    b) Annular rings
    c) Sapwood
    d) Vascular cambium

11. According to the relevant IS code, the weight of the timber is to be reckoned at a moisture content of
    a) Zero
    b) 4%

c) 8%
d) 12%

[IES 1995]

12. The strength of timber is maximum when load applied is
    a) Parallel to grain
    b) Perpendicular to grain
    c) Inclined at 45° to grain
    d) Inclined at 60° to grain

[IES 1995]

13. Assertion (A): Dimensional changes in wood result due to variation in the moisture content of the wood with atmospheric condition.
    Reason (R): The cell walls in wood are highly hygroscopic and when exposed to moisture, absorb large amounts of water and swell.
    a) Both A and R are true and R is the correct explanation of A.
    b) Both A and R are true and R is not the correct explanation of A.
    c) A is true but R is false.
    d) A is false but R is true.

[IES 1995]

14. The nail diameter should not be more than (t = least thickness of the wooden member to be connected)
    a) t/6
    b) t/8
    c) t/10
    d) t/12

[IES 1996]

15. The expansion and shrinkage of ply-woods are comparatively very low as
    a) They are held in position by adhesives
    b) They are glued under pressure
    c) Plies are placed at right angles to each other
    d) They are prepared from veneers

[IES 1996]

16. Seasoning of timber is required to
    a) Soften the timber
    b) Harden the timber
    c) Straighten the timber
    d) Remove sap from the timber

[IES 1996]

17. During the conversion of timber by sawing, in order to obtain strong timber pieces, the cuts should be made by

a) Ordinary sawing
b) Tangential sawing
c) Quarter sawing
d) Radial sawing

[IES 1997]

18. A timber beam of effective span L and of (b × d) cross-section is said to be laterally supported if d/b and L/b are, respectively
    a) Less than 1 and less than 48
    b) Less than 2 and less than 49
    c) Less than 3 and less than 50
    d) Less than 4 and less than 51

[IES 1997]

19. A timber column is made up of two individual members with longitudinal axes parallel, separated at the ends and middle points of their length by blocking, and joined at the ends by timber connectors. Such a column is called a
    a) Built-up column
    b) Composite column
    c) Spaced column
    d) Flitched column

[IES 1997]

20. The moisture content in structural timber should be
    a) Less than 5%
    b) 5 to 10%
    c) 10 to 20%
    d) 15 to 25%

[IES 1998]

21. Match list I with list II and select the correct answer using the codes given below the lists:

	List I		List II
A	The innermost part of core of the stem of a tree	1	Transverse septa (medullary rays)
B	The vascular tissue which encloses the pith	2	Annual rings
C	A cellular tissue and woody fibre arranged in distinct concentric circles	3	The cambium layer
D	The thin layer below the bark not converted into sapwood as yet	4	The outermost cover or skin of the stem
		5	Medulla (pith)

Codes:

	A	B	C	D
a)	2	5	3	4
b)	5	1	2	3
c)	4	3	2	1
d)	5	1	4	3

[IES 1998]

22. Consider the following methods of preservation of timber
    1. Dipping
    2. Brushing or spraying
    3. Pressure impregnation

The correct sequence in decreasing order of the effectiveness of these methods of preservation is

a) 1, 2, 3
b) 2, 1, 3
c) 3, 1, 2
d) 3, 2, 1

[IES 1999]

23. Radial splits in timber originating from "bark" and narrowing towards the "pith" are known as
    a) Heart shakes
    b) Star shakes
    c) Cup shakes
    d) Knots

[IES 1999]

24. The moisture content in a properly seasoned timber will be in the range of
    a) 5–8%
    b) 8–10%
    c) 10–12%
    d) 12–15%

[IES 2000]

25. The strength of timber is maximum in the direction
    a) Perpendicular to the grains
    b) Parallel to the grains
    c) 45° to the grains
    d) At all angles

[IES 2000]

26. On application of external stress on timbers, it behaves like
    a) An elastic material
    b) Non-elastic material
    c) Visco-elastic material
    d) Non-visco-elastic material

[IES 2001]

27. The ratio of tangential shrinkage to radial shrinkage of wood due to reduction in moisture content is
    a) In the range from 3.1 to 5.1
    b) In the range from 2 to 3
    c) In the range from 1 to 2
    d) Less than or equal to 1

[IES 2001]

28. Assertion (A): Trees which have broad leaves and shed in the autumn are classified as hard woods, while trees having needle-like leaves, broadly evergreen are classified as soft woods.

    Reason (R): The term hard wood and soft wood in relation to a species of tree do not necessarily indicate relative hardness or density.
    a) Both A and R are true and R is the correct explanation of A.
    b) Both A and R are true and R is not the correct explanation of A.
    c) A is true but R is false.
    d) A is false but R is true.

[IES 2002]

29. Timber can be made reasonably fire resistant by
    a) Soaking it in ammonium sulphate
    b) Coating with tar paint
    c) Pumping creosote oil into timber under high pressure
    d) Seasoning process

[IES 2002]

30. Which one of the following statements is the correct description of the structure of fibre board?
    a) Thin slices of superior quality of wood are glued and pressed on the surface of inferior wood.
    b) Steamed mass of wood dusts, wood wool and other vegetable fibres are pressed hard to a thickness varying from 3 mm to 12 mm.
    c) Thin and narrow wood shavings are soaked in a refractory binder material and pressed hard.
    d) Wood veneer is backed by fabric mat.

[IES 2002]

31. Consider the following methods of preservation of timber
    1. Pressure application
    2. Brush application
    3. Dipping
    4. Open tank application

The correct sequence of this method in the increasing order of their effectiveness is

a) 1, 3, 4, 2
b) 3, 4, 2, 1
c) 2, 3, 4, 1
d) 4, 2, 1, 3

[IES 2002]

32. Match list I (name of defect) with list II (definition) and select the correct answer using the codes given below the lists:

	List I		List II
A	Cupping	1	Caused by wood limbs encased by the wood of the free trunk
B	Bowing	2	Caused by grain irregularities in the board and can be eliminated by proper stacking
C	Chucks	3	Small cracks appearing at the ends of boards caused by too rapid drying
D	Knots	4	Unequal shrinking in the radial and tangential direction

Codes:

	A	B	C	D
a)	1	2	3	4
b)	4	3	2	1
c)	1	3	2	4
d)	4	2	3	1

[IES 2002]

33. The maximum deflection in timber beams or joints should not be greater than

a) span/300
b) span/325
c) span/360
d) span/380

[IES 2002]

34. Assertion (A): Knots, one of the common features in wood, are associated with the beginning of branches.

    Reason (R): Knots greatly improve the workability.

    a) Both A and R are true and R is the correct explanation of A.
    b) Both A and R are true and R is not the correct explanation of A.
    c) A is true but R is false.
    d) A is false but R is true.

[IES 2002]

35. Assertion (A): Timbers used for engineering construction are derived from deciduous trees.

    Reason (R): Deciduous trees yield hard wood while conifers yield soft wood.

    a) Both A and R are true and R is the correct explanation of A.
    b) Both A and R are true and R is not the correct explanation of A.
    c) A is true but R is false.
    d) A is false but R is true.

[IES 2002]

36. Assertion (A): Dry rot is a disease in wood caused by spores germinating in wood cells.

    Reason(R): Decomposition and putrefaction of tissues of a standing tree are indicators of dry rot.

    a) Both A and R are true and R is the correct explanation of A.
    b) Both A and R are true and R is not the correct explanation of A.
    c) A is true but R is false.
    d) A is false but R is true.

[IES 2002]

37. Dry rot in timber is caused by
    a) Lack of ventilation
    b) Lack of light
    c) Immersion in water
    d) Alternate wet and dry atmosphere

[IES 2003]

38. Wood is impregnated with creosote oil in order to
    a) Change its colour
    b) Protect against fungi
    c) Protect the annular layers
    d) Fill up the pores

[IES 2003]

39. Consider the following statements

    Hardest timber is obtained from the wood grown in

1. The moderately dry climatic regions
2. The Himalayan slopes
3. The open areas
4. The thin jungles

Which of these statements are correct?

a) 1 and 3
b) 1 and 4
c) 2 and 3
d) 2 and 4

[IES 2004]

40. Consider the following statements

Dry rot in timber is due to

1. Stacking wood in open areas
2. Lack of ventilation
3. Decomposition of sap
4. Lack of preservatives

Which of these statements are correct?

a) 1 and 2
b) 1 and 3
c) 2 and 3
d) 3 and 4

[IES 2004]

41. Consider the following statements

The disease of dry rot in timber is caused by

1. Complete submergence in water
2. Alternate wet and dry condition
3. Lack of ventilation

Which of these statements is/are correct?

a) 1 only
b) 3 only
c) 2 only
d) 2 and 3

[IES 2005]

42. Which one of the following is the correct statement?

The strength of timber

a) Is maximum in a direction parallel to the grain
b) Is maximum in a direction perpendicular the grain
c) Is maximum in direction 45° to the grain
d) Remain same in all directions

[IES 2005]

43. Consider the following statements

    Kiln seasoning of timber results in

    1. Reduced density
    2. Reduced life
    3. Dimensional stability

    Which of these statements is/are correct?

    a) 1, 2 and 3
    b) 1 only
    c) 2 and 3
    d) 1 and 3

[IES 2005]

44. In a tree, the cambium layer is situated between
    a) The outer bark and inner bark
    b) The inner bark and sap wood
    c) The sap wood and heart wood
    d) The pith and heart wood

[IES 2005]

45. The timber preservative "creosote" belongs to the group of
    a) Water soluble salts
    b) Organic solvent type
    c) Tar oil type
    d) Inorganic solvent type

[IES 2006]

46. Which one of the following is the most preferred wood for high quality and durable furniture?
    a) Sandalwood
    b) Deodar wood
    c) Teak wood
    d) Shisham wood

[IES 2007]

47. As a natural material, timber is which one of the following?
    a) Isotropic
    b) Anisotropic
    c) Homogeneous
    d) Heterogeneous

[IES 2007]

48. Shear strength of timber depends on which one of the following?
    a) Lignin with fibres
    b) Medullary rays

c) Heartwood
d) Sapwood

[IES 2007]
49. The defect which develops due to uncontrolled and non-uniform loss of moisture from wood is known as which one of the following?
a) Knot
b) Shake
c) Warping
d) Cross grain

[IES 2007]
50. With respect to the moisture content in wood, the fibre saturation point refers to which one of the following?
a) Free water present in the cells
b) Free water present in cell walls and cell
c) No moisture present in cell walls and cell cavities
d) No free water exists in cell cavities but cell walls are saturated

[IES 2008]
51. What is the ratio of the elastic modulus of structural timber in longitudinal direction to that in the transverse direction?
a) 1/2 to 1
b) 1/10 to 1/20
c) 1 to 2
d) 10 to 20

[IES 2009]
52. What is the treatment for making timber resistant?
a) ASCU treatment
b) Abel's process
c) Creosoting
d) Tarring

[IES 2009]
53. How is the process of treatment of wood using a preservative solution and forcing in at a pressure designated?
a) Rueping process
b) Lawry process
c) Full cell process
d) Empty cell process

[IES 2009]
54. Consider the following statements
1. Cambium layer is between sapwood and heartwood.
2. Heartwood is otherwise termed as deadwood.
3. Timber used for construction is obtained from heartwood.

Which of these statements is/are correct?

a) 1, 2 and 3
b) 2 and 3 only
c) 1 and 2 only
d) 2 only

[IES 2010]

55. The advantages in using plywood is that the
    a) Tensile strength is equal in all directions
    b) Higher tensile strength in longer direction
    c) Higher tensile strength in shorter direction
    d) Lower tensile strength in longer direction

[IES 2010]

56. Which one of the following statements is correct as regards tensile strength of wood?
    a) Minimum in the direction parallel to the grains
    b) Maximum in the direction parallel to the grains
    c) Maximum in the direction across the grains
    d) Same in all directions

[IES 2010]

57. Consider the following characteristics regarding timber
    1. Stronger variety
    2. Ability to take very smooth finish
    3. Toughness
    4. Difficult to season

Which of the above characteristics is/are essential for timber to be used as beams?

a) 1 only
b) 2 and 3
c) 3 and 4
d) 1 and 3

[IES 2010]

58. Consider the following statements
    1. Dry rot in sap wood is caused by fungal attack.
    2. Brown rot in coniferous wood is a result of fungal attack.
    3. Alternate wetting and drying of unseasoned timber causes "powdery" form of decay in wood.

Which of these statements are correct?

a) 1, 2 and 3
b) 1 and 2 only
c) 2 and 3 only
d) 1 and 3 only

[IES 2011]

59. Consider the following statements

    Fibre saturation point in wood is reached when

    1. Free water is removed.
    2. Cell water is removed
    3. Shrinkage of wood is rapid
    4. Strength gain is rapid

    Which of these statements are correct?

    a) 1, 2 and 3
    b) 1 and 2 only
    c) 2 and 4 only
    d) 1, 3 and 4

[IES 2011]

60. The radial splits which are wider on the outside of the log and narrower towards the pith are known as

    a) Star shakes
    b) Annular rings
    c) Cup shakes
    d) Heart shakes

[IES 2012]

61. Consider the following distinguishing characteristics of hardwood

    1. They have distinct annular rings.
    2. They are non-resinous.

    Which of these characteristics of hardwood is/are correct?

    a) 1 only
    b) 2 only
    c) Both 1 and 2
    d) Neither 1 nor 2

[IES 2012]

62. Consider the following statements on the specific gravity of wood

    1. It is always greater than 2.
    2. It is less than 1.
    3. It is not dependent upon temperature and equilibrium moisture content.
    4. It is dependent upon type of species.

    Which of these statements are correct?

    a) 1, 2, 3 and 4 only
    b) 1 and 3 only
    c) 2 and 3 only
    d) 2 and 4 only

[IES 2012]

63. The age of a log of timber can be estimated by
    a) Diameter of pith
    b) Thickness of bark
    c) Number of annular rings
    d) Number of medullary rays

[IES 2012]

64. Consider the following statements
    Seasoning of timber results in
    1. Increased strength
    2. Increased durability
    3. Reduced resilience
    4. Increased dimensional stability
        a) 1, 2 and 4
        b) 1, 2 and 3
        c) 1, 3 and 4
        d) 2, 3 and 4

[IES 2012]

65. Statement (I): Planks sawn from trees with twisted fibres are stronger than those cut from trees with normal growth.
    Statement (II): Timber from trees with twisted fibres is used straightway as poles.
    a) Both statement (I) and statement (II) are individually true and statement (II) is the correct explanation of statement (I).
    b) Both statement (I) and statement (II) are individually true and statement (II) is not the correct explanation of statement (I).
    c) Statement (I) is true but statement (II) is false.
    d) Statement (I) is false but statement (II) is true.

[IES 2012]

66. Statement (I): Timber suitable for tension members is obtained from coniferous trees.
    Statement (II): Coniferous trees have distinct annular rings and straight grains.
    a) Both statement (I) and statement (II) are individually true and statement (II) is the correct explanation of statement (I).
    b) Both statement (I) and statement (II) are individually true and statement (II) is not the correct explanation of statement (I).
    c) Statement (I) is true but statement (II) is false.
    d) Statement (I) is false but statement (II) is true.

[IES 2012]

67. Which of the following statements are correct?
    1. Knots in a tree affect the continuity of fibres.

2. Nail knots do not influence the strength of timber.
3. Druxiness of wood is not a defect
4. Tall trees with twisted fibres given good timber for poles
    a) 1, 2 and 3 and 4
    b) 1, 2 and 4 only
    c) 1, 2 and 3 only
    d) 2, 3 and 4 only

[IES 2013]

68. Excrescences in wood are
    1. Defects found in trees
    2. Formed due to injuries inflicted on trees when they are growing
    3. Defects which render wood suitable as firewood only
        a) 1, 2 and 3
        b) 1 and 2 only
        c) 1 and 3 only
        d) 2 and 3 only

[IES 2013]

69. Deterioration of structure of timber due to dry rot is
    1. Caused by fungi
    2. Due to dry-spell after heavy rains
    3. Due to attack of termites
    4. Indicated by surface stripes on scantlings
        a) 1 and 2
        b) 3 and 4
        c) 2 and 3
        d) 1 and 4

[IES 2013]

70. Alternate wetting and drying of timber
    1. Results in shrinkage and swelling
    2. Brings about wet rot onset
    3. Increases the durability
    4. Causes transmission of spores from germination
        a) 1, 2 3 and 4
        b) 1, 2 and 4 only
        c) 1, 2 and 3 only
        d) 2, 3 and 4 only

[IES 2013]

71. Consider the following with regard to "the application of preservation of timber"
    1. Increase in the life span of the member
    2. Increase in the strength of the timber

3. Removal of moisture
4. Prevention of growth of fungi by killing them

Which of the above are correct?

a) 1, 2, 3 and 4
b) 2 and 4 only
c) 1 and 4 only
d) 2 and 3 only

[IES 2014]

72. Assertion (A): Timber suitable for furniture is obtained from conifers only.
Reason (R): Woods with distinct annual rings are conifers.
a) Both A and R are true and R is the correct explanation of A.
b) Both A and R are true and R is not the correct explanation of A.
c) A is true but R is false.
d) A is false but R is true.

[IES 2015]

73. Assertion (A): Seasoning of timber gives dimensional stability, safety against attack by fungi and improved workability.
Reason (R): Seasoning of timber removes moisture in the form of sap from timber.
a) Both A and R are true and R is the correct explanation of A.
b) Both A and R are true and R is not the correct explanation of A.
c) A is true but R is false.
d) A is false but R is true.

[IES 2015]

74. Consider the following statements
1. There will be no defects in select grade timbers.
2. The codal values for strength of grade II timber without defects may be reduced by 37.5%.
3. For timber used as columns, the permissible stress in ungraded timbers is adopted with a multiplying factor of 0.50.
4. In case of wind force and earthquakes, a modification factor of 1.33 is adopted.

Which of the above statements are correct?

a) 1 and 3 only
b) 1 and 4 only
c) 2 and 4 only
d) 2 and 3 only

[IES 2016]

75. Consider the following statements regarding timber

1. The strength of timber increases by kiln seasoning.
2. Cutting of wood is to be done prior to treatment.
3. Water seasoning is good for prevention of warping.
4. ASCU treatment enhances the strength of wood.

Which of the above statements are correct?

a) 1, 2 and 3 only
b) 2, 3 and 4 only
c) 1, 3 and 4 only
d) 1, 2, 3 and 4

[IES 2016]

76. Gase(s) emitted during rotting or decomposition of timber is/are mainly
    a) Methane and hydrogen
    b) Hydrogen sulphide
    c) Carbonic acid and hydrogen
    d) Ammonia

[IES 2016]

77. In the cross-section of a timber, cambium layer can occur in
    a) Inner bark and sap wood
    b) Pith and heart wood
    c) Sap wood and heart wood
    d) Outer bark and sap wood

[IES 2016]

78. Statement (I): Splitting of fibres is a type of seasoning defect in wood.

    Statement (II): Seasoning of timber is a general requirement for structural purposes.
    a) Both statement (I) and statement (II) are individually true and statement (II) is the correct explanation of statement (I).
    b) Both statement (I) and statement (II) are individually true and statement (II) is not the correct explanation of statement (I).
    c) Statement (I) is true but statement (II) is false.
    d) Statement (I) is false but statement (II) is true.

[IES 2016]

79. Statement (I): "Compreg" timbers have higher specific gravity of up to 1.30 and are stronger than other timbers.

    Statement (II): Impregnation of resins and special curing methods are adopted to develop "Compreg" timbers.
    a) Both statement (I) and statement (II) are individually true and statement (II) is the correct explanation of statement (I).
    b) Both statement (I) and statement (II) are individually true and statement (II) is not the correct explanation of statement (I).

c) Statement (I) is true but statement (II) is false.
d) Statement (I) is false but statement (II) is true.

[IES 2017]

80. Statement (I): Air seasoning of structural timber renders it more durable, tough and elastic.

Statement (II): Air seasoning of timber is the most economical and eco-friendly method of treatment when time is not a constraining criterion.

a) Both statement (I) and statement (II) are individually true and statement (II) is the correct explanation of statement (I).
b) Both statement (I) and statement (II) are individually true and statement (II) is not the correct explanation of statement (I).
c) Statement (I) is true but statement (II) is false.
d) Statement (I) is false but statement (II) is true.

[IES 2018]

81. Seasoning of timber is done for removing
a) Knots from timber
b) Sap from timber
c) Roughness of timber
d) None of the above

[SSC JE 2007]

82. Generally, wooden moulds are made from
a) Plywood
b) Shisham wood
c) Deodar wood
d) Teak wood

[SSC JE 2007]

83. Plywood is made from
a) Common timber
b) Bamboo fibre
c) Leak wood only
d) Asbestos sheets

[SSC JE 2008]

84. The age of a tree can be known by examining
a) Cambium layer
b) Annular rings
c) Medullary rays
d) Heart wood

[SSC JE 2009]

85. Dry rot
a) Cracks the timber

b) Reduces the timber to powder
c) Reduces the strength of timber
d) Shrinks the timber

[SSC JE 2010]

86. Age of a tree may be ascertained by the
a) Radius of its stem
b) Circumference of its stem
c) Number of branches
d) Number of annual rings

[SSC JE 2010]

87. The moisture content in a well-seasoned timber is
a) 5–10%
b) 10–12%
c) 12–15%
d) 30–50%

[SSC JE 2010]

88. Plywood is obtained by gluing wooden sheets at a pressure of
a) 100–150 N/cm^2
b) 100–130 N/cm^2
c) Both (a) and (b)
d) Neither (a) nor (b)

[SSC JE 2010]

89. The disease of dry rot in timber is caused by
a) Complete submergence in water
b) Alternative wet and dry conditions
c) Lack of ventilation
d) None of these

[SSC JE 2012]

90. Which of the following is the hardest wood?
a) Babul
b) Chir
c) Teak
d) Shisham

[SSC JE 2012]

91. The life of teakwood doors and windows is usually taken to be
a) 80 years
b) 60 years
c) 40 years
d) 20 years

[SSC JE 2017]

92. How does the seasoning of timber help?
    A. It increases the weight of timber.
    B. It improves the strength properties of timber.

    a) Only A
    b) Only B
    c) Both A and B
    d) None of these

[SSC JE 2017]

93. Pick up the correct statement from the following method of sawing timber
    a) Tangentially to annual rings is known as tangential method.
    b) In four quarters such that each board cuts annual rings at angles not less than 45° is known as quarter sawing method.
    c) Cut out of quarter logs, parallel to the medullary rays, and perpendicular to annual rings is known as radial sawing.
    d) All options are correct.

[SSC JE 2017]

94. For the manufacture of plywood, veneers are
    a) At right angles
    b) Parallel
    c) Inclined at 45°
    d) Inclined at 60°

[SSC JE 2017]

95. The solution of salts from the soil absorbed by the trees which becomes a viscous solution due to loss of moisture and the action of carbon dioxide is known as
    a) Pith
    b) Cambium
    c) Bark
    d) Sap

[SSC JE 2017]

96. The most valuable timber may be obtained from
    a) Chir
    b) Shisham
    c) Sal
    d) Teak

[SSC JE 2017]

97. The timber having maximum resistance against white ants is obtained from
    a) Chir
    b) Shisham

c) Sal
d) Teak

[SSC JE 2017]
98. Due to attack of dry rot, the timber
a) Cracks
b) Shrinks
c) Reduces to powder
d) None of these

[SSC JE 2017]
99. The defect in timber that arises due to the swelling caused by growth of layers of sap wood over the wounds after branch is cut off is called as
a) Checks
b) Knots
c) Shakes
d) Rind gall

[SSC JE 2018]
100. Which of the following qualities of timber can improved using Abel's process?
a) Durability
b) Fire resistance
c) Chemical resistance
d) Strength

[SSC JE 2018]
101. According to the IS code, at what moisture content weight of timber is noted?
a) 0.05
b) 0.12
c) 0.23
d) 0.3

[SSC JE 2018]
102. When timber is burned in the wood fire over depth of about 15 mm the process of treatment is known as
a) Charring
b) Rueping process
c) Bethal process
d) Boucherie process

[SSC JE 2018]
103. For which of the following process Boucherie process is used?
a) Manufacturing of bricks
b) Manufacturing of cement

c) Production of clay tiles
d) Treatment of green timber

[SSC JE 2018]

104. Which of the timber is used to make the goods that are used in the sports industry?
a) Alder
b) Asanfona
c) Mulberry
d) Balsa

[SSC JE 2018]

105. In the air drying process, the practical limit of moisture content is
a) 0.05
b) 0.15
c) 0.25
d) 0.35

[SSC JE 2018]

106. Which one of the following treatments is used to make the timber fire resistant?
a) Abel's process
b) Empty cell process
c) Envelope treatment
d) Tarring

[SSC JE 2018]

107. Saw dust can be rendered chemically inert by boiling it in water containing
a) Ferrous sulphate
b) Potassium chloride
c) Ammonia
d) None of these

[SSC JE 2018]

108. Due to attack of dry rob, the timber
a) Cracks
b) Twist
c) Shrinks
d) Reduce to powder

[AE 2013]

109. A well-seasoned timber may contain moisture up to
a) 1%
b) 2%

c) 5%
d) 12%

[AE 2013]

110. The central part of a tree is called
a) Heart wood
b) Pith
c) Sap wood
d) Cambium layer

[AE 2013]

111. The optimum age for the felling of trees varies between .... decades.
a) 5–10
b) 10–15
c) 15–20
d) 20–25

112. Name of the seasoning method for timber specifically preferred in Kerala due to less cost is
a) River seasoning
b) Water seasoning
c) Both a and b
d) None of the above

113. The method of application of creosote oil on the timber section is known as
a) Soathing
b) Creosoting
c) Pressure treatment
d) None of the above

114. Timber section is attacked by fungi if the moisture content in it is higher than .... % and also there is the availability of air surrounding that section.
a) 2
b) 10
c) 15
d) 20

115. Radial fibres that extend from the pith to the cambium layer and hold the annual rings of heartwood and sapwood in position are known as
a) Medullary rays
b) Pith rays
c) Cambium rays
d) None of the above

116. Assume unidirectional loading condition. What is the most preferred orientation angle of fibre in FRP with respect to loading?
   a) 0
   b) 45
   c) 90
   d) None of the above

1.	d	2.	a	3.	b	4.	d	5.	b	6.	d	7.	d	8.	b	9.	d	10.	d
11.	d	12.	a	13.	a	14.	a	15.	c	16.	d	17.	d	18.	c	19.	c	20.	c
21.	b	22.	c	23.	b	24.	c	25.	b	26.	a	27.	a	28.	b	29.	a	30.	b
31.	c	32.	d	33.	c	34.	c	35.	a	36.	c	37.	a	38.	c	39.	a	40.	c
41.	b	42.	a	43.	d	44.	b	45.	c	46.	c	47.	b	48.	a	49.	c	50.	d
51.	d	52.	b	53.	a	54.	b	55.	a	56.	b	57.	d	58.	a	59.	d	60.	a
61.	b	62.	d	63.	c	64.	a	65.	d	66.	d	67.	b	68.	b	69.	a	70.	b
71.	a	72.	d	73.	a	74.	b	75.	a	76.	c	77.	a	78.	b	79.	a	80.	b
81.	b	82.	b	83.	a	84.	b	85.		86.	d	87.	b	88.	c	89.	c	90.	d
91.	d	92.	b	93.	d	94.	a	95.	d	96.	d	97.	d	98.	c	99.	d	100.	b
101.	b	102.	a	103.	d	104.	c	105.	b	106.	a	107.	a	108.	d	109.	d	110.	B
111.	a	112.	c	113.	b	114.	d	115.	a	116.	a								

REFERENCES

IS 2408. (2015). *Methods of static tests of timbers in structural sizes.* Bureau of Indian Standards. New Delhi.

Graduate Aptitude Test in Engineering (GATE). (2020). *GATE Office*, Chennai, India. http://gate.iitm.ac.in/gate2019/previousqp18.php (accessed September 18, 2020).

Engineering Service Exam (ESE). (2020). *Union Public Service Commission*, New Delhi, India. https://www.upsc.gov.in/examinations/previous-question-papers (accessed September 18, 2020).

Assistant Engineer (AE). (2020). *Tamil Nadu Public Service Commission (TNPSC)*, Tamil Nadu, India. http://www.tnpsc.gov.in/previous-questions.html (accessed September 18, 2020).

Junior Engineer (JE), *Staff Selection Commission (SSC)*, New Delhi, India. https://ssc.nic.in/Portal/SchemeExamination (accessed September 18, 2020).

9 Paints and Varnishes

9.1 INTRODUCTION

Paints and varnishes are used as a protective membrane on many surfaces. They primarily protect the surface from deterioration and aggressive agents. Deterioration includes corrosion in the case of structures made up of steel and other metals. In the case of concrete structures, the surface should be protected from external agents such as temperature, air, water, chloride, etc. Besides, the paint is also used as a decorative material. Moreover, due to recent developments in colloidal materials, paints that can self-cleanse have been introduced. Still, there have been a considerable number of developments and innovations in paints with respect to their vast applications. These materials that are being used in construction and many more areas require a primary check in order to assess their quality. This unit introduces some necessary tests to check the quality of coating materials.

9.2 TESTING PROCEDURES

9.2.1 Paint and Varnishes – Pull-off Test for Adhesion

This test method is used to compare the adhesion behaviour of different coatings.

Test Summary

This test consists of applying a coating on the required surface (deformable or rigid or dolly) and allowing it to cure for the specified period. This is followed by conducting a pull-off test in order to determine the maximum perpendicular force the surface can bear before failure (detachment of plug of material). This failure can be either an adhesive failure (the weakest interface) or a cohesive failure (the weakest component) or a mix of adhesion and cohesion failure (ASTM D7234-19, 2019).

Use and Significance

- This test method is used to understand the quality of a coating applied on a substrate. Based on the magnitude of the required pull-off force, the weakest component can be identified. Moreover, a comparable understanding of the suitability of a coating for a different practical application can be decided.

Required Apparatus

- Tensile tester (capable of applying tensile force in the perpendicular direction that can be increased at a substantially uniform rate ($\ngtr$1 MPa/ s)).
- Test dollies
- Centring devices

- Cutting device
- Adhesives (preliminary screening should be done in order to understand the suitability of adhesives for the test. The strength of the adhesives should be more than coating under test), e.g. cyanoacrylate, solvent less epoxide and per-oxide catalysed polyester adhesives can be used
- Substrate selected from ISO 1514. In most of the cases, the same type of material is used. The substrate should be clean and free from distortion

Pre-treatment and Coating

- Pre-treatment and conditioning of substrate and coating are carried out similar to the original surfaces where the coating will be applied. (Pre-treatment method should be mentioned in student remarks.)
- Coating thickness should be according to the requirement. ISO 2808 can be used to determine the thickness of the coating in μm.
- Drying and conditioning are carried out for a specified time period before testing.
- The coating is dried for a minimum period of 16 hours at 23 ± 2°C and relative humidity of 50 ± 5%.

Procedure

Step 1. At least six specimens should be tested at a temperature and relative humidity of 23 ± 2°C and 50 ± 5%, respectively.

Step 2. To an uncoated, freshly cleaned dolly surface, prepared adhesives are coated uniformly. The adhesive-coated dolly face is placed in contact with the coating, as shown in Figure 9.1.

Step 3. The time period of placing the adhesive-coated dolly to the coating surface should be equal to the curing time of adhesive.

Step 4. At the end of the curing time, use the cutting device to cut a circumference around the dolly along with the coating surface and the substrate to which the coating was given.

Step 5. Place the outer ring and test, as indicated in Figure 9.1.

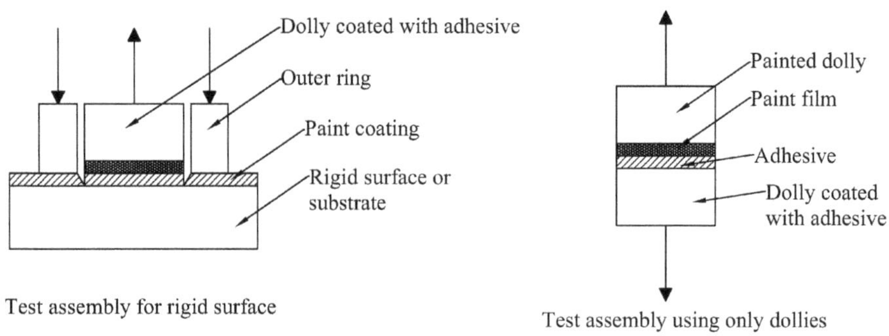

FIGURE 9.1 Test assembly for rigid surface and dollies only.

Step 6. Similar to the above method, two dollies can be used to test, as shown in Figure 9.1. In this case, an adhesive is coated evenly in one of the dolly's cleaned surface. This dolly is placed on the surface of the other dolly, which is coated with the material to be tested.

Step 7. Now, the test assembly is assembled by use of the centring device. The assemble time should be equal to the curing time of the adhesive, as shown in the figure.

Step 8. After assembling and centring the test setup, the specimen is pulled-off by pull-off adhesion tester. Care should be taken to apply uniform force across the test area.

Step 9. Tensile stress is applied at a rate not more than 1 MPa/s perpendicular to the coated surface or dolly such that the failure occurs within a period of 90 seconds.

Step 10. Tensile stress required for failure is noted down. Repeat the same procedure for other prepared test assemblies.

Note: Adhesives are prepared and applied uniformly to form a thin film between different components according to manufacturer instructions. Excess adhesives are removed.

Observations and Calculations

After visual inspection, the nature of the fracture can be reported as follows:

In the case of cohesion failure of substrate – A
In the case of cohesion failure of coat – B (in case of multi-coat this is considered as a first coat)
In the case of adhesion failure between the substrate and the first coat – A/B
In the case of adhesion failure between the first and the second coat – B/C
In case of a multi-coat system of "n" coats, the adhesion failure of the nth coat is – n
In case of adhesion failure between the final coat and the adhesive – */n
Cohesion failure of adhesive is given by – n
Adhesive failure between adhesive and dolly – n/d

Breakage strength is given by $\sigma = \frac{F}{A}$

F – breaking force in newtons

A – is the area of the dolly in mm^2

Report

Average breaking strength .... MPa
Nature of failure ....

Chosen pre-treatment of the surface ....

Thickness of the coating system or thickness of the individual layer .... µm

Student Remark

....

....

....

....

$$\frac{\textit{Marks obtained}}{\textit{Total marks}} = -$$

Instructor signature

Instructor Remark

....

....

....

....

9.2.2 Paint and Varnishes – Determination of the Degree of Settling of Paint

Test Summary

This test method is used to determine the degree of pigment suspension and ease of re-mixing a paint into a homogeneous suspension that can be used for the intended purpose (ASTM 869–85, 2015).

Use and Significance

- Under-processed paint can result in excessive settling. Excessive settling decreases the quality of the paint.

Required Apparatus

- A container of 500 ml capacity (friction-top can paint container) diameter – 85.5 ± 1.5 mm and height – 98.5 ± 1.5 mm
- Spatula weighing 45 ± 1 g with square end blade 125 mm length and 20 mm in width

Procedure

Step 1. Place the specimen to be tested for settling in the container (filling the can to within the 13 mm of the top).

Step 2. The container is closed firmly and placed in an undisturbed place for the shelf-ageing period, as provided by the manufacturer.

Step 3. After the prescribed time, the container is examined without removing the supernatant vehicle, while taking care not to disturb the container (without shaking or agitation).

Step 4. Use the spatula to determine the quantity of paint separated during the shelf-life period.

Step 5. Place the spatula in such a way that the bottom of the spatula is at the top of the container (perpendicular and at the centre area of the paint). Drop the spatula and rate the condition as mentioned below for degree of settling:

a. 10 – Perfect suspension. No change from the original condition of the paint
b. 9 – A definite feel of settling and a slight deposit brought up on spatula. No significant resistance to the sidewise motion of spatula
c. 6 – Definite cake of settled pigment. Spatula drops through the cake to the bottom of the container under its weight. Definite resistance to sidewise motion of spatula. Coherent portions of the cake may be removed on the spatula
d. 4 – Spatula does not fall to the bottom of the container under its own weight. Difficult to move spatula through cake sidewise and slight edgewise resistance is noticed. Paint can be remixed readily to a homogeneous state
e. 2 – When spatula has been forced through the settled layer, it is complicated to move spatula sidewise. Definite edgewise movement of the spatula. The paint can be remixed to a homogeneous state
f. 0 – Very firm cake that cannot be again incorporated with the liquid to form a smooth paint by stirring manually

Step 6. After examining per step 5, if a portion was separated out from the paint as a firm cake, then remove the supernatant liquid separately in a container. Mix the firm cake by reincorporating the supernatant liquid slowly until the pigment has reincorporated to form homogeneous paint for the intended purpose, or the cake cannot be remixed by hand stirring. Condition of the paint is rated according to the step 5.

Report

Rating ....

Description of paint condition ....

Student Remark

....

....

....

....

$$\frac{\textit{Marks obtained}}{\textit{Total marks}} = -$$

Instructor signature

Instructor Remark

....
....
....
....
....

9.2.3 Paint and Varnishes – Measurement of Yield Stress of Paints, Inks and Related Liquid Materials

Test Summary

This test method is used to determine the yield stress of the paint. Yield stress is defined as the minimum force required to deform the solid-state and allow it to flow (fluid-like behaviour). In this test, a small quantity of the material to be investigated is taken in a container or in the available setup of rheometer or viscometer and sheared at different shear rates or shear stress (ASTM D7836-13, 2013).

Use and Significance

- Determining the yield stress of a material is useful in many ways. Paint needs to spread at a specific rate to cover an entire area. While spreading the paint, if the yield stress is higher, then the ability of paint to spread and form a uniform layer will be reduced.
- Flowing and levelling of coating is inversely proportional to the yield stress. Sag resistance is directly proportional to the yield stress.

Required Apparatus

- Rheometer (stress-controlled or strain-controlled) with required measuring system (vane system, parallel plate or cone plate)
- Viscometer

Procedure

Test Method A:

Step 1. Prepare the specimen according to the requirement. The required specimen can be decided based on the type of measuring system used for the measurement.

Step 2. Specimen is taken in the respective measuring system and equilibrated for the required temperature at which the test to be conducted.

Step 3. Minimise the disturbance to the specimen as much as possible. This will aid in measuring accurate yield stress of the specimen under investigation. If disturbed, then allow the specimen to equilibrate for a specific time period. In this period, particles that are in the suspension can reform their structure by internal interactions.

Step 4. Now, the specimen is sheared at different shear rates and the corresponding shear stress is noted down. A plot is made between shear stress and shear rate, and the plot is extrapolated to the zero shear rate to identify the respective stress of the specimen.

Step 5. The identified stress is the minimum stress required by the specimen to deform or move from its solid state to the liquid state.

Test Method B:

Step 1. Alternatively, the specimen is subjected to shear stress and yielding point can be determined in a stress-controlled rheometer.

Step 2. In this method, a cone plate or a parallel plate or a van type or a coaxial cylinder measuring system is used.

Step 3. Pre-set stress ramp applied to the specimen rises from zero to some specific value. The stress at which the measuring system moves is considered to be the yield stress of the specimen under investigation.

Step 4. In this method, the rate of application of stress will affect the determined yield stress. Either a correction needs to be incorporated or the rate of application of stress should be made visualising the practical application which the specimen undergoes.

Report

1. Rheometer model and the type of measuring system used
2. Type of product under investigation
3. Test method used
4. Shear rate or the profile used
5. Measured yield stress
6. Temperature of the specimen

Student Remark

....

....

....

....

$$\frac{\textbf{\textit{Marks obtained}}}{\textbf{\textit{Total marks}}} = -$$

Instructor signature

Instructor Remark

....
....
....
....

9.3 PRACTICE QUESTIONS AND ANSWERS FROM COMPETITIVE EXAMS

1. During a pull-off test, the failure can be
   a) Adhesion
   b) Cohesion
   c) Both a and b
   d) None of the above

2. What is adhesion?
   a) Attraction between two same materials
   b) Attraction between two different materials
   c) Repulsion between two different materials
   d) Cannot be defined

3. Modifier EVA consists of ....% vinyl acetate.
   a) 70
   b) 75
   c) 80
   d) 85

4. What is the purpose of a pull-off test?
   a) Suitability of a particular coating with a substrate can be identified
   b) Excessive settling in the paint can be identified
   c) Sag of the paint can be identified
   d) None of the above

5. Paint is a
   a) Newtonian material
   b) Non-Newtonian material
   c) Bingham material
   d) Herschel-Bulkley material

6. The stress at which the paint starts to change from solid to liquid state is called
   a) Plastic stress
   b) Apparent stress
   c) Yield stress
   d) None of the above

7. What will happen if the yield stress of the paint is high?
   a) It will spread very easily
   b) It won't spread at all
   c) It will sag very easily
   d) None of the above

8. Sag of paint is directly proportional to
   a) Apparent viscosity
   b) Penetration resistance
   c) Yield stress
   d) None of the above

9. Yield stress of the material can be measured by
   a) Capillary meter
   b) Rheometer
   c) Viscometer
   d) All the above

10. If the temperature of the paint increases, what happens to the yield stress?
    a) Increases
    b) Decreases
    c) Remains the same
    d) None of the above

11. Assertion (A): While painting on flush doors of plywood, putty-filling is done after prime coat.
    Reason (R): This reduces the quantity of paint and the effort involved in the regular coats of the paints.
    a) Both A and R are true, and R is the correct explanation of A.
    b) Both A and R are true, and R is not the correct explanation of A.
    c) A is true, but R is false.
    d) A is false, but R is true.

[IES 1996]

12. In paints, linseed oil is used as
    a) A thinner
    b) A drier
    c) A vehicle
    d) A water-proofing base

[IES 1998]

13. The most commonly used base for timber painting is
    a) Red lead
    b) Zinc white
    c) White lead
    d) Titanium white

[SSC JE 2004]

14. The most durable varnish is
    a) Water varnish
    b) Spirit varnish
    c) Turpentine varnish
    d) Oil varnish

[SSC JE 2008]

15. In paints, the pigments are responsible for
    a) Durability
    b) Colour
    c) Smoothness
    d) Glassy face

[SSC JE 2009]

16. The commonly used thinner in oil paints is
    a) Naptha
    b) Turpentine
    c) Both (a) and (b)
    d) Neither (a) nor (b)

[SSC JE 2010]

17. The base material for distemper is
    a) Chalk
    b) Lime
    c) Clay
    d) Lime putty

[SSC JE 2013]

18. In paints, methylated spirit, naphtha and turpentine are used as
    a) Base
    b) Binder
    c) Solvent
    d) Extender

[SSC JE 2014]

19. Turpentine oil is used in paint as a
    (a) Base
    (b) Carrier
    (c) Drier
    (d) Thinner

[SSC JE 2015]

20. The paints that are most resistant to fire are
    (a) Enamel paints

(b) Aluminium paints
(c) Asbestos paints
(d) Cement paints

[SSC JE 2017]
21. Terra cotta, in buildings, is used for
(a) Insulation
(b) Ornamental work
(c) Sewage lines
(d) Sanitary services

[SSC JE 2017]
22. Bullet proof glass is made of thick glass sheet and a sandwiched layer of
(a) Steel
(b) Stainless steel
(c) High strength plastic
(d) Chromium plate

[SSC JE 2017]
23. The detachment of the paint film from the surface is known as
(a) Chalking
(b) Cracking
(c) Flaking
(d) Wrinkling

[SSC JE 2018]
24. Which of the following is used as the vehicle in the enamel paints?
(a) Linseed oil
(b) Mustard
(c) Varnish
(d) Water

[SSC JE 2018]
25. Which one of the following is used as a carrier in paint?
(a) Almond oil
(b) Linseed oil
(c) Mustard oil
(d) Olive oil

[SSC JE 2018]
26. The metallic oxide used in the form of powder in paint is called
(a) Extender
(b) Base
(c) Vehicle
(d) Solvent

[TN AE 2007]

27. The painting work is generally specified by
    (a) Weight of the paint applied
    (b) Labour used in the painting
    (c) Area of the painted surface
    (d) Number of coating applied

[TN AE 2008]

1.	c	2.	b	3.	c	4.	a	5.	d	6.	c	7.	b	8.	c	9.	d	10.	A
11.	a	12.	c	13.	c	14.	d	15.	b	16.	d	17.	a	18.	c	19.	d	20.	c
21.	b	22.	c	23.	c	24.	c	25.	b	26.	b	27.	c						

REFERENCES

Assistant Engineer (AE). 2020. Tamil Nadu Public Service Commission (TNPSC), Tamil Nadu, India. http://www.tnpsc.gov.in/previous-questions.html (accessed September 18, 2020).

ASTM 869-85. (2015). *Standard test method for evaluating degree of settling of paint.* American Society for Testing and Materials. Pennsylvania, United States.

ASTM D7234-19. (2019). *Standard test method for pull-off adhesion strength of coatings on concrete using portable pull-off adhesion testers.* American Society for Testing and Materials. Pennsylvania, United States.

ASTM D7836-13. (2013). *Standard test methods for measurement of yield stress of paints, inks and related liquid materials.* American Society for Testing and Materials. Pennsylvania, United States.

Engineering Service Exam (ESE). 2020. Union Public Service Commission, New Delhi, India. https://www.upsc.gov.in/examinations/previous-question-papers (accessed September 18, 2020).

Graduate Aptitude Test in Engineering (GATE). 2020. GATE Office, Chennai, India. http://gate.iitm.ac.in/gate2019/previousqp18.php (accessed September 18, 2020).

Junior Engineer (JE), Staff Selection Commission (SSC), New Delhi, India. https://ssc.nic.in/Portal/SchemeExamination (accessed September 18, 2020).

10 Steel

10.1 INTRODUCTION

Steel is one of the widely used materials for construction. Steel generally has a high strength to mass ratio.

10.2 TESTING PROCEDURES

10.2.1 Tensile Strength of Steel

Modern construction applications, especially reinforced cement concrete (RCC), use steel bars. Steel bars are used as a construction material to increase the load-bearing capacity of the structural members. Notably, in those structural members, tensile strength is considered to be very important. It is of utmost importance to check the tensile behaviour of steel bars. (IS 1608 (Part 1))

Test Summary

This test procedure is used to determine the behaviour of steel bars when subjected to tension.

Use and Significance

- In RCC, structural failure can happen because of different means. Commonly used concrete is good in compression but weak in tension. During the transfer of load in a structural member especially, in case of beams, concrete is reinforced with steel bars to ensure that the members are capable of withstanding tensile stresses.

Required Apparatus

- Universal testing machine (UTM)
- Steel specimen under consideration with markings at regular interval
- Extensometer

Testing

Step 1. Steel reinforcement is placed in the universal testing machine, with an extensometer holding the specimen between the gauge length. The gauge length is denoted as L_0 and the initial diameter of the specimen is denoted as D_0.

Step 2. Once the sample is placed in the position and required adjustments are given in the system, the UTM is started. Based on the load requirement, which depends upon the tensile strength of the steel reinforcement, an appropriate UTM is chosen.

Step 3. The rate of loading is chosen in such a way that a proper reading can be recorded from the extensometer and the UTM.

Step 4. Once the specimen fails and breaks into half, the broken half is fitted and the "gauge length" as well as the broken diameter is measured. This length is denoted as L_f, and the broken diameter is denoted as D_f.

Step 5. Draw a plot between stress and strain.
- Stress vs. strain graph until the fracture
- An offset point of 0.2% is drawn and the yield point is denoted

Step 6. The following characteristics can be obtained from the stress vs. strain plot
- Modulus of elasticity
- Yield stress (0.2% offset)
- Ultimate tensile stress
- Percentage reduction in area
- Modulus of resilience
- Modulus of toughness

Report

Readings are reported, as shown in Table 10.1.

TABLE 10.1
Properties of Steel under Investigation

Parameter	Unit	Observations
Initial gauge length (L_0)	mm	
Final gauge length (L_f)	mm	
Initial diameter (D_o)	mm	
Final diameter (D_f)	mm	
Modulus of elasticity	MPa	
Yield strength	MPa	
Ultimate tensile strength	MPa	
Elongation	$\frac{L_f - L_0}{L_0} \times 100\%$	
Reduction of area	$\frac{A_f - A_0}{A_0} \times 100\ \%$	
Modulus of resilience	MPa	
Modulus of toughness	MPa	

Student Remark

....

....

....
....

$$\frac{\textit{Marks obtained}}{\textit{Total marks}} = -$$

Instructor signature

Instructor Remark

....
....
....
....

10.2.2 Bend and Re-bend Test for Steel

Steel is considered to be one of the vital construction materials. The process of construction involves the bending of steel to a particular form. Consequently, it is important to check the ductility of the steel, which is the ability to deform under applied pressure.

Test Summary

This test procedure is used to check the ductility of the steel bar under applied pressure. During the construction of reinforced cement concrete (RCC) structures, bending and re-bending the steel during the process of construction is inevitable. To ensure that the material is safe to deform without fracture, a bend test is performed at two different angles (90° and 180°). Moreover, strain ageing may take place which may result in a reduction in ductility, as well as an increase in the ultimate yield stress of the steel bar. Reduction in ductility can be checked by the re-bend test.

Use and Significance

- Bending test can ensure that the rebars are deformable without any fracture while bending the rebar into different shapes.
- Re-bend test ensures that the load-carrying capacity is not affected; in other words, the ductility of the material is not reduced (ultimate yield strength of the material is not increased due to strain ageing) by retesting the rebar, which has already tested and entered into its inelastic range.

Required Apparatus

- Rebar bending test fixtures (clamps and mandrels)

Sampling

Bend and re-bend test is done for every one specimen for a lot of 20 tonnes of material.

Testing

Bend test

Step 1. Based on the diameter of the chosen steel specimen, the mandrel is selected per Table 10.2.

Step 2. One end of the specimen is clamped; on the other end, force is applied and bent initially to 90°. After bending to 90°, the bent end is examined for cracks.

Step 3. Now, the specimen is bent to an additional 90° and a total bent angle of 180° is made. Once again, the specimen is observed for cracks in the bent portion.

Step 4. If no cracks are observed during the bend test, the specimen has passed the test, otherwise, it has failed.

Re-bend test

Step 1. The mandrel is selected per Table 10.3.

Step 2. One end of the test specimen is clamped, and the other end of the test specimen is bent to 45°.

Step 3. Now, the specimen is placed in boiling water at a temperature of 100°C for a period of 30 minutes.

Step 4. The test specimen is removed from the boiling water and allowed to cool at room temperature.

TABLE 10.2
Bend Test – Mandrel Diameter

	Diameter (∅) of a mandrel for different grades		
Steel rebar nominal size	**Fe415**	**Fe500**	**Fe550**
Until 22 mm	3∅	4∅	5∅
Above 22 mm	4∅	5∅	6∅

TABLE 10.3
Re-bend Test – Mandrel Diameter

	Diameter (∅) of the mandrel for different grades	
Steel rebar nominal size	**Fe 415 and Fe 500**	**Fe 550**
Until 10 mm	5∅	7∅
Above 10 mm	7∅	8∅

Step 5. The test specimen is fixed in the appropriate mandrel and bent back to $22\frac{1}{2}^{\circ}$. The bent portion is examined for cracks.

Step 6. If no cracks are observed during the re-bend test, the specimen has passed the test; else it has failed.

Report

- Bend test – Passed/Failed
- Re-bend test – Passed/Failed

Student Remark

....
....
....
....

$$\frac{\textit{Marks obtained}}{\textit{Total marks}} = -$$

Instructor signature

Instructor Remark

....
....
....
....

10.2.3 Corrosion Testing of Steel Rebar

This test method is used to understand the corrosion activity of reinforced steel by estimating the electrical corrosion potential (ASTM C876-15, 2016).

Use and Significance

- This test method can be used directly in the field as well as in the laboratory to study the corrosion of reinforcement. Regardless of the depth of cover concrete, this test method is applicable to estimate the electrical corrosion potential.
- This test method can be used at any time during the service life of the concrete.

Limitations

- This test method cannot be used for concrete with high resistivity. Variation in relative corrosion activity of adjacent reinforcements may not be discriminated if the cover concrete is above 75 mm.
- The parameters obtained from this test method cannot be used to determine the properties of rebar in the concrete.

- In addition to corrosion potential obtained from this method, the following parameters are additionally required to understand the probable effect of corrosion activity on the service life of the structure. They are chloride content, carbonation depth, corrosion rate and environmental exposure.

Required Apparatus

- Reference electrode: Capable of providing a stable and reproducible potential for the measurement of corrosion potential within a temperature range of 0–49°C.
- In general, the reference electrode is a copper–copper sulphate electrode (Figure 10.1). Depending upon the requirement, other electrodes can also be used. In those cases, the measured potential needs to be converted with respect to the copper–copper sulphate electrode.
- This experiment is similar to a half cell potential reaction, where copper releases two electrons and converts to copper 2+. At 22.2°C, the potential of the Cu–$CuSO_4$ electrode with reference to H_2 electrode is -0.30 V. The above value highly depends upon the temperature. For every 1°F increase of temperature, potential increases negatively by 0.0005 V within a temperature range of 32–120°F (0–49°C).
- Electrical junction device is a pre-wetted sponge immersed in a low electrical resistance contact solution. This sponge will be attached at the end of the reference electrode. This sponge acts as a medium to provide electrical conductivity between the porous electrode and the concrete member. The minimum area of contact should be three times the diameter of the coarse aggregates used in the concrete.
- Electrical contact solution: Solution that is used for pre-wetting constitutes of 95 mL of commercially available wetting agent or liquid detergent thoroughly mixed with potable water of 19 L. Approximately, 15% of isopropyl alcohol or denatured alcohol is added if the working temperature is below 10°C (50°F). This is to avoid the clouding of electrical charges during the contact. This clouding may result in a reduction of water penetrating the

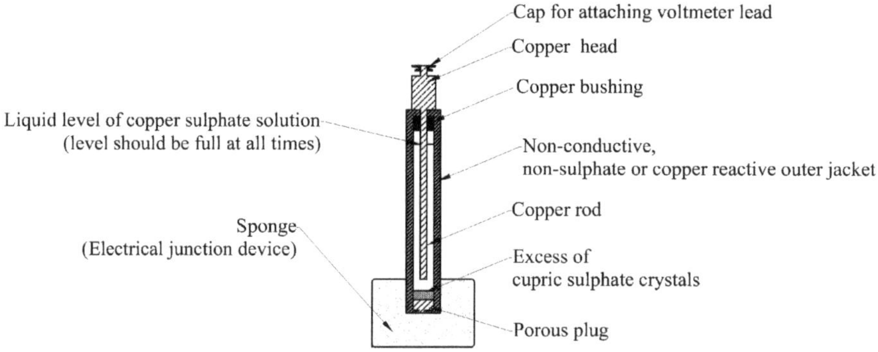

FIGURE 10.1 Sectional view of the copper–copper sulphate reference electrode.

concrete to be tested. Corrosion potential drifts can be reduced by applying conductive gels at the junctions. While measurements are taken in large structures, such as bridges, where the length of longitudinal reinforcement is long, then preliminary surface cleaning of concrete may be helpful.
- The voltmeter should be battery operated and capable of measuring DC voltages with adequate input impedance.
- Errors from high-resistive concrete can be avoided by pre-determining the precision of voltmeter with a variable input impedance voltmeter with a range of 10–200 MΩ. The initial reading is taken at 10 MΩ, and successive readings are taken at higher impedance values until two successive readings remain constant. If a constant reading is not obtained until 200 MΩ, then a galvanometer of 1 GΩ can be used for this purpose.
- Electromagnetic interferences such as AC power or radiofrequency waves within the proximity need to be avoided.
- Electrical lead wire of not more than 150 m length with very less resistivity (less than 0.0001 V) direct burial type insulation is used.

Cautions to Be Observed before Performing the Test

- Care should be taken while storing, calibrating and maintaining the electrodes. Electrodes should not be dried out or contaminated. If the electrodes are not used for a longer time, then the electrodes should be covered with the sponge plugs to avoid drying.
- Voltmeter and reference standards should be calibrated as and when required.

Procedure

Step 1. Measurement spacing should be maintained correctly. If the predefined spacing is not available for measurements, then a spacing chart should be made prior to the measurement. The spacing and the measurement should be consistent. A cover meter can be used to match the reinforcements.

Step 2. For large horizontal structures, 1.2 m is considered as appropriate. In general, larger spacing should be avoided. This may result in missing regions of localised corrosion. The spacing between two readings needs to be decreased in the case of adjacent value readings more than 50 mV. In the same way, in case of structural cracks, cold joints and the places where the dynamic activity of the structure is high, then the readings may shoot more than few 100 mV within a distance of 300 mm. In the above cases, care must be taken not to miss out any localised corrosion.

Step 3. To establish a connection between the voltmeter and the reinforcing steel bar, a temporary or a permanent connection can be made. The connection is generally made by clamping or welding a protruding bar with the reinforcing steel bar. In order to ensure low electrical resistivity, the clamp bar and the reinforcing bar should be cleaned to ensure a bright metal-to-bright metal contact. In some cases, reinforcements are exposed by the removal of concrete cover. Connect the reinforcing steel to the positive terminal of the voltmeter. In the case of pre-stressing steel, only a mechanical connection needs to be made. Care should be taken to ensure that no damage is done to the pre-stressed bars.

Step 4. In case exposing steel is connected directly to the reinforcing steel, then the exposing steel can be used for connection with the voltmeter.

Step 5. Electrical continuity should be checked by surveying opposite ends of the diagonal areas.

Step 6. The reference electrode is connected to the negative (ground) terminal of the voltmeter by the lead wire. Figure 10.2 shows the circuit to identify corrosion potential.

Step 7. Pre-wetting needs to be done by checking the following conditions.

- Place the reference electrode on the surface of the concrete and ensure that there is no deviation in the voltmeter. The reading should be stable (±0.02 V) for at least a period of 5 minutes. If the above condition is established, then pre-wetting of the concrete surface is not necessary. As the resistivity of the concrete is less.
- On the other hand, if the resistivity is high, then pre-wetting needs to be carried out. Pre-wetting needs to be adopted until the reading is stable (±0.02 V) for a period of 5 minutes. If the reading is not stabilised after 5 minutes, then the electrical

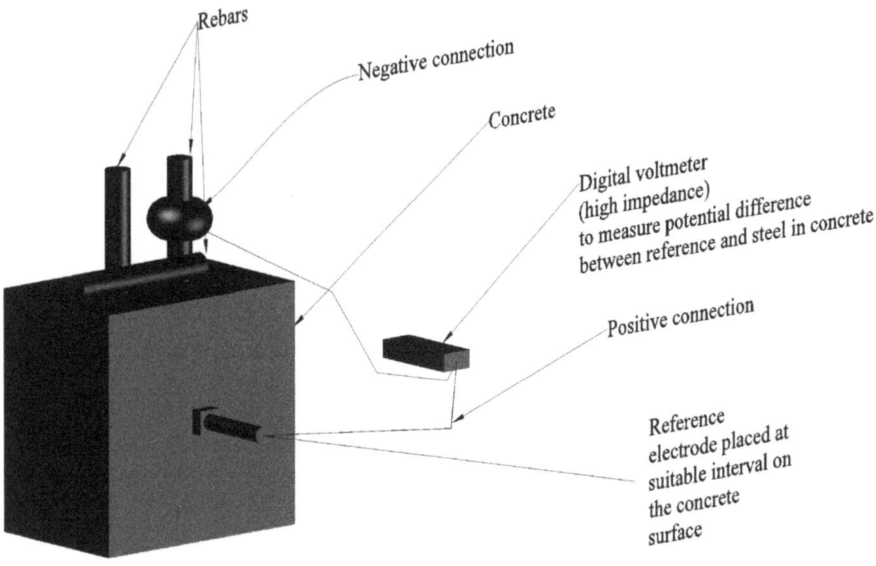

FIGURE 10.2 Circuit to identify corrosion potential.

resistivity of the circuit is too high. Therefore, a valid corrosion potential cannot be measured by this method.

Step 8. In case the stable condition as described in (1) can be established by a small amount of pre-wetting, then this can be accomplished by spraying or wetting the entire surface of concrete under investigation with the pre-wetting solution prepared. During the measurement, no free water should be available on the surface.

Step 9. On the other hand, an alternative measurement consists of placing a pre-wetted sponge on the surface of the concrete for a period until a stable condition as described in (1) can be achieved. Until the reference electrode measurement and corrosion potential measurements are made, do not remove the pre-wetted sponges from the placed places.

Step 10. Potential readings can only indicate the existence of corrosion activity at a particular location, whereas the precise location of the corrosion activity cannot be detected by these readings.

Step 11. If the readings are to be carried out underwater or on horizontal or vertical members, caution needs to be taken, particularly in case of underwater measurements and the measurements made in the tunnel. In the above two cases, the availability of oxygen will be very less near the surface of the reinforcements, thus resulting in a shift of corrosion potential to very high negative values. In most of the cases, while reading underwater corrosion, it is challenging to locate underwater corrosion points. Care should be taken while using reference electrodes underwater. Subsequently, the reference electrode may get contaminated. Therefore, apart from the porous Cu–$CuSO_4$ tip, other parts of the reference electrode should not be immersed during the underwater investigation.

Step 12. Corrosion potential is recorded to the nearest of 0.01 V. Temperature correction is adopted if the reference electrode temperature is outside the range of 22.2 ± 5.5°C (72 ± 10°F).

Representation of Data and Interpretation of Results

- Data is represented as an equipotential contour map. The equipotential contour map is drawn for the complete member based on the corrosion potential obtained. Corrosion potentials of the same values are matched and drawn as a contour. The interval between two contours should be 0.10 V.
- Numeric magnitude technique or potential difference technique are used to interpret corrosion potentials.
    - Numeric magnitude technique is used as an indication of corrosion in reinforcement in un-carbonated and atmospherically exposed concrete. This technique cannot be used for underwater concrete structures, water-saturated or near-saturated members such as tunnels, water tanks and basements when submerged. Moreover, this technique cannot give the rate of corrosion of the reinforcements.

- Potential difference technique can be used to identify localised corrosion. A major difference in corrosion potential may arise within a few millimetres. This method can also give a sense of the magnitude of corrosion.
- In case of numeric magnitude technique, if the measured potential over an area is more positive than -0.20 V, then there is more than 90% probability that no reinforcing steel is corroded in that region.
- If the potential is in the range of -0.20 V and -0.35 V, corrosion activity of the reinforcing steel over those zones are uncertain.
- If the potential is more than -0.35 V, then there is a 90% probability that the corrosion is in progress in those regions.
- The above criteria should not be used in the following cases
    - Carbonation has occurred in concrete until a depth of embedded steel.
    - Interior concrete where drying is predominant. If the concrete is subjected to frequent wetting condition or not allowed to dry after casting, then the above criteria can be used.
    - When outdoor reinforced concrete which has relatively high moisture variation or oxygen content, or both, needs to be compared.
    - In case of rehabilitated structures, where the provided treatment has changed the moisture content or oxygen content, or both.

Report

- Type of cell used (including calibration details)...................
- The average temperature during testing ....
- Pre-wetting method and method of attaching the voltmeter to the reinforcing steel ....
- An equipotential map showing the location of reinforcements

Percentage of total corrosion potential is more than – 0.35 V and less than – 0.20 V.

Student Remark

....
....
....
....
....

$$\frac{\textit{Marks obtained}}{\textit{Total marks}} = -$$

Instructor signature

Instructor Remark

....
....

....

....

10.3 PRACTICE QUESTIONS AND ANSWERS FROM COMPETITIVE EXAMS

1. The ratio of the yield stress of Grade 1 steel to the yield stress of Fe550 is
   a) 0.4545
   b) 0.5045
   c) 0.4955
   d) 0.4595

2. When pH is less than ……………………………………, corrosion is initiated.
   a) 13
   b) 14
   c) 12
   d) 11

3. Tensile test of steel reinforcement (Fe 550) was performed. 10% more yield strength was observed compared to expected standard yield strength. What is the ratio of yield strength of Grade 1 steel to the tested steel reinforcement?
   a) 0.413
   b) 0.513
   c) 0.423
   d) 0.523

4. Why is steel used as a member in RCC?
   a) Co-efficient of thermal expansion is the same as that of concrete
   b) Steel can take tensile stress
   c) Both a and b
   d) None of the above

5. Which of the following is not advised to be used in the marine environment as well as in sites where the workmanship is poor?
   a) Cold twisted bar
   b) Thermo-mechanically treated bar
   c) Epoxy coated bar
   d) Quenched rebar

6. Usage of epoxy coated rebar requires a considerable amount of care. If they are used in the sites without proper handling, what will happen?
   a) Holiday/Pinholes will form that results in enhanced corrosion activity.
   b) No problem will be there.
   c) Enhanced protection against corrosion will be there.
   d) None of the above.

7. Pitting corrosion is also referred to as
   a) Chloride corrosion
   b) Corrosion due to carbon-di-oxide
   c) Both a and b
   d) None of the above

8. How is yield stress calculated?
   a) Taking offset of stress vs. strain plot at 0.2% strain
   b) Taking offset of stress vs. strain plot at 0.02% strain
   c) Taking offset of stress vs. strain plot at 0.002% strain
   d) Taking offset of stress vs. strain plot at 2% strain

9. Why is an electrical contact solution necessary during corrosion testing of the steel bar?
   a) To avoid clouding of electrical charges during the contact
   b) To reduce the resistivity of concrete
   c) Both a and b
   d) To enhance the clouding of electrical charges during the contact

10. A sponge is attached at the end of the reference electrode. Why?
    a) To increase the electrical conductivity
    b) To reduce the resistivity
    c) Both a and b
    d) None of the above

11. The property of materials due to which it can be transformed into the thin state by heating is known as
    a) Malleability
    b) Ductility
    c) Toughness
    d) Hardness

[TN AE 2007]

---

1. a 2. d 3. a 4. c 5. c 6. a 7. a 8. a 9. c 10. c
11. a

REFERENCES

Assistant Engineer (AE). 2020. Tamil Nadu Public Service Commission (TNPSC), Tamil Nadu, India. http://www.tnpsc.gov.in/previous-questions.html (accessed September 18, 2020).

ASTM C876-15 (2016). *Standard test method for corrosion potentials of uncoated reinforcing steel in concrete*. American Society for Testing and Materials. West Conshohocken, Pennsylvania.

Engineering Service Exam (ESE). 2020. Union Public Service Commission, New Delhi, India. https://www.upsc.gov.in/examinations/previous-question-papers (accessed September 18, 2020).

Graduate Aptitude Test in Engineering (GATE). 2020. GATE Office, Chennai, India. http://gate.iitm.ac.in/gate2019/previousqp18.php (accessed September 18, 2020).

IS 1608. Part 1 (2018). *Metallic materials – Tensile testing*. Bureau of Indian Standards. New Delhi.

Junior Engineer (JE), Staff Selection Commission (SSC), New Delhi, India. https://ssc.nic.in/Portal/SchemeExamination (accessed September 18, 2020).

11 Sophisticated Analytical Techniques for Investigation of Building Materials

11.1 SCANNING ELECTRON MICROSCOPE (SEM)

11.1.1 Introduction

Early microscopes were developed to view small objects (i.e. to magnify small objects) and to examine it clearly (with high resolution). Due to mechanization in the early eighteenth century, simple microscopes that use light as a source were invented that can give a resolution of 1 micron (μm) (William, 2006). Still, looking at a magnification less than a micron was not widely used until the concept of using an electron as a source for magnification instead of light was developed. This contributed towards the development of the electron microscopy. The wide use of electron microscopes is due to the fact that their magnification and resolution are immensely high, and they can be used to see details in and below the size of an atom. Since the size of the electron is much smaller than the size of an atom, electrons as a source for magnification cannot be observed directly by naked eyes. Therefore, electron microscopes have a viewing screen that can convert electron intensity into light intensity for photographing or to look at. In recent times, due to technological advancements, imaging can be done by television detectors.

All the materials in nature are made up of atoms; the atom consists of a nucleus and electrons. During the process of imaging a material by electron microscopy, the electrons will interact with the material. The interaction will be in such a way that the electrons revolving around the nucleus in a particular shell can be knocked out. This process results in the emission of different electron signals from the material under investigation. These electron signals can be used as a source to characterize and understand the material. Different kinds of signals emitted during the interaction are as follows:

1. Backscattered electrons (BSE)
2. Auger electrons

3. Secondary electrons (SE)
4. Characteristic X-rays
5. Visible light
6. Elastically scattered electrons
7. Bremsstrahlung X-rays
8. Inelastically scattered electrons

One of the essential secondary signals produced during the interaction is characteristic X-rays that are produced from each element in the sample. These X-rays have a specific wavelength characteristic to a particular element. These can be used to quantify and analyze materials by using energy dispersive spectroscopy (EDS).

11.1.2 Apparatus Required and Their Usage

The SEM is a mapping rather than an imaging device. The sample to be analyzed is placed and impinged by a source of electrons through an electron column. Electrons scan the surface of the sample and provide images of the sample. The electron column consists of an electron gun (cathode and associated electrodes), which is placed under a high vacuum and connected to an external high output voltage source (30–40 kV) (Anwar, 2018). Acceleration and magnitude of electrons emerging can be varied by the potential difference in the electron gun. The force at which the electrons emitted from the electron gun impinge the surface of the sample depends upon the accelerating voltage (typically varied from 2 kV to 30 kV). The performance of the electrons emitted from the electron gun depends upon its brightness and current density. As the electron source moves towards the sample inside the electron column, the current density of the source electron varies. This is due to the blockage of various components inside the electron column that reduces the current density. On the other hand, brightness can also be considered as a measure of the performance of the electron gun. Brightness can be defined as the number of electrons emitted per unit area of the source and the solid angle subtended by the emitted electrons. Hence, the angular speed of the beam during its movement is also crucial. Therefore, the brightness remains constant throughout the electron column. Moreover, the brightness is directly proportional to the applied voltage at the output source. Therefore, the brightness of the source (filament in the electron gun) increases with an increase in applied voltage. The description of the brightness is vital in this context. Since the quantity of current (i.e. the electron) emitted from the probe of a given diameter highly depends upon the brightness, the brightness determines the amount of electrons that can be impinged at a given point of the sample in 1 second. This directly affects the quality of the image obtained.

One more important factor that shall affect the quality of the image obtained is the diameter of the source (i.e. the diameter of the impinging electron released from the probe (tip of the electron gun)). If the tip of the electron gun is small, then the released source diameter will be less; this results in high spatial resolution. Therefore, the source of the electron from where it is released should have a smaller diameter as possible. This can reduce the effort in de-magnifying the electron source. Therefore, a high current will be present in the probe to image the sample, which ultimately results in excellent imaging. In the same way, a number of electromagnetic lenses (made up of copper coils and iron shields) and the applied voltage to them should be controlled precisely in order to get the right focal length. Therefore, the focus of the electron on the sample can be precisely controlled for better brightness and proper resolution by a strong condenser lens, which has larger demagnification. Analogous to the lenses, apertures are a small rectangular strip made up of molybdenum. They are used to control the convergence angle of electrons passing through the column. Figure 11.1 shows a scanning electron microscope.

11.1.3 Working Principle

The SEM consists of an electron source; the best resolution with the SEM can be controlled by adjusting the diameter of the final electron beam emitted from the source. The electron source is usually a cathode. The most commonly used cathode is made up of thermionic tungsten. Because of its relatively less life time, tungsten was updated by a sintered polycrystalline material made up of lanthanum hexaboride (LaB_6), which has 10 times more brightness and long-lasting life. The source of electrons needs to be magnified and de-magnified by a set of electron lens systems. In general, when the electrons are emitted from the cathode or electron gun, their diameter will be of 10–50 μm. This needs to be de-magnified to a size of 1 nm to 1 μm. Reducing lenses or condenser lenses (basically a magnetic field outside the electron gun) are used for de-magnifying the electron beam from the source so that the final spot covered in the sample is 1–2 nm. An objective lens is used as a final lens in controlling the movement of the probe to focus the image of the sample.

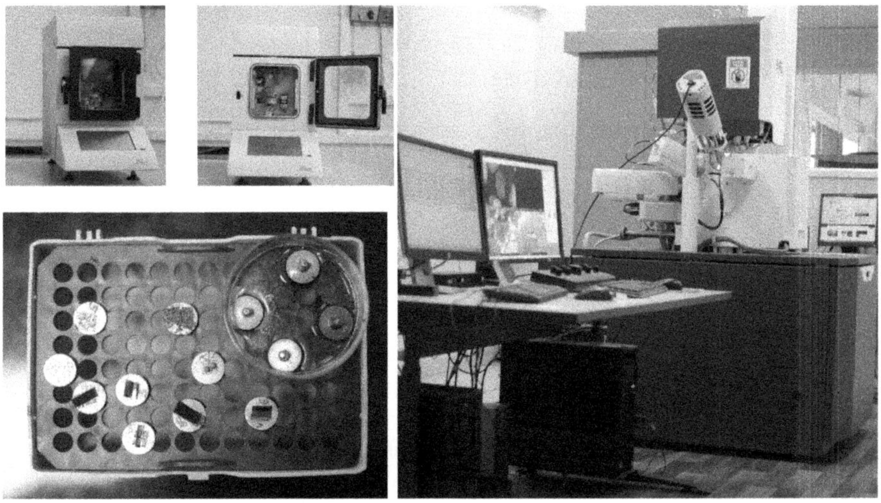

FIGURE 11.1 Scanning electron microscope.

During the collision of the electron beam on the surface of the sample, different electrons are emitted from the sample. Two of the important electrons that escape from the sample are backscattered electrons (BSEs) and secondary electrons (SEs). BSEs are emitted due to elastic collision, whereas SEs are emitted due to inelastic collision. Selective detection and collection of these electrons are of utmost importance. In order to do that, the SEM consists of different electron collectors, which are used to collect BSEs and SEs emitted from the sample. These electrons are the source of SEM images. BSEs are highly energetic in nature and escape from a few micrometres underneath the sample. They move in straight lines and consist of very important information, such as the average atomic number and chemical background of the sample under investigation. Similarly, SEs consists majorly of topographical features of the sample. Once the SEs and BSEs are collected, they are used individually or combined to create a signal and the signal is converted into an image.

Backscattered Electrons (BSE)

These are electrons emitted from few micrometres beneath the surface of the sample. The primary information carried by these electrons is the average atomic number of the elements of the sample under investigation (i.e. compositional image). Solid-state photodiodes are used as a detector to collect BSEs. The intensity of the image increases with an increase in the mean atomic number.

Secondary Electrons (SE)

Secondary electrons are detected and collected by electron collectors (Everhart-Thornley (ET-detectors)). Based on the SEs collected image of the sample is obtained. They are emitted from a shallow depth of the sample under investigation. Morphological characteristics of the sample can be obtained using SE mode.

11.1.4 Steps Involved in Sample Preparation Specific to Cement-Based Materials

Step 1. Cast samples (hardened cement paste/mortar/concrete) are cut to the appropriate size. In order to stop the hydration, the prepared samples are placed inside iso-propanol. The solution of iso-propanol is changed for every 1 hour initially for a time period of 5 hours. After this the solution is changed once every 24 hours. Usually, the samples are placed for one day in order to arrest the hydration. Few studies like the microstructure development with respect to time; then the samples are placed in iso-propanol after the specified time in which the microstructure needs to be studied.

Step 2. Coarse grinding and polishing of the surface of the sample to be studied are among the important and essential requirements during the preparation of the sample for SEM analysis.

Step 3. In order to prepare the sample, a silicon carbide sheet is used for initial manual grinding. Grinding needs to be done for a sufficient time period to achieve a glossy or shiny surface when the sample is exposed against the light.

Step 4. After grinding the surface to the required level, compressed air is used to clean the surface to remove the leftover powders. After confirmation of the surface to the required level, it is placed in the desiccator. The pressure is lowered down so that no air remains inside the desiccator.

Step 5. For the SEM analysis, the sample needs to be impregnated in epoxy resin. Impregnation is done by a small mould custom made with the help of a polyvinyl chloride (PVC) pipe or a silicone mould or mould, which can solve the impregnation purpose effectively.

Step 6. Place the sample in a clean and flat position. Centre and place the mould around the sample. As prescribed by the manufacturer pour the mixed resin in the mould, ensuring that the sample is in the centre. Utmost care is taken to ensure that no resin is passing underneath the sample.

Step 7. Ensure that no air bubbles are there during impregnation of the sample into the resin; this can be achieved by placing the impregnated sample in a desiccator under reduced pressure for 4–8 hours. The time period depends upon the air bubbles to be removed during the process.

Step 8. The pre-grinding in steps 3 and 4 is done just to ensure that the surface of the sample is flat and smooth enough for further grinding and polishing.

Step 9. After the successful impregnation of the sample in the epoxy resin, the samples are taken for further fine grinding and polishing. A grinding and a polishing machine are used to grind and smoothen the surface of the sample to be investigated. The samples are initially ground to remove the surface of the resin covering the sample.

Step 10. After removal of the resin from the surface of the sample (can be ensured by exposing the surface to light and checking for glossy or shiny surface) by grinding, polishing is started. The required amount of polishing can be done by viewing the sample through an optical microscope.

Step 11. Polishing is done by a special diamond grinder. Petrol or iso-propanol or any other solvent that is not reactive with the sample is used as a lubricant. Ensure that the used solvent is not hazardous or reacts with the sample, aiding in the formation of new or additional hydration products.

Step 12. After grinding and polishing to the required level, the sample is coated with a conductive medium before it can be used for the SEM experiment and analysis.

Step 13. The above type of sample preparation described from step 2 to step 11 is preferred only if the interest is to collect the SE images,

BSE images and FDXM (characteristic X-ray of the excited elements).

Step 14. Following steps can be used for a preliminary understanding of the images obtained from the SEM experiment.

1. Chemical composition should be analyzed in more than one place. For C-S-H, at least 200 places should be analyzed to ensure correct quantification.
2. To find the variation among different phases in the system which has quiet similar chemical compositions (for example: C_3S, C_2S), elemental ratios can be calculated and used. For example, in C_3S each silicate has 3 calcium and C_2S each silicate has 2 calcium, each can be arrived at by the difference in amount obtained during imaging.
3. A comprehensive understanding can be obtained by examining the phases by ratio. For example: Al/Ca ratio vs. Ca/Si ratio chart can reveal a lot of information about the formed phases.
4. One of the major drawbacks of the above method is that it cannot be used for detecting less-heavy metals like hydrogen.
5. In the case of calcium hydroxide ($Ca(OH)_2$) and free lime (CaO) differentiation can be made by identifying the time of hydration and compositional ratios.
   i. In such cases, the oxygen carried by $Ca(OH)_2$ will be more compared to the oxygen in free lime.
   ii. Time of hydration of the sample: If the hydration time is three days then more or less all the free lime might have got consumed by then due to its higher reactivity.
6. SEM images can be used for phase identification and to identify the percentage crystallinity of the material after certain processing in order to understand the reactivity of the material.
7. In most of the cases, the combined effect of BSEs and SEs can be beneficial in identifying the phases as well as the morphology of the hydration products.
8. SEM images can be classified into point detection and phase mapping when using FDXM for compositional analysis.
9. Mapping of the phases got by this method is more beneficial than the results obtained from other techniques.

Note: During sample preparation, water needs to be removed from the sample in order to avoid the build-up of static charge on the surface of the sample, which may result in distorted images.

11.1.5 Experimentation and Interpretation of Obtained Results

Electrons emitted from the electron gun raster the surface of the sample. During this process, electrons interact at an atomic scale with the sample. Atoms are the building blocks of elements, and elements combine to form different materials. Cement is also one such type of material formed by processing different compounds, as discussed in the Chapter 1. Cement, upon mixing with water, starts hydrating and forms hydration products. When a sample of cement is prepared by the prescribed steps as described in sample preparation and impinged with a source of the electron under vacuum in the SEM, a high contrast image can be obtained by capturing BSEs and SEs. Majorly, there are two types of scattering that take place due to the impinging of electron on the surface of the sample. They are elastic scattering and inelastic scattering.

Elastic Scattering and Inelastic Scattering

When the electrons raster through the sample, if the kinetic energy of the electrons (negatively charged) do not change much due to the deflection of the nucleus (positively charged) then the scattering is called as elastic scattering. This type of electrons that are emanating from the sample is called as backscattered electrons (BSE). These are source electrons from the initial beam (electron gun). This type of scattering has the most probability in case of materials with heavy elements (due to their strong positively charged nucleus). When the energy of the electron increases, the tendency of the electron to scatter elastically decreases. This is due to the fact that the kinetic energy of the electrons will be strong enough to overcome the attractive forces of the nucleus, thus resulting in less elastic scattering. The images formed by BSEs are called as backscattered electron images. These images are useful in observing compositional or atomic number contrast.

On the other hand, inelastic scattering happens if the impinging electrons lose kinetic energy due to the attraction of atomic nuclei. In this case, the electrons knock out the loosely bound orbital electron in the atom, creating a vacancy in the orbiting shell of the atom. They are deflected at small angles (as their accelerating energy is higher). The electrons knocked out of the orbital shells are called secondary electrons (SE). These belong to the atoms of the sample under investigation. The images formed from these electrons are called secondary electron images. These images are useful in observing morphological details.

The foremost difference between BSEs and SEs is their energy level. SEs exhibits a very low energy level (50 eV or less). This energy level is used as a boundary line separating SEs from BSEs.

Backscattered Electron Image or Compositional Contrast or Z-Contrast

Compositional or atomic number contrast produced during the collection of BSEs is due to the variation in the atomic number of the elements present in the sample under investigation. In case of high atomic number elements in the sample, elastic collision will be predominant. This results in a higher number of BSE. This can be given by a coefficient called as backscatter coefficient. This coefficient is a result of the ratio between the numbers of backscattered electrons and the number of incident beam electrons. In the case of high atomic number elements in the sample, the backscattered coefficient ratio will be the highest. This is attributed due to the fact that the attractive forces exhibited by the nucleus of high atomic number elements will be higher. This results in the highest number of elastic scattering of electrons. Moreover, these electrons can be retracted from the surface of the sample without significant loss in their kinetic energy. This contributes to the increase in signal strength due to an increase in backscatter coefficient and as the energy carried by the BSEs is the highest. Therefore, the resulting contrast will be the brightest for the element with high atomic number (high density) whereas the darkest for the element with low atomic number (low density) in the sample. Consequently, whenever the source of electron passes through the voids, there will be a dark contrast in the compositional image obtained. This is true in case of a cement sample as well. On the other hand, at low beam energy (from the source less than 5 keV) backscatter coefficient will be vice versa (i. e.) backscatter coefficient increases for elements with low atomic number (less than 30) and decreases for elements with high atomic number (more than 30). This results in a complication, while the compositional images obtained from BSEs contrast. From the above discussion, it can be understood that BSE images can be used to understand the compositional variation of the sample under investigation.

Overall, to summarize, the compositional contrast will be brighter (i.e. strong contrast) for the phases with high atomic number and will be dark (i.e. weak contrast) for the phases with low atomic number. Moreover, to differentiate two phases with low atomic numbers, compositional contrast (z-contrast) can be used if the difference in obtained contrast is more than 10%. Since SEM can differentiate a compositional contrast equal to or more than 10%. On the other hand, using compositional contrast to discern between phases of high atomic number is difficult. As the percentage difference in their compositional contrast is very low (less than 1%), it cannot be differentiated by SEM. Backscattering coefficient can be calculated by the following empirical equation

$$\eta = -0.0254 + 0.016Z - 1.86\times10^{-4}Z^2 + 8.3\times10^{-7}Z^3$$

Where, η is the backscattering coefficient, Z is the atomic number of the element under consideration.

The contrast between two locations in a sample is called as z-contrast (C_z). Z-contrast between two locations in a sample can be defined as follows.

$$C_z = \frac{\eta_2 - \eta_1}{\eta_2}$$

Where, η_1 and η_2 are backscattering coefficients at different locations

Other than the compositional contrast obtained by the BSEs, it can also give details about the topographical contrast by directionality component of BSEs. Depending upon the roughness of the sample surface, the directionality component can vary, resulting in different trajectories of the emitted BSEs. Therefore, the directionality component can be separated from z-contrast to obtain topographical images separately.

Secondary Electron Image or Topographical Contrast

Similar to BSEs, SEs also have secondary electron coefficient (δ). This is a ratio between the numbers of secondary electrons emitted from the sample to that of the number of incident beam electrons. In general, the emanation depth of the SEs is less when compared to that of BSEs. This is due to the fact that the secondary electrons emanate from the surface of the sample by inelastic collision. Therefore, SEs do not carry very high energy as compared to BSEs, which are source electrons from an electron gun. As the SE are emanated from the surface of the sample by inelastic collision, the energy level of theses electrons will be very low. With deficient energy level (<50 eV) their ability to leave the surface will be difficult, since this will not be enough to exceed the work potential of the sample surface. Similarly, the depth from which these electrons emanate is of few nanometres (nm) when compared to BSEs whose depth can be of few μm.

One of the prominent reasons behind the emanation from lesser depth is the fact that the energy level of electrons from the inner shells of the atoms are difficult to remove. If those inner electrons are removed, they do not possess enough amount of energy to exit from the surface of the sample due to several inelastic collisions with other electrons. Thus resulting SEs are released from few nanometres from the surface of the sample. Notably, the depth of emanation will be further reduced for conductive samples when compared to insulating samples. SEs will have a very low kinetic energy less than 50 eV. About 90% of the SEs will have energy less than 10 eV.

During the emanation of SEs, different SEs emanate from the surface of the sample: SE1 from the immediate vicinity of the source electron (where the electron beam from the gun interacts) end SE2 from the much farther distance from the source electron. SE2 are predominantly emitted due to the inelastic scattering of BSEs which can collide with some electrons in the orbiting shells of the atoms. This results in the emanation of SE2 electrons. The strength of SE2 signal is much higher when compared to SE1 signal. However, the current density distribution of SE1 electrons are higher; thus, the resolution at high magnification is determined by SE1 electrons, whereas SE2 contributes to the background. In light elements, SE1 will be predominant; this is due to the fact that the inelastic scattering due to BSEs will be less as fewer BSEs will be available. In contrast, in heavier elements, SE2 will be dominant due to the higher number of inelastic scattering by BSEs. In heavier elements, attraction by the nucleus will be higher, resulting in a higher number of BSEs, which in turn can result in higher inelastic scattering, thereby increasing emanation of SE2 electrons from the surface of the sample with heavier elements. At low beam energy, penetration of the source electron will be shallow. Therefore, the secondary electron coefficient will be higher. This is due to the fact that these electrons are emanated at a low energy from the shallow depth of the surface. SEs emanating from the surface of the sample are independent of the atomic number of the elements in the sample, though any contaminations in the surface of the sample profoundly affects the emanation of SEs.

SEs are used to observe the size, shape and surface texture of the samples (i.e. topographical features). Therefore, secondary electron imaging is used to image the surface characteristics of the sample under consideration. Everhart-Thornley (ET-detectors) detectors are used to collect SEs when the ET-detectors are biased positively (+200–300 V). A significant effect called edge effect results in an increase in brightness at certain parts of the sample. This is attributed due to the higher secondary electrons emanated from those parts of the sample. This generally takes place in raised areas, steep surfaces, protrusions, edges of thin surfaces and holes. It will appear less bright on other flat areas compared to the above areas. Similarly, spherical particles will have very bright light due to the grazing effect. When the beam strikes the surface of the spherical particles at an inclined angle, more SEs escapes from closer to the surface. Whereas, at a certain depth below the surface, the number of SEs emanated will be less, this also happens in case of the beams striking the surface at 90° angle. As the electron beam travels farther inside the sample, it loses energy. When the energy of the electrons reaches a few eV, they finally stop. The distance travelled by the electrons inside the sample depends upon many factors: density of the specimen, atomic number of the atoms, incident beam energy and the stopping power of the sample under investigation. Approximate reduction in the electron energy is 1–10 eV per nanometre (nm) of the distance travelled by the electrons. In general, heavy elements have more power to stop the electrons quicker when compared to lighter elements (i.e. to obtain a greater depth of penetration in heavy elements, higher-energy electrons are necessary compared to lighter elements). However, in lighter elements, due to high presence of electrons per unit volume, electron source losses energy more rapidly due to higher inelastic collision with the electrons in the sample (higher SEs are emitted from the surface of the sample). For this reason, it is clear that, when the source of electron hits the sample, it is not converged into a specific volume and that the electrons do not travel in straight lines. Instead, they cover a larger width and penetrate to a greater depth. As a general rule, SEs are emanated from the surface until a depth of 100 nm, BSEs are until a depth of 1 μm and characteristic X-rays until a depth of 5 μm.

11.2 THERMO-GRAVIMETRIC ANALYSIS (TGA)

11.2.1 Introduction

Minerals formed during a process of a reaction may undergo different thermal reactions such as dehydration, oxidation, dehydroxylation, decomposition, de-carbonation, etc. The changes brought about by thermal reactions result in mass changes and subsequent changes in the amount of heat released or taken by the system. Therefore, thermo-gravimetric analysis (TGA) is used as a method to analyze the change in mass with respect to the increase in temperature, whereas, change in mass at a particular temperature corresponds to a specific mineral that undergoes one of the above thermal reactions. The thermo-gravimetric analysis is one among the widely used techniques in the field of cement chemistry to follow the reactivity of cementitious materials (Lothenbach, Durdziński, & Weerdt, 2016). This method is used widely to supplement interesting facts obtained from other methods such as SEM and XRD. Primarily, in the case of cementitious materials, this method is used to study the bound water and the amount of portlandite ($Ca(OH)_2$) formed or consumed during the reaction. Often, this is used as a method to understand the reactivity of supplementary cementitious materials (SCM's). In a typical TGA, the sample is heated, and the weight loss of the sample is recorded. Figure 11.2 shows a thermo-gravimetric analyser.

FIGURE 11.2 Thermo-gravimetric analyzer.

11.2.2 Factors That Influence TGA

The type of device used profoundly influences the thermal analysis. Following are some of the crucial factors that affect the thermal analysis. They are the rate of heating, type of vessel used, gas used for purging (O_2, N_2, argon, helium, etc.), size of the particles, purging rate, amount of sample, sample pre-treatment method, etc. Therefore, it is generally not preferred to compare the measurement obtained from one device with the other. Consequently, if a particular measurement protocol is followed for one sample, it is advisable to follow the same protocol throughout the entire investigation. Important parameters that affect the obtained results are discussed in the following section.

Rate of Heating

The heating rate affects the peaks observed during the thermal analysis. Higher rate of heating results in narrower and better-defined peaks, although they may result in the faster build-up of vapour pressure inside the chamber. As the chambers are closed, it results in the shift of dehydration and dehydroxylation temperature to higher magnitude. Vice versa, at the low heating rate the developed vapour pressure is less, which may result in dehydration at a much lower magnitude of temperature.

For example: In most of the general cases, dehydration of gypsum ($CaSO_4.2H_2O$) to anhydride ($CaSO_4$) takes place approximately at 140°C (Hudson-Lamb, Strydom, & Potgieter, 1996). At a higher rate of heating, by reason of increased vapour pressure in the closed chamber, the dehydration of gypsum ($CaSO_4.2H_2O$) to hemihydrate ($CaSO_4.1/2H_2O$) then hemihydrate ($CaSO_4.1/2H_2O$) to anhydride ($CaSO_4$.) can be differentiated very clearly. It takes place at a higher temperature than normal (Paulik, Paulik, & Arnold, 1992).

Rate of Gas Purging

Similar to the heating rate, the flow of gas affects the temperature at which a particular mineral is decomposed. In general, a higher flow rate may result in a shift of temperature to a lower magnitude. Moreover, a good differentiation between different peaks cannot be observed. Likewise, the type and the size of the pan also affect the vapour pressure produced; therefore the magnitude of the temperature at which the dehydration or dehydroxylation is observed (Paulik et al., 1992).

Weight of the Sample

Weight of the sample taken for thermal analysis is one among the critical factor that affects the results obtained. Higher the weight of the sample, higher is the magnitude of the temperature at which the dehydration takes place. The width of the peak obtained will be broader as well. For that reason, in order to have a proper comparison between different samples, the results obtained are normalized to 100% of the initial weight of the sample.

For example: If a small amount of gypsum is present in the sample, then the main peak where the mass loss happens occurs at the initial stage around 115°C. Dehydration of anhydride takes place beyond 135°C. If the presence of gypsum in the sample is higher, then the peak will be shifted to a higher temperature. The primary peak dehydration from gypsum to hemi-hydride will be around 145°C, followed by hemi-hydride to anhydride around 180°C. This is due to the fact that a higher quantity of gypsum results in the higher release of water vapour during dehydration. It is thus causing higher vapour pressure in the closed chamber, therefore resulting in the occurrence of a peak at the higher temperature (Paulik et al., 1992).

11.2.3 Methods Involved in Pre-treating the Sample for TGA and Their Details Specific to Cement-Based Materials

1. Pre-treatment of samples is an essential factor that affects the thermal analysis. Typically, pre-treatment involves stopping of hydration at the prescribed time period by quenching in liquid nitrogen or solvent exchange (Zhang & Scherer, 2011). This is followed by freeze-drying, vacuum drying or other methods to remove the water (free water). Nevertheless, each treatment has its own influence on the results obtained during thermal analysis.
2. Solvent exchange method with methanol, ethanol and iso-propyl alcohol was found to interact with hydration products such as C-S-H, portlandite, ettringite and AFm phases. Prolonged exposure of the sample during the solvent exchange method results in the exchange of water from interlayers of ettringite and AFm phases by solvents. Particularly, monosulfate phases are dehydrated during the solvent exchange process. Moreover, intercalated

solvents are difficult to be removed from the interlayers even by vacuum drying methods (Khoshnazar, Beaudoin, Raki, & Alizadeh, 2013a, 2013b).
3. The action of methanol as a solvent is very vigorous in destabilizing the ettringite and AFm phases formed during the process of hydration. Besides, methanol and acetone also result in the interaction with portlandite during TGA, forming calcium carbonates. This can be evidenced by increased mass loss above 600°C. Above mentioned interactions result in a change in the mass loss signal during TGA.
4. To avoid the sorption of organic solvents by the hydrated phases, other methods such as immersing in liquid nitrogen or less sorption organic solvents such as diethyl ether can be used. Stopping of hydration reaction by diethyl ether is a 2-step process (Deschner et al., 2012; Schöler, Lothenbach, Winnefeld, & Zajac, 2015).
    a. The first step involves immersing in iso-propanol where the pore water is replaced by iso-propanol followed by immersing in diethyl ether.
    b. This is trailed by drying at 40°C to evaporate the diethyl ether remaining in the pores.
    c. Following the above 2-step procedure may result in decreased disturbance of portlandite or no evidenced carbonation of portlandite compared to freeze-drying method.

Freeze-drying can also be used as an alternative method to stop hydration without much influence of other reactions. In the first step, the required amount of sample is quenched in liquid nitrogen. This step is followed by vacuum drying in a CO_2-free environment where sublimation of ice takes place, resulting in the direct conversion of ice into water vapour. However, this method requires certain care when the point of focus is ettringite, since on freezing and drying water in ettringite may evaporate resulting in suppression of associated signal in TGA. Other hydrates may also lose their loosely bound water from their interlayers.

11.2.4 Typically Observed TGA Data for Cementitious Materials

Table 11.1 gives a broad idea about the decomposition temperature of different hydration products obtained in TGA.

11.2.5 Experimentation and Interpretation of Obtained Results

Experimentation is generally performed with following boundary conditions.

1. Aluminium crucibles are used which have a volume of 150 µL (which can sufficiently accommodate 50 mg of paste sample). Direct testing of other powder sample is also adopted.
2. Heating rate is generally chosen at 10–20°C per minute.
3. Temperature ranges from 40°C to 1,000°C.

TABLE 11.1
Temperature Range at which the Cementitious Materials Are Decomposed by Thermal Reactions (Lothenbach et al., 2016)

Cementitious materials or hydration products	Particulars	Decomposition or dehydration or dehydroxylation temperature range (°C)
Gypsum	Two-step dehydration process	Closed vessel with a small hole should be used for a distinct difference
	Gypsum to hemihydrate	100–140
	Hemihydrate to anhydride	140–150
Portlandite	Dehydroxylation [$CaO+H_2O$]	460
Magnesium hydroxide or Brucite	Dehydroxylation [$MgO+H_2O$]	420
Calcium carbonate or Calcite	De-carbonation [$CaO+CO_2$]	600–800
Coarser calcite	Un-hydrated cement	720
Mono or hemi-carbonates	From carbonation of portlandite or C-S-H	600–650
Magnesium carbonate or Magnesite	De-carbonation [$MgO+CO_2$]	500–600
Aragonite and Vaterite (polymorphs of calcium carbonate)	Re-crystallization without any weight change	450 (approx.)
Amorphous calcium carbonate	De-carbonates partially [CaO + Calcite]	400–600
Dolomite (carbonates of calcium and magnesium with a trace of iron)	MgO + Calcite	650
	Calcite decomposes to $CaO+CO_2$	600–800
C-S-H	loss of water over a wide range of temperature (dehydration or dehydroxylation)	50–600
	C-S-H to wollastonite	800
AFt	Lose of water	100
AFt	Water loses from dehydroxylation of aluminium hydroxide	200–400
AFm	Mono-carbonates (loss of 5 inter-layer water)	60– 200
AFm	Mono-carbonates (loss of 6 inter-layer water)	200–300
AFm	Loss of CO_2	650
AFm	Mono-sulphates (loss of water from interlayers)	250–350
AFm	Chlorides (Friedel's salt) (loss of water from interlayers)	140
Calcium aluminium hydrates	Loses its water	270
Calcium aluminium hydrates	Katoite	320
Calcium aluminium hydrates	Hydrogarnets	340

4. The type of purging gas and rate of purging: N_2 is used as purging gas at a rate of 30–50 mL/minute.
5. Before each series of experiment, a blank curve is determined with the selected crucible at a similar heating rate with similar purging gas and purging rate.
6. After the required time period of hydration, the samples are crushed and ground in a mortar and pestle. Ensure all precautionary measure to reduce the carbonation that can take place during the above process.
7. Based on the requirement of analysis, a pre-treatment method is chosen to stop the hydration. Either, a solvent exchange method can be used or freeze-drying can be used.
    a. In the case of solvent exchange method, use iso-propanol. About 5 grams of grounded cement paste is taken and placed in a container with iso-propanol for 10–15 minutes. This time period is sufficed for the free water to be replaced by iso-propanol.
    b. The suspension is filtered in a Büchner funnel and a vacuum pump to remove the excess iso-propanol in the suspension.
    c. Now, the filtered powder is washed again with diethyl ether (5–10 mL) to remove the excess iso-propanol and vacuum dried until the sample becomes light in colour. This can be identified by indication from the pressure gauge of the vacuum pump (pressure increases).
    d. Now the obtained powder is put in a petri dish and dried for 8–10 minutes in an air oven at 40°C or in the vacuum desiccator.
    e. Analyze the sample immediately if possible. Otherwise, store the sample in a closed vessel for a few days under a light vacuum.
    f. Alternatively, freeze-drying can also be used. Immerse the required amount of sample in liquid nitrogen for 15 minutes.
    g. The frozen sample is dried in a freeze dryer. The dried sample is crushed in a mortar and pestle.

8. Once the sample is ready, weigh approximately 50 mg of the sample and place it in the crucible. Detect the blank before the start of the experiment.
9. In general, the amount of gypsum, portlandite and calcite can be quantified by the measured weight loss during TGA as follows. Around 400–500°C portlandite decomposes to CaO and H_2O. Weight loss at this temperature is an indication that the portlandite decomposition has taken place. Based on the weight loss (WL ($Ca(OH)_2$)) at this temperature, the amount of portlandite present in the sample can be measured. We know that the molecular weight of portlandite is 74 g/mol and water is 18 g/mol. Therefore, the amount of portlandite present is determined as follows

$$Ca(OH)_{2\,\text{measured}} = WL(Ca(OH)_2) \times \frac{\text{molecular weight of portlandite}}{\text{molecular weight of water}}$$

$$= WL(Ca(OH)_2) \times \frac{74}{18}$$

10. Similarly, calcium carbonate decomposes above 600°C into CaO and CO_2. The weight loss at this temperature can be used as an indicator to measure the amount of calcium hydroxide carbonated at a specific time period as follows

$$Ca(CO)_{3\,measured} = WL(CaCO_3) \times \frac{\text{molecular weight of } CaCO_3}{\text{molecular weight of } CO_2}$$

$$= WL(CaCO_3) \times \frac{100}{44}$$

11. During the complete process of TGA, the mass of the sample is continuously changing. Therefore, an alteration needs to be done in order to account for the continuous change in the mass of the sample during the process of analysis. Consequently, for 100 grams of paste, the following modification needs to be made on the measured quantity of portlandite as well as calcite

$$Ca(OH)_{2\ paste} = \frac{Ca(OH)_{2\ measured}}{1 - H_2O_{bound} \times \left(1 + \frac{Water}{Cement}\right)}$$

$$Ca(OH)_{2\ paste} = \frac{Ca(OH)_{2\ measured}}{\text{weight at } 600°C \times \left(1 + \frac{Water}{Cement}\right)}$$

$$Ca(OH)_{2\ anhydrous} = \frac{Ca(OH)_{2\ measured}}{\text{weight at } 600°C}$$

12. Quantification of other compounds can also be carried out based on the weight loss at respective temperatures. Nonetheless, due to the high overlap of TG curves between two or more compounds results in de-convolution of TGA outputs difficult.
13. Typically, two methods are used to calculate the area underweight loss in TGA curves. Choice of these methods depends upon the software being used. Quantification of calcite is comparably more straightforward as the de-carbonation takes place above 600°C. Above this temperature, the influence of other hydration products is minimal.
14. Still, there exists a problem for quantification of portlandite, since the weight loss from C-S-H overlaps the dehydration portion of portlandite. For that reason, to erase those effects, the tangential method or step-wise method is used. In general, the step-wise method over-estimates portlandite content as the curve may include the weight loss due to C-S-H also. On the other hand, the tangential method assumes a correction for other hydrates, especially for C-S-H, a linear increase in weight loss in the prescribed temperature range is deduced. For better quantification, manual integration of the area under the curve can also be used.

In case blast furnace slags are used in the cement during TGA, loss in weight is observed due to the release of H_2S gas. This happens at a temperature approximately below 700°C. On the other hand, above 700°C, weight gain can be observed. This is due to the fact that oxidation of sulphur to sulphate (approximately, 1.996 g per g of S^{2-}) (Montes-Morán et al., 2012). Besides, oxidation of metallic iron and manganese could also lead to an increase in the weight of slag samples. The formation of a oxidizable product is profoundly affected by the type of gas present in the chamber as well as the rate of heating

11.3 X-RAY DIFFRACTION TECHNIQUE (XRD)

11.3.1 Introduction

X-rays were discovered by a German physicist Rontgen. However, their use to characterize materials by diffraction was found in 1912, after the discovery of wave-particle duality. X-rays, as a part of electromagnetic (EM) radiation, have a short wavelength in the range of 0.1–100Å. They are emitted when fast-moving electrons towards a source (copper (Cu) or molybdenum (Mo)) are decelerated by applying a potential difference. In a conventional type crystallographic technique, a cathode (tungsten filament (W)) is heated to emit electrons. These electrons impinge on an anode which can be rotated along its axis. On impinging on the anode, these electrons undergo multiple collisions; as a result, a huge amount of heat is generated. About 10% of the source electrons are transformed to produce X-rays that can be used for crystallographic investigation. A schematic diagram of the sealed X-ray tube is shown in Figure 11.3.

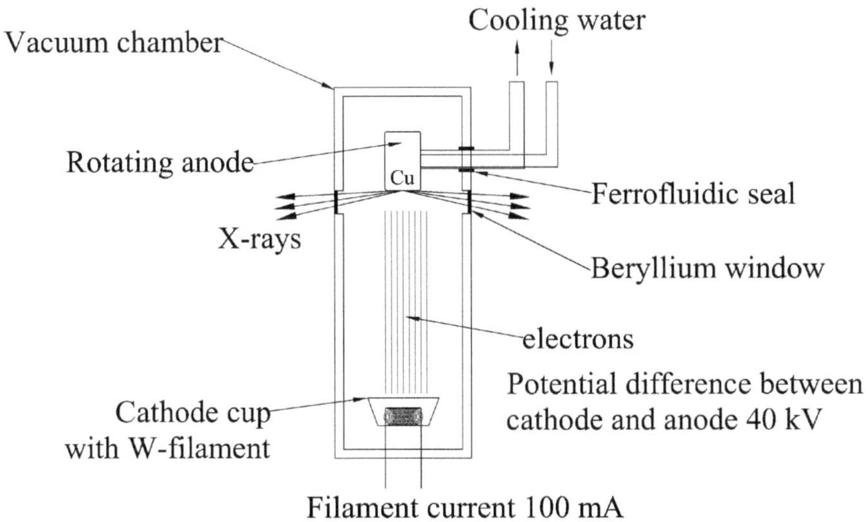

FIGURE 11.3 Schematic of a sealed X-ray tube.

The type and the quality of X-rays emitted from the anode highly depend upon the accelerating voltage (Suryanarayana & Norton, 1998). Higher the accelerating voltage, lesser will be the continuous distribution of X-ray wavelengths (i.e. less will be the white light emitted from the source). Therefore, higher the accelerating voltage, greater is the intensity of radiation emitted from the source as well as shorter will be the wavelength. Characteristic X-rays are emitted from the source (anode) when the electrons impinge on the anode. This electron impingement results in the removal of electrons from different shells of the atom in anode resulting in the transition of electrons from the outer shells to the inner shells. This transition of electrons from higher energy level (outer shell) to lower energy level (inner shell) in the anode results in the emission of characteristic X-rays from the anode. The wavelengths of the X-ray highly depend on the energy levels of the electrons involved in the transition (L – shell to K –shell or M – shell to K – shell). If E_1 and E_2 are two energy levels of the electrons, then the wavelength of the emitted X-rays can be given by

$$\lambda = \frac{hc}{\left|E_2 - E_1\right|}$$

Where, h is the Plank's constant, c is the velocity of the light, E_1 and E_2 are the energy levels of the electrons corresponding to the respective shells. Moreover, on electron transition from L – shell to K – shell or M – shell to K – shell, two distinct wavelengths are observed. In case of X-rays emitted, considering copper (Cu) as the anode (most commonly used source for X-rays in XRD), during L to K – shell transition two distinguishable wavelengths are $K_{\alpha 1}$ and $K_{\alpha 2}$ with an average value of 1.54184 Å (with a wavelength intensity ratio of 2:1). In the same way, for an M to K – shell transition two distinguishable wavelengths ($K_{\beta 1}$ and $K_{\beta 2}$) with an average value of 1.38851 Å (with a wavelength intensity ratio of 2:1). In case if the Bragg angle is larger (i.e. at a higher value of scattering), the averaged wavelengths need to be resolved.

Choosing the correct anode (the source from where the X-rays are emitted) is of utmost importance. When X-rays transmitted through a material, based on the atomic number of the element X-rays are attenuated (attenuation will be higher if the atomic number is higher and vice versa). Therefore, the transmitted intensity of the X-rays exponentially depends upon a coefficient called as a linear absorption coefficient of the material denoted as μ and the path length through which the X-rays are transmitted (L). With respect to the wavelength of X-rays, the linear absorption coefficient of the material varies. For any material, the linear absorption coefficient decreases directly proportional to $\lambda^{5/2}$ where, λ is the wavelength. Therefore, greater will be the depth of X-ray penetration because of the increase in energy of the emitted radiation. Consequently, the wavelength of the X-ray decreases, there will be a point at which the emitted radiation is strong enough to knock down an electron from the atom of a particular energy level (from K, L or M shell). The linear absorption coefficient dramatically increases at this point and is identified as an absorption edge or resonance level. Further decrease in wavelength results in a decrease in the linear absorption coefficient of the material. So what does this have to do with the selection of an anode? Based on the identified absorption edge or resonance level, practical applicability of the particular radiation will differ. For example, X-rays radiated from copper as an anode cannot be used for materials with a higher percentage of iron in it. This is due to the fact that the K absorption edge or resonance level for iron is 1.7433 Å. For that reason, the emitted X-radiation (from Cu) will be absorbed by atoms of iron and will be remitted as a characteristic K spectrum of iron. In such cases, anode needs to be chosen cautiously (Suryanarayana & Norton, 1998).

Continuous emission of X-rays from the anode with a specific wavelength is of utmost importance in order to obtain high-quality results. These emitted X-rays can be used as a source to investigate the crystalline structure of the material under investigation. In the case of cement-based materials, this analytical technique can be used for qualitative as well as quantitative phase identification.

11.3.2 Working Principle

Crystalline materials are arranged in some specific pattern. This arrangement is distinguished from one another by defining the repetitive pattern occurring in the material. The basic unit of the arrangement is called a unit cell. Each unit cell has a definite arrangement, and whenever X-rays passes through the material based on the arrangement of the unit cell in the material, a characteristic diffraction pattern can be obtained. These patterns are used to distinguish one unit cell from another. When a source of X-ray is diffracted by a specimen, then the diffracted rays obey Bragg's law as given by

$$n\lambda = 2d\sin\theta$$

Where, n is an integral number, λ is the wavelength of the X-rays emitted from the source, d is the distance between two inter-planes in a crystal, and θ is the angle of diffraction.

When the X-rays diffract a crystal based on the arrangement of atoms in the crystal, two or more rays can interfere. Interference is the result of the superposition of two or more waves. Superposition results in either a constructive interference or destructive interference based on the amplitude of the diffracted waves under consideration. When constructive interference takes place, then the intensity of the superposition will be the maximum and vice versa. Bragg's law forms the basis for indexing X-ray diffraction pattern of any specimen under investigation. Therefore, for any crystal structure based on the arrangement of the atoms in the inter-plane, the diffraction pattern will vary, resulting in a constructive as well as destructive interference. This is given by a factor called a structure factor, F, which describes the effect of crystal structure on the intensity of the diffracted beam.

11.3.3 Apparatus Required and Their Usage

X-ray diffractometer comprises of three essential components. They are: (1) X-ray source, (2) Specimen chamber and (3) X-ray detector, all of which lie on the circumference of a circle. This is also called focusing circle. One more circle can be drawn with the sample as the centre and with a radius of the X-ray source or the X-ray detector. This circle is referred as goniometer circle.

X-ray source consists of an X-ray tube, as shown in Figure 11.3. A tungsten (W) filament on heating emits electrons, and these electrons bombard the anode placed in a rotating table. Anode emits characteristic X-rays based on the transition of the electrons from a higher energy shell to a lower energy shell. The complete setup inside the X-ray tube is placed under high vacuum. The specimen to be investigated is placed in the specimen mounting table in the goniometer circle. Usually, a non-diffracting material is used to place the specimen. Non-diffracting material is usually a glass slide. The quantity of specimen (powder) is very small (1 mg and slightly above), should be less than 50 μm in size. If the size of grains is too large, then preferential orientation can take place. Preferred orientation may result in relatively high-intensity reflections. Similarly, if the size of the grains is too small, then the diffraction pattern will be broader that may overlap with other peaks.

The angle at which the source X-rays impinge the specimen is called as the Bragg angle (θ). Therefore, the angle between the source and the detector is 2θ (0–170°). The type of diffraction patterns produced from the above geometry is called as θ–2θ scans. In this particular geometry, the source of the electron is fixed. Only the detector moves on the circumference of the circle at different angles. As the angle of the detector increases, the radius of the focusing circle decreases and vice versa. The range of scan angle highly depends upon the choice of crystal under investigation. To define the collimation (parallel beams) of the X-ray beam soller slits are used. These soller slits are made up of high atomic elements because of their high absorption capacity. In the same way, to control the divergence of the X-ray beam (width), divergence slits are used. As soon as the X-ray beam diffracts the specimen, it passes through a set of slits before received by the detector for processing. The background radiation is reduced by an anti-scatter slit followed by a receiving slit. X-ray beam converges on passing through the receiving slit. Before received by the detector, the beam passes through one more soller slit. A monochromatic beam can be achieved only if other X-ray beams (K_β) are removed. This can be achieved by placing a filter (monochromator) or most recently by a graphite crystal after the diffracted beam in front of the detector. If the monochromator is used, then the set of soller slits placed after the specimen may not be necessary, since the width of the diffracted beam can be controlled by monochromator itself. A modern X-ray diffractometer is shown in Figure 11.4.

11.3.4 Steps Involved in Sample Preparation Specific to Cement-Based Materials

1. Based on the requirement, the sample is dried by a necessary method. Either the samples are placed in the oven at an appropriate temperature or they are dried, or organic solvent exchange methods are used to stop the hydration at respective days.

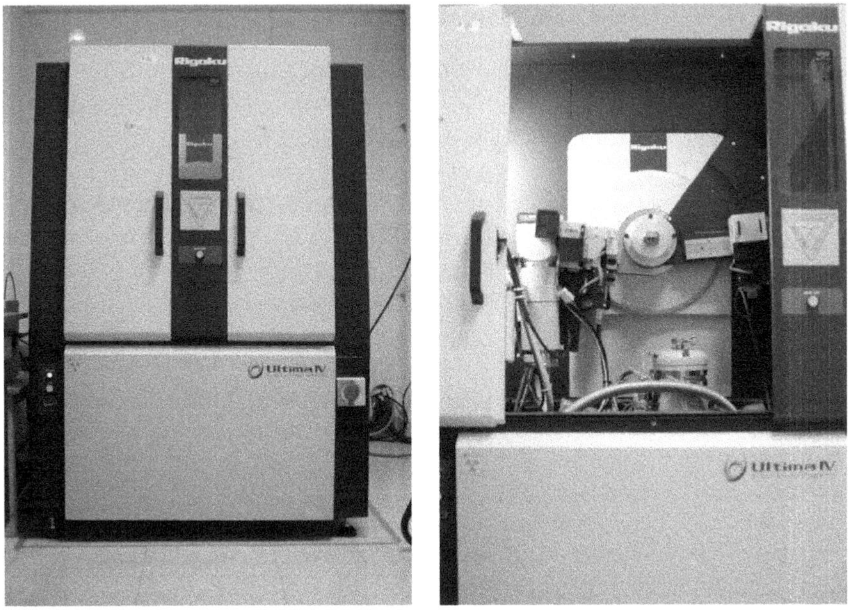

FIGURE 11.4 Modern X-ray diffractometer.

2. All the above methods have their own advantages and disadvantages. The method chosen should be based on the requirement and a wide range of materials to be tested.
3. Once the sample to be investigated is treated with one of the above methods, fill the sample in the mould in two layers with compaction given by spatula.
4. Then the open end is closed with a lid and removed from the holder with care.
5. The mould is inverted and the surface now exposed is used for the investigation. Since this surface is not exposed to the regular hand levelling, up to 90% preferential orientation can be avoided.
6. Cleaning of the mould and holders are done preferably by iso-propanol.
7. Preferential orientation is mainly due to the crystals like $Ca(OH)_2$ in hydrated samples and sometimes by C-S-H crystals.
8. $Ca(OH)_2$ and C-S-H aspect ratios are much higher when compared to other crystals that formed during the progress of hydration.
9. If measurements were done on in-situ hydrating samples, care should be taken to avoid carbonation. Carbonation can have a higher effect on diffracted patterns. In order to avoid a holder with a close lid can be used to avoid the interaction of a cement sample with atmospheric CO_2.

11.3.5 Experimentation and Interpretation of Obtained Results

Analysis of the X-Ray Pattern: Search-Match Procedure

The analysis is carried out on the basis of spacing and relative intensities of the peaks in the XRD pattern in comparison with patterns of known components. This comparison can be a collection of patterns of pure phases or powder diffraction files maintained by the International centre for diffraction data (ICDD). The automatic suggestions obtained based on X'Pert high score plus or any such similar programme can be taken as a suggestion. Ultimately, to determine the availability of all phases, a preliminary chemical investigation needs to be carried out by other methods. Further, analysis obtained using different techniques such as scanning electron microscopy (SEM) and thermo-gravimetric analysis (TGA) can also be used for confirmation. Quantitative analysis through the Rietveld method, controlled phase identification by selective dissolution, and density or magnetic separation can be of great help in identifying the concentration of trace phases.

For quantitative analysis about the composition of multiphase mixtures, the relationship between the intensity of the reflection of phase (S_α) in a mixture (m) and its concentration in the mixture (W_α) is given by

$$S_\alpha = K_e \frac{W_\alpha}{\rho V^2 \mu_m}$$

Where, K_e is the constant depending on the diffractometer and the component α, ρ_α and V_α are the density and cell volume of component α, and μ_m is the mass absorption coefficient of the mixture.

Quantification can be done by two methods: one related to the extracted intensities of distinct lines in the XRD pattern; different approaches in this group include addition/dilution method, use of standards or the direct reference intensity ratio (RIR) method. The second group of methods (Rietveld method) is a full pattern-fitting. In the Rietveld approach, diffraction pattern can be calculated using crystal structure data; then using a set parameter, calculated parameters can be compared with the experimental pattern.

Refinement is calculated using the sum of the weighted square differences between observed and calculated intensities at every step in the digital powder pattern. This method requires knowledge of approximate crystal structures such as space group symmetry, the atomic positions and site occupancies, and the lattice parameters (Snellings, 2016).

Caution to be Observed while Interpreting the Data from XRD:

1. Measurement made in low angles needs to be taken care off. Since at low angles, both source and detector will be at 180°, resulting in radiation being exposed directly from source to detector resulting in higher peak intensities. This can be avoided by using back slits that separate the source from the detector.

2. The main drawback is that elements that can be identified only at low angle measurements cannot be detected. E.g., kaolinite in clay; Friedel salt; bicarbonates and ettringite.
3. Similar to low angles, high-intensity problems will be there at high angle measurements also.
4. The chosen type of slit highly depends upon the power that is under investigation. An adjustment needs to be made on resolution or sample size based on the analysis requirement. While a finer slit can give quiet higher resolution, it takes a longer time; a coarser slit gives lower resolution, but it can complete the analysis in a shorter time. Preferably, a medium-sized slit is used for all analysis. If homogenization is present and the application is related to some specific tasks, then finer slits are used.
5. The type of source used is more critical since the same element will give different kinds of intensities and diffraction pattern for copper (Cu) and cobalt (Co). When presenting data using XRD, the type of metal source need to be indicated.
6. Scanning above 80° is not generally preferred. During the quantitative analysis of the cementitious system, using Rietveld method measurements made above 65° won't give any practically useful information. As the angle increases signal is very weak to be captured. NIST has a list of files available for cement-based material analysis.

11.4 PRACTICE QUESTIONS AND ANSWERS

1. Convergence angle of the electrons passing through the column is controlled by an aperture made up of
   a) Copper
   b) Cobalt
   c) Molybdenum
   d) Tungsten
2. During the collision of an electron beam on the surface of the sample, different electrons are emitted from the sample. Backscattered electrons (BSEs) are emitted due to
   a) Inelastic collision
   b) Elastic collision
   c) Partial scattering
   d) None of the above
3. Both BSEs and SEs are collected to create SEM images. From where the SEs are emitted
   a) From the source electron (thermionic tungsten or lanthanum hexaboride)
   b) From the material under investigation
   c) Escaped from the shell of material under investigation
   d) All the above

4. What type of image is obtained from BSEs?
   a) Morphological-based images
   b) Composition-based images
   c) Combination of a and b
   d) None of the above
5. Why petrol or iso-propanol is used as a solvent to aid in polishing of the sample instead of water?
   a) It acts as a lubricant
   b) It avoids the formation of new hydration products
   c) Both a and b
   d) None of the above
6. Why is the surface of the sample coated before SEM analysis?
   a) To allow the electrons to impinge the surface of the sample
   b) To make the surface of the sample conductive
   c) Both a and b
   d) None of the above
7. During an inelastic collision, which of the following are emitted?
   a) BSEs and SEs
   b) Characteristic X-rays and BSEs
   c) Characteristic X-rays and SEs
   d) All the above
8. The probability of emanation of BSEs is higher in which of the following cases?
   a) Light elements
   b) Heavy elements
   c) Hydrogen
   d) Water
9. In a compositional image of cement, how will you differentiate a void from other compounds?
   a) Location of the void will be dark
   b) Location of the void will be bright
   c) Location of other compounds will be dark
   d) None of the above
10. How is calcium hydroxide identified in SEs image?
    a) Tubular structure
    b) Hexagonal pyramid
    c) Hexagonal prism
    d) All the above
11. What is the principle behind thermo-gravimetric analysis (TGA)?
    a) Thermal reaction results in a change of weight
    b) Thermal reaction results in a change of volume
    c) Thermal reaction results in a change of length
    d) None of the above

12. How does the rate of heating affect TGA?
    a) Shifts the decomposition temperature to higher range
    b) Shifts the decomposition temperature to lower range
    c) Narrows down the range of peak and shifts the peak to the higher temperature
    d) Broadens the range of peak and shifts the peak to the higher temperature
13. A higher flow rate of gas may result in a shift of temperature to a lower magnitude. This is attributed because
    a) Increase in flow rate decreases the vapour pressure inside the chamber
    b) Increase in flow rate increases the vapour pressure inside the chamber
    c) Decrease in flow rate decreases the vapour pressure inside the chamber
    d) All the above
14. How does the weight of the sample taken affect the peak values?
    a) Increase in weight of the sample increases the peak temperature range.
    b) Increase in weight of the sample decreases the peak temperature range.
    c) Decrease in weight of the sample increases the peak temperature range.
    d) None of the above.
15. Compared to methanol and iso-propanol, which of the following can be used as an alternative solvent and why?
    a) Ethanol and high sorptivity
    b) Benzene and less sorptivity
    c) Nitro-benzene and high sorptivity
    d) Diethyl ether and less sorptivity
16. What are the decomposition ranges or temperature of calcium hydroxide, calcite and C-S-H (in °C)?
    a) 460, 600–800, 50–600
    b) 600–800, 460, 50–600
    c) 460, 50–600, 600–800
    d) 50–600, 600–800, 460
17. Which of the following method over-estimates portlandite?
    a) Tangential method
    b) Step-wise method
    c) Manual integration method
    d) All the above
18. Reactivity of any supplementary cementitious material (SCM) can be identified by knowing the mass of unreacted
    a) C-S-H
    b) $Ca(OH)_2$
    c) AFm
    d) $CaCO_3$
19. Decomposition temperature of hydro garnets ……… °C
    a) 140
    b) 270
    c) 320
    d) 340

20. Usage of freeze-drying method should be practised with caution. Why?
    a) It can dehydrate some of the hydration products during the drying process.
    b) It decomposes most of the hydration products.
    c) It can vigorously destabilize ettringite and AFm phases.
    d) None of the above.
21. How are X-rays emitted?
    a) An elastic collision of source electrons
    b) An inelastic collision of source electrons
    c) Both a and b
    d) None of the above
22. Why X-rays radiated from copper cannot be used for investigation of material with higher percentage of iron in it?
    a) K absorption edge of copper and iron are similar.
    b) Their resonance levels are similar.
    c) Their linear absorption coefficients are similar.
    d) All the above.
23. What will happen if the size of the specimen is too large?
    a) Preferential orientation may take place.
    b) Diffraction pattern may overlap.
    c) Both a and b.
    d) None of the above.
24. Characteristic X-rays emitted from anode highly depend upon the applied voltage. If the applied voltage is less, what will happen?
    a) Wavelength of X-rays will be unique and short.
    b) Wavelength of X-rays will be continuous and short.
    c) Wavelength of X-rays will be continuous and large.
    d) Wavelength of X-rays will be unique and large.
25. X-rays are collimated by allowing the X-ray beams to pass through
    a) Soller slits
    b) Anti-scatter slits
    c) Receiving slits
    d) Monochromator
26. Monochromator is used to
    a) Collimate the beam
    b) Receive the beam for processing
    c) Convert the beam to a single wavelength
    d) None of the above
27. What is the crystal structure of calcium carbonate?
    a) Hexagonal
    b) Trigonal
    c) Cubical
    d) Tetragonal

1.	c	2.	b	3.	c	4.	b	5.	b	6.	c	7.	c	8.	b	9.	a	10.	c
11.	a	12.	c	13.	a	14.	a	15.	d	16.	a	17.	b	18.	b	19.	d	20.	a
21.	b	22.	d	23.	a	24.	c	25.	a	26.	c	27.	b						

REFERENCES

Anwar, U.-H. (2018). *A beginners' guide to scanning electron microscopy* (1st ed.). Cham. Switzerland: Springer Nature.

Deschner, F., Winnefeld, F., Lothenbach, B., Seufert, S., Schwesig, P., Dittrich, S., … Neubauer, J. (2012). Hydration of a Portland cement with high replacement by siliceous fly ash. *Cement and Concrete Research*, *42*, 1389–1400.

Hudson-Lamb, D. L., Strydom, C. A., & Potgieter, J. H. (1996). The thermal dehydration of natural gypsum and pure calcium sulphate dihydrate (gypsum). *Thermochimica Acta*, 282–283, 483–492.

Khoshnazar, R., Beaudoin, J. J., Raki, L., & Alizadeh, R. (2013a). Solvent exchange in sulfoaluminate phases; Part II: Monosulfate. *Advances in Cement Research*, *25*, 322–331.

Khoshnazar, R., Beaudoin, J. J., Raki, L., & Alizadeh, R. (2013b). Solvent exchange in sulphoaluminate phases; Part I: Ettringite. *Advances in Cement Research*, *25*, 314–321.

Lothenbach, B., Durdziński, P., & Weerdt, K. De. (2016). Thermogravimetric analysis. In K. Scrivener, R. Snellings, & B. Lothenbach (Eds.), *A practical guide to microstructural analysis of cementitious materials* (1st ed., pp. 177–212). Boca Raton, FL: Taylor & Francis.

Montes-Morán, M. A., Concheso, A., Canals-Batlle, C., Aguirre, N. V., Ania, C. O., Martín, M. J., & Masaguer, V. (2012). Linz–Donawitz steel slag for the removal of hydrogen sulfide at room temperature. *Environmental Science & Technology*, *46*, 8992–8997.

Paulik, F., Paulik, J., & Arnold, M. (1992). Thermal decomposition of gypsum. *Thermochimica Acta*, *200*, 195–204.

Schöler, A., Lothenbach, B., Winnefeld, F., & Zajac, M. (2015). Hydrate formation in quaternary Portland cement blends containing blast-furnace slag, siliceous fly ash and limestone powder. *Cement and Concrete Composites*, *55*, 374–382.

Snellings, R. (2016). X-ray powder diffraction applied to cement. In K. Scrivener, R. Snellings, & B. Lothenbach (Eds.), *A practical guide to microstructural analysis of cementitious materials* (1st ed., pp. 107–162). Boca Raton, FL: Taylor & Francis.

Suryanarayana, C., & Norton, M. G. (1998). *X-ray diffraction a practical approach* (1st ed.). New York: Plenum Publishing Corporation.

William, J. C. (2006). *Under the microscope a brief history of microscopy* (1st ed.). Singapore: World Scientific Publishing. Pte. Ltd.

Zhang, J., & Scherer, G. W. (2011). Comparison of methods for arresting hydration of cement. *Cement and Concrete Research*, *41*, 1024–1036.

Index

For Product Safety Concerns and Information please contact our EU
representative GPSR@taylorandfrancis.com
Taylor & Francis Verlag GmbH, Kaufingerstraße 24, 80331 München, Germany

www.ingramcontent.com/pod-product-compliance
Ingram Content Group UK Ltd.
Pitfield, Milton Keynes, MK11 3LW, UK
UKHW022212190926
475702UK00003B/326
9780367644956